Birkhäuser

Progress in Nonlinear Differential Equations and Their Applications

PNLDE Subseries in Control

Volume 94

More information about this subseries at http://www.springer.com/series/15137

Tatsien Li · Bopeng Rao

Boundary Synchronization for Hyperbolic Systems

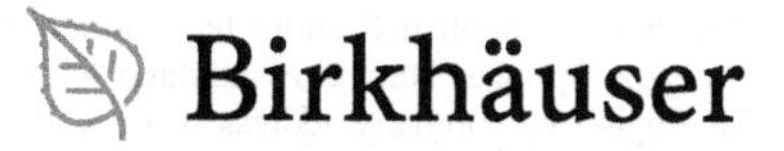

Tatsien Li
School of Mathematical Sciences
Fudan University
Shanghai, China

Bopeng Rao
Institut de Recherche Mathématique
Avancée
Université de Strasbourg
Strasbourg, France

ISSN 1421-1750 ISSN 2374-0280 (electronic)
Progress in Nonlinear Differential Equations and Their Applications
PNLDE Subseries in Control
ISBN 978-3-030-32851-1 ISBN 978-3-030-32849-8 (eBook)
https://doi.org/10.1007/978-3-030-32849-8

This book is published under the imprint Birkhäuser, www.birkhauser-science.com by the registered company Springer Nature Switzerland AG
The registered company address is: Gewerbestrasse 11, 6330 Cham, Switzerland

Contents

Chapter 1
Introduction and Overview

An introduction and overview of the whole book can be found in this chapter.

1.1 Introduction

Synchronization is a widespread natural phenomenon. Thousands of fireflies may twinkle at the same time; audiences in the theater can applaud with a rhythmic beat; pacemaker cells of the heart function simultaneously; and field crickets give out a unanimous cry. All these are phenomena of synchronization (cf. [71, 74]).

In principle, synchronization happens when different individuals possess likeness in nature, that is, they conform essentially to the same governing equation, and meanwhile, the individuals should bear a certain coupled relation.

The phenomenon of synchronization was first observed by Huygens [22] in 1665. The research on synchronization from a mathematical point of view dates back to Wiener [81] in the 1950s.

The previous studies focused on systems described by ordinary differential equations (ODEs), such as

$$X_i' = f(t, X_i) + \sum_{j=1}^{N} A_{ij} X_j \quad (i = 1, \cdots, N), \tag{1.1}$$

where $X_i (i = 1, \cdots, N)$ are n-dimensional state vectors, " $'$ " stands for the time derivative, $A_{ij}(i, j = 1, \cdots, N)$ are $n \times n$ matrices, and $f(t, X)$ is an n-dimensional vector function independent of $i = 1, \cdots, N$ (cf. [71, 74]). The right-hand side of (1.1) shows that every $X_i (i = 1, \cdots, N)$ possesses two basic features: satisfying a fundamental governing equation and bearing a coupled relation among

T. Li and B. Rao, *Boundary Synchronization for Hyperbolic Systems*,
Progress in Nonlinear Differential Equations and Their Applications 94,
https://doi.org/10.1007/978-3-030-32849-8_1

one another. If for any given initial data

$$t = 0: \quad X_i = X_i^{(0)} \ (i = 1, \cdots, N), \tag{1.2}$$

the solution $X = (X_1, \cdots, X_N)^T = X(t)$ to the system satisfies

$$X_i(t) - X_j(t) \to 0 \ \ (i, j = 1, \cdots, N) \ \text{as } t \to +\infty, \tag{1.3}$$

namely, all the states $X_i(t)$ $(i = 1, \cdots, N)$ tend to coincide with each other as $t \to +\infty$, then we say that the system possesses the synchronization in the **consensus sense**, or, in particular, if the solution $X = X(t)$ satisfies

$$X_i(t) - a(t) \to 0 \ \ (i = 1, \cdots, N) \ \text{as } t \to +\infty, \tag{1.4}$$

where $a(t)$ is a state vector which is a priori unknown, then we say that the system possesses the synchronization in the **pinning sense**. Obviously, the synchronization in the pinning sense implies that in the consensus sense. These kinds of synchronizations are all called the **asymptotic synchronization**, which should be realized on the infinite time interval $[0, +\infty)$.

What the authors of this monograph have been doing in the recent years is to extend, in both concept and method, the universal phenomena of synchronization from finite-dimensional dynamical systems of ordinary differential equations to infinite-dimensional dynamical systems of partial differential equations (PDEs). This should be the first attempt in this regard, and the earliest relevant paper was published by the authors in [42] in a special issue of Chinese Annals of Mathematics in honor of the scientific heritage of Lions, the results of which were announced in 2012 in CRAS [41].

For fixing the idea, in this chapter we will only consider the following coupled system of wave equations with Dirichlet boundary conditions (cf. Part 1 and Part 2 of this monograph. The corresponding consideration with Neumann or coupled Robin boundary conditions will be given in Part 3 and Part 4 or in Part 5 and Part 6):

$$\begin{cases} U'' - \Delta U + AU = 0 & \text{in } (0, +\infty) \times \Omega, \\ U = 0 & \text{on } (0, +\infty) \times \Gamma_0, \\ U = DH & \text{on } (0, +\infty) \times \Gamma_1 \end{cases} \tag{1.5}$$

with the initial data

$$t = 0: \quad U = \widehat{U}_0, \ U' = \widehat{U}_1 \text{ in } \Omega, \tag{1.6}$$

where Ω is a bounded domain with smooth boundary $\Gamma = \Gamma_0 \cup \Gamma_1$ such that $\overline{\Gamma}_0 \cap \overline{\Gamma}_1 = \emptyset$ and $\text{mes}(\Gamma_1) > 0$, $U = (u^{(1)}, \cdots, u^{(N)})^T$ is the state variable, $\Delta = \sum_{i=1}^{n} \frac{\partial^2}{\partial x_i^2}$ is the n-dimensional Laplacian operator, $A \in \mathbb{M}^N(\mathbb{R})$ is an $N \times N$ **coupling matrix** with constant elements D, called the **boundary control matrix**, is an $N \times M$ full column-rank matrix ($M \leqslant N$) with constant elements, $H = (h^{(1)}, \cdots, h^{(M)})^T$ stands for the boundary control.

Here, a boundary control matrix D is added to the boundary condition on Γ_1. This approach is more flexible: we will see in what follows that the introduction of D enables us to simplify the statement and the discussion to a great extent.

For systems governed by PDEs, we can similarly consider the asymptotic synchronization on an infinite time interval as in the case of systems governed by ODEs, namely, we may ask the following questions: under what conditions do the system states with any given initial data possess the asymptotic synchronization in the consensus sense:

$$u^{(i)}(t, \cdot) - u^{(j)}(t, \cdot) \to 0 \quad (i, j = 1, \dots, N) \quad \text{as } t \to +\infty, \tag{1.7}$$

or, in particular, if the system states with any given initial data possess the asymptotic synchronization in the pinning sense:

$$u^{(i)}(t, \cdot) - u(t, \cdot) \to 0 \quad (i = 1, \cdots, N) \quad \text{as } t \to +\infty, \tag{1.8}$$

where $u = u(t, \cdot)$ is called the **asymptotically synchronizable state**, which is a priori unknown? if the answer of this question is positive, these conclusions should be realized spontaneously on an infinite time interval $[0, +\infty)$, and is a naturally developed result decided by the nature of the system itself.

But for systems governed by partial differential equations, as there are boundary conditions, another possibility exists, i.e., to give artificial intervention to the evolution of state variables through appropriate boundary controls, which combines synchronization with controllability and introduces the study of synchronization to the field of control. This is also a new perspective on the investigation of synchronization for systems of partial differential equations.

Here, the boundary control comes from the boundary condition $U = DH$ on Γ_1. The elements in H are adjustable boundary controls, the number of which is $M(\leqslant N)$. To put the boundary control matrix D before H will provide many possibilities for combining boundary controls.

On the other hand, precisely due to the artificial intervention of control, we can make a higher demand, i.e., to meet the requirement of synchronization within a limited time, instead of waiting until $t \to +\infty$.

The corresponding question is whether there is a suitably large $T > 0$, such that for any given initial data $(\widehat{U}_0, \widehat{U}_1)$, through proper boundary controls with compact support in $[0, T]$ (that is, to exert the boundary control at the time interval $[0, T]$, and abandon the control from the time $t = T$), the solution $U = U(t, x)$ to the corresponding problem (1.5)–(1.6) satisfies, as $t \geqslant T$,

$$u^{(1)}(t, \cdot) \equiv u^{(2)}(t, \cdot) \equiv \cdots \equiv u^{(N)}(t, \cdot) := u(t, \cdot), \tag{1.9}$$

that is, all state variables tend to be the same since the time $t = T$, while $u = u(t, x)$ is called the corresponding **exactly synchronizable state** which is unknown beforehand. If the above is satisfied, we say that the system possesses the **exact boundary**

synchronization. Here, “exact” means that the synchronization of state variables is exact without error, and the so-called “boundary” indicates the means or method of control, i.e., to realize the synchronization through boundary controls.

In the above definition of synchronization, through boundary controls on the time interval $[0, T]$, we not only demand synchronization at the time $t = T$, but also require synchronization to continue when $t \geqslant T$, i.e., after all boundary controls are eliminated. This kind of synchronization is not a short-lived one, but exists once and for all, as is needed in applications.

We have to point out that in Part 1 and Part 2 of this monograph we always assume that the initial value $(\widehat{U}_0, \widehat{U}_1) \in (L^2(\Omega))^N \times (H^{-1}(\Omega))^N$, and the solution to problem (1.5)–(1.6) belongs to the corresponding function space, which will not be specified one by one in this chapter. As to the synchronization considered in the framework of classical solutions in the one-space-dimensional case, see [20, 21, 34, 57, 58].

1.2 Exact Boundary Null Controllability

The exact boundary synchronization on a finite time interval is closely related to the **exact boundary null controllability**.

If there exists $T > 0$, such that for any given initial data $(\widehat{U}_0, \widehat{U}_1)$, through boundary controls with compact support in $[0, T]$, the solution $U = U(t, x)$ to the corresponding problem (1.5)–(1.6) satisfies, as $t \geqslant T$,

$$u^{(1)}(t, x) \equiv u^{(2)}(t, x) \equiv \cdots \equiv u^{(N)}(t, x) \equiv 0, \tag{1.10}$$

then we say that the system possesses the so-called exact boundary null controllability in control theory. This is of course a very special case of the abovementioned exact boundary synchronization.

For a single wave equation, the exact boundary null controllability can be proved by the HUM method proposed by Lions [61, 62]. For a coupled system of wave equations, since for the purpose of studying the synchronization, the coupling matrix A should be an arbitrarily given matrix, the proof of the exact boundary null controllability cannot be simply reduced to the case of a single wave equation, however, using a compact perturbation result in [69], it is possible to establish a corresponding observability inequality for the corresponding adjoint system, and then the HUM method can be still applied.

As a result, we can get the following conclusions (cf. Chap. 3):

(1) Assume that $M = N$ and let the domain Ω satisfy the usual multiplier geometrical condition (cf. [7, 26, 61, 62]): there exists $x_0 \in \mathbb{R}^n$, such that, setting $m = x - x_0$, we have

$$(m, \nu) > 0, \ \forall x \in \Gamma_1; \quad (m, \nu) \leqslant 0, \ \forall x \in \Gamma_0, \tag{1.11}$$

where ν is the unit outward normal vector, and $(\cdot, \cdot)$ denotes the inner product in $\mathbb{R}^n$.

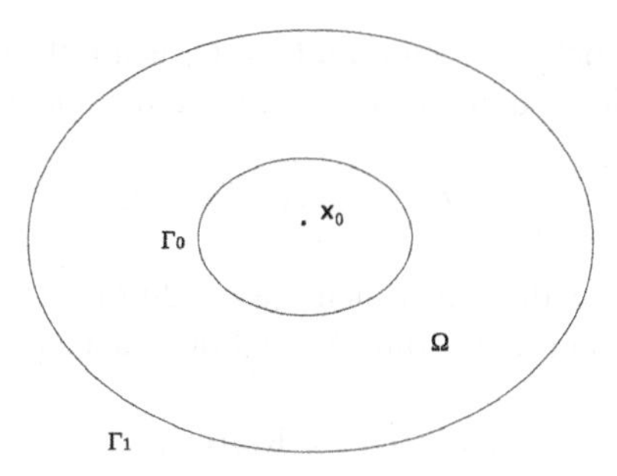

Then through N boundary controls, when $T > 0$ is suitably large, the exact boundary null controllability can surely be realized for all $(\widehat{U}_0, \widehat{U}_1) \in (L^2(\Omega))^N \times (H^{-1}(\Omega))^N$.

(2) Assume that $M < N$, that is, if the number of boundary controls is fewer than N, then no matter how large $T > 0$ is, the exact boundary null controllability cannot be achieved for all $(\widehat{U}_0, \widehat{U}_1) \in (L^2(\Omega))^N \times (H^{-1}(\Omega))^N$.

Thus, in the case of partial lack of boundary controls, which kind of controllability in a weaker sense can be realized by means of fewer boundary controls? It is a significant problem from both theoretical and practical points of view, and can be discussed in many different cases as will be shown in the sequel.

1.3 Exact Boundary Synchronization

Really meaningful synchronization should exclude the trivial situation of null controllability, and thus we get the following results (cf. Chap. 4):

Assume that the system under consideration is exactly synchronizable, but not exactly null controllable, that is to say, assume that the system is exactly synchronizable with $\text{rank}(D) < N$. Then the coupling matrix $A = (a_{ij})$ should satisfy the following condition of compatibility:

$$\sum_{p=1}^{N} a_{kp} = a \quad (k = 1, \cdots, N), \tag{1.12}$$

where a is a constant independent of $k = 1, \cdots, N$, namely, the sum of all elements in every row of A is the same (the **row-sum condition**).

Let $e_1 = (1, \ldots, 1)^T$. The condition of compatibility (1.12) is equivalent to that e_1 is an eigenvector of A, corresponding to the eigenvalue a.

Let

$$C_1 = \begin{pmatrix} 1 & -1 & & & \\ & 1 & -1 & & \\ & & \ddots & \ddots & \\ & & & 1 & -1 \end{pmatrix}_{(N-1)\times N} \tag{1.13}$$

be the corresponding **synchronization matrix**. C_1 is a full row-rank matrix. Obviously, the synchronization requirement (1.9) can be written as

$$t \geqslant T: \quad C_1 U(t, x) \equiv 0. \tag{1.14}$$

Moreover, $\mathrm{Ker}(C_1) = \mathrm{Span}\{e_1\}$, and then the condition of compatibility (1.12) is equivalent to that $\mathrm{Ker}(C_1)$ is a one-dimensional invariant subspace of A:

$$A\mathrm{Ker}(C_1) \subseteq \mathrm{Ker}(C_1). \tag{1.15}$$

Then, the condition of compatibility (1.12), often called the **condition of C_1-compatibility** in what follows, is also equivalent to that there exists a unique matrix $\overline{A}_1$ of order $(N-1)$, such that

$$C_1 A = \overline{A}_1 C_1. \tag{1.16}$$

Such matrix $\overline{A}_1$ is called the **reduced matrix of A by C_1**.

Under the condition of C_1-compatibility, let

$$W_1 = (w^{(1)}, \cdots, w^{(N-1)})^{\mathrm{T}} = C_1 U. \tag{1.17}$$

It is easy to see that the original system (1.5) for the variable U can be reduced to the following self-closed system for the variable W_1:

$$\begin{cases} W_1'' - \Delta W_1 + \overline{A}_1 W_1 = 0 & \text{in } (0, +\infty) \times \Omega, \\ W_1 = 0 & \text{on } (0, +\infty) \times \Gamma_0, \\ W_1 = C_1 D H & \text{on } (0, +\infty) \times \Gamma_1. \end{cases} \tag{1.18}$$

Obviously, under the condition of C_1-compatibility, the exact boundary synchronization of the original system (1.5) for U is equivalent to the exact boundary null controllability of the reduced system (1.18) for W. Hence, we have:

Assume that the condition of C_1-compatibility is satisfied and the domain Ω satisfies the usual multiplier geometrical condition, then there exists a suitably large $T > 0$, such that for any boundary control matrix D with rank $(C_1 D) = N - 1$, the exact boundary synchronization of system (1.5) can be realized at the time $t = T$. On the contrary, if $\mathrm{rank}(C_1 D) < N - 1$, in particular, if $\mathrm{rank}(D) < N - 1$, i.e., the number of boundary controls is fewer than $(N-1)$, then no matter how large $T > 0$ is, the exact boundary synchronization can never be achieved for all initial data $(\widehat{U}_0, \widehat{U}_1) \in (L^2(\Omega))^N \times (H^{-1}(\Omega))^N$.

Therefore, the above condition of C_1-compatibility is not only sufficient but also necessary to ensure the exact boundary synchronization. Under this condition, for the boundary control matrix D such that $\mathrm{rank}(D) = \mathrm{rank}(C_1 D) = N - 1$, appropriately chosen $(N-1)$ boundary controls suffice to meet the requirement.

We point out that in the study of synchronization for systems governed by ODEs, the row-sum condition (1.12) is imposed on the system according to physical meanings as a reasonable sufficient condition. However, for our systems governed by PDEs and for the synchronization on a finite time interval, it is actually a necessary condition, which makes the theory of synchronization more complete for systems governed by PDEs.

In the case that system (1.5) possesses the exact boundary synchronization at the time $T > 0$, as $t \geqslant T$, the exactly synchronizable state $u = u(t, x)$ satisfies

$$\begin{cases} u'' - \Delta u + au = 0 \text{ in } (T, +\infty) \times \Omega, \\ u = 0 \qquad\qquad\qquad \text{on } (T, +\infty) \times \Gamma, \end{cases} \tag{1.19}$$

where a is given by the row-sum condition (1.12). Thus, the evolution of the exactly synchronizable state $u = u(t, x)$ with respect to t can be uniquely determined by its initial data:

$$t = T: \quad u = \widehat{u}_0, \ u' = \widehat{u}_1. \tag{1.20}$$

However, generally speaking, the value $(\widehat{u}_0, \widehat{u}_1)$ of (u, u') at $t = T$ should depend on the original initial data $(\widehat{U}_0, \widehat{U}_1)$ as well as on the boundary controls H, which realize the exact boundary synchronization. Moreover, the value of $(\widehat{u}_0, \widehat{u}_1)$ at $t = T$ for a given initial data $(\widehat{U}_0, \widehat{U}_1)$ can be determined, and in some special cases it is independent of boundary controls H, which realize the exact boundary synchronization.

The attainable set of all possible values of $(\widehat{u}_0, \widehat{u}_1)$ at $t = T$ is the whole space $L^2(\Omega) \times H^{-1}(\Omega)$, when the original initial data $(\widehat{U}_0, \widehat{U}_1)$ vary in the space $(L^2(\Omega))^N \times (H^{-1}(\Omega))^N$. That is to say, any given $(\widehat{u}_0, \widehat{u}_1)$ in $L^2(\Omega) \times H^{-1}(\Omega)$ can be the value of an exactly synchronizable state (u, u') at $t = T$ (cf. Chap. 5).

1.4 Exact Boundary Synchronization by *p*-Groups

How will be the situation if the number of boundary controls is further reduced? Then we can only further lower the standard to be reached.

One possible way is not to require synchronization of all state variables, but to divide them into several groups, e.g., p groups ($p \geqslant 1$), and then demand synchronization of state variables within every group, whereby we get the concept of **exact boundary synchronization by p-groups**.

Precisely speaking, let $p \geqslant 1$ be an integer and let

$$0 = n_0 < n_1 < n_2 < \cdots < n_p = N \tag{1.21}$$

be integers such that $n_r - n_{r-1} \geqslant 2$ for all $1 \leqslant r \leqslant p$. We rearrange the components of U into p groups

$$(u^{(1)}, \cdots, u^{(n_1)}), (u^{(n_1+1)}, \cdots, u^{(n_2)}), \cdots, (u^{(n_{p-1}+1)}, \cdots, u^{(n_p)}). \tag{1.22}$$

System (1.5) is exactly synchronizable by p-groups at the time $T > 0$, if for any given initial data $(\widehat{U}_0, \widehat{U}_1) \in (L^2(\Omega))^N \times (H^{-1}(\Omega))^N$, there exist suitable boundary controls $H \in L^2_{loc}(0, +\infty; (L^2(\Gamma_1))^M)$ with compact support in $[0, T]$, such that the corresponding solution $U = U(t, x)$ satisfies the following final conditions:

$$t \geqslant T: \quad \begin{cases} u^{(1)} \equiv \cdots \equiv u^{(n_1)} := u_1, \\ u^{(n_1+1)} \equiv \cdots \equiv u^{(n_2)} := u_2, \\ \cdots \\ u^{(n_{p-1}+1)} \equiv \cdots \equiv u^{(n_p)} := u_p, \end{cases} \tag{1.23}$$

where $u = (u_1, \cdots, u_p)^T$ is called the **exactly synchronizable state by p-groups,** which is *a priori* unknown.

Let S_r be the following $(n_r - n_{r-1} - 1) \times (n_r - n_{r-1})$ matrix:

$$S_r = \begin{pmatrix} 1 & -1 & 0 & \cdots & 0 \\ 0 & 1 & -1 & \cdots & 0 \\ \vdots & \vdots & \ddots & \ddots & \vdots \\ 0 & 0 & \cdots & 1 & -1 \end{pmatrix} \tag{1.24}$$

and let C_p be the following $(N - p) \times N$ full row-rank **matrix of synchronization by p-groups**:

$$C_p = \begin{pmatrix} S_1 & & & \\ & S_2 & & \\ & & \ddots & \\ & & & S_p \end{pmatrix}. \tag{1.25}$$

The exact boundary synchronization by p-groups (1.23) is equivalent to

$$t \geqslant T: \quad C_p U \equiv 0. \tag{1.26}$$

We can establish all the previous results in a similar way after having overcome some technical difficulties (see Chaps. 6 and 7)). For instance, if the exact boundary synchronization by p-groups can be realized by means of $(N - p)$ boundary controls, we can get the corresponding **condition of C_p-compatibility**: There exists a unique reduced matrix $\overline{A}_p$ of order $(N - p)$, such that

$$C_p A = \overline{A}_p C_p, \tag{1.27}$$

which actually means that the coupling matrix A satisfies the **row-sum condition by blocks.**

Table 1.1 The exact boundary synchronization by p-groups

	Condition of C_p-compatibility	Minimal number of boundary controls
Exact boundary null controllability		N
Exact boundary synchronization	$C_1A = \overline{A}_1C_1$	$N-1$
Exact boundary synchronization by 2-groups	$C_2A = \overline{A}_2C_2$	$N-2$
....		
Exact boundary synchronization by p-groups	$C_pA = \overline{A}_pC_p$	$N-p$

Correspondingly, under the condition of C_p-compatibility, we assume that the domain Ω satisfies the usual multiplier geometrical condition, then there exists a suitably large $T > 0$, such that for any given boundary control matrix D with rank $(C_pD) = N - p$, the exact boundary synchronization by p-groups of system (1.5) can be realized at the time $t = T$. On the contrary if $\text{rank}(C_pD) < N - p$, in particular, if $\text{rank}(D) < N - p$, then no matter how large $T > 0$ is, the exact boundary synchronization by p-groups can never be achieved for all initial data $(\widehat{U}_0, \widehat{U}_1) \in (L^2(\Omega))^N \times (H^{-1}(\Omega))^N$.

In summary, we get the following table (Table 1.1).

What can we do when the number of boundary controls that can be chosen further decreases?

The aforementioned controllability and synchronization should both be established in the exact sense, however, from the practical point of view, some minor error will not affect the general situation, so it also makes sense that these requirements are tenable only in the approximate sense. What we are considering here is the situation that no matter how small the error given beforehand is, we can always find suitable boundary controls so that the controllability or synchronization can be realized within the permitted range of accuracy. Since the error can be chosen progressively smaller, this corresponds actually to a limit process, which thus allows the analytical method to be applied more effectively. The corresponding controllability and synchronization are called approximate boundary null controllability and approximate boundary synchronization.

1.5 Approximate Boundary Null Controllability

We first consider the approximate boundary null controllability given by the following definition:

System (1.5) possesses the **approximate boundary null controllability** at the time $T > 0$ if for any given initial data $(\widehat{U}_0, \widehat{U}_1) \in (L^2(\Omega))^N \times (H^{-1}(\Omega))^N$, there exists a sequence $\{H_n\}$ of boundary controls, $H_n \in L^2_{loc}(0, +\infty; (L^2(\Gamma_1))^M)$ with compact support in $[0, T]$, such that the corresponding sequence $\{U_n\}$ of solutions to problem (1.5)–(1.6) satisfies the following condition:

$$U_n \to 0 \quad \text{as } n \to +\infty \tag{1.28}$$

in the space

$$C^0_{\text{loc}}([T, +\infty); (L^2(\Omega))^N) \cap C^1_{\text{loc}}([T, +\infty); (H^{-1}(\Omega))^N). \tag{1.29}$$

Obviously, the exact boundary null controllability leads to the approximate boundary null controllability. However, since from the previous definition we cannot get the convergence of the sequence $\{H_n\}$ of boundary controls, the approximate boundary null controllability cannot lead to the exact boundary null controllability in general.

Consider the corresponding adjoint problem

$$\begin{cases} \Phi'' - \Delta\Phi + A^{\mathrm{T}}\Phi = 0 & \text{in } (0, +\infty) \times \Omega, \\ \Phi = 0 & \text{on } (0, +\infty) \times \Gamma, \\ t = 0: \quad \Phi = \widehat{\Phi}_0, \ \Phi' = \widehat{\Phi}_1 & \text{in } \Omega, \end{cases} \tag{1.30}$$

where A^T denotes the transpose of A. We give the following definition:

The adjoint problem (1.30) is D-observable on the interval $[0, T]$, if

$$D^{\mathrm{T}}\partial_\nu\Phi \equiv 0 \quad \text{on } [0, T] \times \Gamma_1 \Longrightarrow (\widehat{\Phi}_0, \widehat{\Phi}_1) \equiv 0, \ i.e., \ \Phi \equiv 0, \tag{1.31}$$

where ∂_ν denotes the outward normal derivative on the boundary.

The D-observability is only a weak observability in the sense of uniqueness, which cannot guarantee the exact boundary null controllability. However, we can prove the following result (cf. Chap. 8):

System (1.5) is approximately null controllable at the time $T > 0$ if and only if the adjoint problem (1.30) is D-observable on the interval $[0, T]$.

Obviously, if $M = N$, then by Holmgren's uniqueness theorem (cf. [18, 62, 75]), system (1.5) is always approximately null controllable by means of N boundary controls for $T > 0$ large enough, even without the multiplier geometrical condition.

However, it should be pointed out that the approximate boundary null controllability can be realized even if $M < N$, namely, by means of fewer boundary controls.

To transform the approximate boundary null controllability of the original system equivalently to the D-observability of the adjoint problem cannot solve specifically the problem of judging whether a system possesses the approximate boundary null controllability, nor is it clear to what extent the weakened concept may reduce the number of boundary controls needed. But we can thereby prove the following result (cf. Chap. 8):

Assuming that system (1.5) is approximately null controllable at the time $T > 0$, then we necessarily have that the following enlarged matrix composed of A and D is of full rank:

$$\text{rank}(D, AD, \cdots, A^{N-1}D) = N. \tag{1.32}$$

This is a necessary condition which helps to conveniently eliminate a set of systems that do not meet the requirements. Equation (1.32) is nothing but the so-called Kalman's criterion for guaranteeing the exact controllability for the following system of ODEs (cf. [23, 73]):

$$X' = AX + Du, \tag{1.33}$$

where u stands for the vector of control variables, however here we get it from a different point of view.

Since condition (1.32) is independent of T, it is not a sufficient condition of the D-observability for the adjoint problem (1.30) in general, otherwise the D-observability should be realized almost immediately, but it is not the case because of the finite speed of wave prorogation.

However, under certain assumptions on A, condition (1.32) is also sufficient for guaranteeing the approximate boundary null controllability for $T > 0$ large enough for some special kinds of systems, for instance, some one-space-dimensional systems, some 2×2 systems, the cascade system, and more generally, the nilpotent system under the multiplier geometrical condition, *etc.*

For the exact boundary null controllability of system (1.5), the number $M = \text{rank}(D)$, namely, the number of boundary controls, should be equal to N, the number of state variables. However, the approximate boundary null controllability of system (1.5) could be realized if the number $M = \text{rank}(D)$ is substantially small, even if $M = \text{rank}(D) = 1$. Nevertheless, even if the rank of D might be small, but because of the existence and influence of the coupling matrix A, in order to realize the approximate boundary null controllability, the rank of the enlarged matrix $(D, AD, \cdots, A^{N-1}D)$ should be still equal to N, the number of state variables. From this point of view, we may say that the rank M of D is the number of "direct" boundary controls acting on Γ_1, and $\text{rank}(D, AD, \cdots, A^{N-1}D)$ denotes the number of "total" controls. Differently from the exact boundary null controllability, for the approximate boundary null controllability, we should consider not only the **number of direct boundary controls**, but also the **number of total controls**.

1.6 Approximate Boundary Synchronization

Similar to the approximate boundary null controllability, we give the following definition:

System (1.5) possesses the **approximate boundary synchronization** at the time $T > 0$ if for any given initial data $(\widehat{U}_0, \widehat{U}_1) \in (L^2(\Omega))^N \times (H^{-1}(\Omega))^N$, there exist a sequence $\{H_n\}$ of boundary controls, $H_n \in L^2_{loc}(0, +\infty; (L^2(\Gamma_1))^M)$ with compact support in $[0, T]$, such that the corresponding sequence $\{U_n\} = \{(u_n^{(1)}, \cdots, u_n^{(N)})^T\}$ of solutions to problem (1.5)–(1.6) satisfies

$$u_n^{(k)} - u_n^{(l)} \to 0 \quad \text{as } n \to +\infty \tag{1.34}$$

for all $1 \leqslant k, l \leqslant N$ in the space

$$C^0_{loc}([T, +\infty); L^2(\Omega)) \cap C^1_{loc}([T, +\infty); H^{-1}(\Omega)). \tag{1.35}$$

Obviously, if system (1.5) is exactly synchronizable, then it must be approximately synchronizable; however, the inverse is not true in general.

Moreover, the approximate boundary null controllability obviously leads to the approximate boundary synchronization. We should exclude this trivial situation in advance.

Assume that system (1.5) is approximately synchronizable, but not approximately null controllable. Then, as in the case of exact boundary synchronization, the coupling matrix A should satisfy the same condition of C_1-compatibility (1.12) (cf. Chap. 9).

Then, under the condition of C_1-compatibility, setting $W_1 = C_1 U$ as in (1.17), we get again the reduced system (1.18) and its adjoint problem (the **reduced adjoint problem**):

$$\begin{cases} \Psi_1'' - \Delta\Psi_1 + \overline{A}_1^{\mathrm{T}}\Psi_1 = 0 & \text{in } (0, +\infty) \times \Omega, \\ \Psi_1 = 0 & \text{on } (0, +\infty) \times \Gamma, \\ t = 0: \quad \Psi_1 = \widehat{\Psi}_0, \ \Psi_1' = \widehat{\Psi}_1 \ \text{in } \Omega. \end{cases} \tag{1.36}$$

Similarly to the D-observability, we say that the reduced adjoint problem (1.36) is C_1D-**observable** on the interval $[0, T]$, if

$$(C_1 D)^{\mathrm{T}} \partial_\nu \Psi_1 \equiv 0 \quad \text{on } [0, T] \times \Gamma_1 \Longrightarrow (\widehat{\Psi}_0, \widehat{\Psi}_1) \equiv 0, \ i.e., \ \Psi \equiv 0. \tag{1.37}$$

We can prove that (cf. Chap. 9):

Under the condition of C_1-compatibility, system (1.5) is approximately synchronizable at the time $T > 0$ if and only if the reduced adjoint problem (1.36) is C_1D-observable on the interval $[0, T]$.

Then, it is easy to see that under the condition of C_1-compatibility if $\text{rank}(C_1 D) = N - 1$ (which implies $M \geqslant N - 1$), then, system (1.5) is always approximately synchronizable, even without the multiplier geometrical condition.

We should point out that even if $\text{rank}(C_1 D) < N - 1$, and in particular, if we essentially use fewer than $(N - 1)$ boundary controls, it is still possible to realize the approximate boundary synchronization.

Moreover, as in the case of approximate boundary null controllability, under the condition of C_1-compatibility, we have:

Assume that system (1.5) is approximately synchronizable at the time $T > 0$, then we necessarily have

$$\text{rank}(C_1 D, C_1 AD, \cdots, C_1 A^{N-1} D) = N - 1. \tag{1.38}$$

Condition (1.38) is not sufficient in general for the approximate boundary synchronization, however, it is still sufficient for $T > 0$ large enough for some special systems under certain additional assumptions on A.

On the other hand, we can prove:

Assume that system (1.5) is approximately synchronizable under the action of a boundary control matrix D. No matter whether the condition of C_1-compatibility is satisfied or not, we necessarily have

$$\text{rank}(D, AD, \cdots, A^{N-1} D) \geqslant N - 1, \tag{1.39}$$

namely, at least $(N - 1)$ total controls are needed in order to realize the approximate boundary synchronization of system (1.5).

Furthermore, assume that system (1.5) is approximately synchronizable under the minimal rank condition

$$\text{rank}(D, AD, \cdots, A^{N-1} D) = N - 1. \tag{1.40}$$

Then the matrix A should satisfy the condition of C_1-compatibility and some algebraic properties, and there exists a scalar function u as the **approximately synchronizable state**, such that

$$u_n^{(k)} \to u \quad \text{as } n \to +\infty \tag{1.41}$$

in the space (1.35) for all $1 \leqslant k \leqslant N$. Moreover, the approximately synchronizable state u is independent of the sequence $\{H_n\}$ of applied boundary controls. Thus the original approximate boundary synchronization in the consensus sense reduces to that in the pinning sense.

1.7 Approximate Boundary Synchronization by *p*-Groups

Generally speaking, we can define the **approximate boundary synchronization by *p*-groups** ($p \geqslant 1$). Let us re-group the components of the state variable as in (1.22).

We say that system (1.5) is approximately synchronizable by p-groups at the time $T > 0$ if for any given initial data $(\widehat{U}_0, \widehat{U}_1) \in (L^2(\Omega))^N \times (H^{-1}(\Omega))^N$, there exists a sequence $\{H_n\}$ of boundary controls in $L^2_{loc}(0, +\infty; (L^2(\Gamma_1))^M)$ with compact support in $[0, T]$, such that the sequence $\{U_n\}$ of the corresponding solutions satisfies the following conditions:

$$u_n^{(k)} - u_n^{(l)} \to 0 \quad \text{as } n \to +\infty \tag{1.42}$$

in the space (1.35) for all $n_{r-1} + 1 \leqslant k, l \leqslant n_r$ and $1 \leqslant r \leqslant p$, or equivalently

$$C_p U_n \to 0 \quad \text{as } n \to +\infty \tag{1.43}$$

in the space

$$C^0_{loc}([T, +\infty); (L^2(\Omega))^{N-p}) \cap C^1_{loc}([T, +\infty); (H^{-1}(\Omega))^{N-p}), \tag{1.44}$$

where C_p is given by (1.24)–(1.25).

As in the case of exact boundary synchronization by p-groups, we can get the corresponding condition of C_p-compatibility (cf. Chap. 10): There exists a unique reduced matrix $\overline{A}_p$ of order $(N - p)$, such that (1.27) holds. Under the condition of C_p-compatibility, we necessarily get the following Kalman's criterion:

$$\text{rank}(C_p D, C_p AD, \cdots, C_p A^{N-1} D) = N - p. \tag{1.45}$$

On the other hand, we can prove:

Assume that system (1.5) is approximately synchronizable by p-groups under the action of a boundary control matrix D. Then, no matter whether the condition of C_p-compatibility is satisfied or not, we necessarily have

$$\text{rank}(D, AD, \cdots, A^{N-1} D) \geqslant N - p, \tag{1.46}$$

namely, at least $(N - p)$ total controls are needed in order to realize the approximate boundary synchronization by p-groups of system (1.5).

Furthermore, assume that system (1.5) is approximately synchronizable by p-groups under the minimal rank condition

$$\text{rank}(D, AD, \cdots, A^{N-1} D) = N - p, \tag{1.47}$$

then the matrix A should satisfy the condition of C_p-compatibility and $\text{Ker}(C_p)$ admits a supplement V, which is also invariant for A. Moreover, there exist some linearly independent scalar functions $u_1, \cdots, u_p$ such that the approximately synchronizable state $u = (u_1, \cdots, u_p)^T$ is independent of the sequence $\{H_n\}$ of applied boundary controls. Thus the original approximate synchronization by p-groups in the consensus sense reduces to that in the pinning sense.

In summary, we get the following table (Table 1.2).

Table 1.2 The approximate boundary synchronization by p-groups

	Condition of C_p-compatibility	Minimal number of total controls
Approximate boundary null controllability		N
Approximate boundary synchronization	$C_1A = \overline{A}_1C_1$	$N-1$
Approximate boundary synchronization by 2-groups	$C_2A = \overline{A}_2C_2$	$N-2$
....		
Approximate boundary synchronization by p-groups	$C_pA = \overline{A}_pC_p$	$N-p$

1.8 Induced Approximate Boundary Synchronization

We cannot always realize the approximate boundary synchronization by p-groups under the minimal rank condition (1.47). In fact, let $\mathbb{D}_p$ be the set of all boundary control matrices D which realize the approximate boundary synchronization by p-groups for system (1.5). Define the minimal number N_p of total controls by

$$N_p = \inf_{D \in \mathbb{D}_p} \text{rank}(D, AD, \cdots, A^{N-1}D). \tag{1.48}$$

Under a suitable condition on the matrix A, we can prove

$$N_p = N - q, \tag{1.49}$$

where $q \leqslant p$ is the dimension of the largest subspace W, which is contained in $\text{Ker}(C_p)$ and admits a supplement V, both W and V are invariant for the matrix A. So, generally speaking, $(N-q)$ total controls are necessary for the approximate boundary synchronization by p-groups of system (1.5), while (1.43) provides only the convergence of $(N-p)$ components of state variables. Hence, there is a loss of $(p-q)$ information hidden in the minimal rank condition

$$\text{rank}(D, AD, \cdots, A^{N-1}D) = N - q. \tag{1.50}$$

Let C_q^* be the **induced extension matrix** defined by $\text{Ker}(C_q^*) = W$. $\text{Ker}(C_q^*)$ is the biggest subspace of A, which is contained in $\text{Ker}(C_p)$ and admits a supplement V, both W and V are invariant for A. Moreover, we have the following rank condition:

$$\text{rank}(C_q^*D, C_q^*AD, \cdots, C_q^*A^{N-1}D) = N - q, \tag{1.51}$$

which is necessary for the approximate boundary null controllability of the corresponding reduced system:

$$\begin{cases} W'' - \Delta W + A_q^* W = 0 & \text{in } (0, +\infty) \times \Omega, \\ W = 0 & \text{on } (0, +\infty) \times \Gamma_0, \\ W = C_q^* D H & \text{on } (0, +\infty) \times \Gamma_1 \end{cases} \tag{1.52}$$

with the initial data

$$t = 0: \quad W = C_q^* \widehat{U}_0, \; W' = C_q^* \widehat{U}_1 \quad \text{in } \Omega, \tag{1.53}$$

where $W = C_q^* U$ and A_q^* is given by $C_q^* A = A_q^* C_q^*$.

In some specific situations discussed in Chap. 11, the above rank condition (1.51) is indeed sufficient for the approximate boundary null controllability of the reduced system (1.52). On this occasion, we have

$$C_q^* U_n \to 0 \quad \text{as } n \to +\infty \tag{1.54}$$

in the space

$$C^0_{loc}([T, +\infty); (L^2(\Omega))^{N-q}) \cap C^1_{loc}([T, +\infty); (H^{-1}(\Omega))^{N-q}). \tag{1.55}$$

In this case, we say that system (1.5) is **induced approximately synchronizable**.

Since (1.54) provides the convergence of $(N - q)$ components of state variables, by this way, we recover the $(p - q)$ missed information in (1.43). Moreover, there exist some scalar functions $u_1, \cdots, u_p$ independent of the sequence $\{H_n\}$ of applied boundary controls, such that the sequence $\{U_n\}$ of the corresponding solutions satisfies the following conditions:

$$u_n^{(k)} \to u_r \quad \text{as } n \to +\infty \tag{1.56}$$

for all $n_{r-1} + 1 \leqslant k \leqslant n_r$ and $1 \leqslant r \leqslant p$ in the space (1.35). Thus, the approximate boundary synchronization by p-groups in the consensus sense is in fact in the pinning sense. Nevertheless, unlike the case $N_p = N - p$, these functions $u_1, \cdots, u_p$ are linearly dependent (cf. Chap. 11).

1.9 Organization

The organization of this monograph is as follows.

Some preliminaries on linear algebra which are necessary in the whole book are presented in Chap. 2

We consider the exact boundary synchronization in Part 1 (from Chaps. 3 to 7) and the approximate boundary synchronization in Part 2 (from Chaps. 8 to 11) for a coupled system of wave equations with Dirichlet boundary controls.

In Part 3 (from Chaps. 12 to 15) and Part 4 (from Chaps. 16 to 18), we study the same subjects, respectively, for the coupled system of wave equations with the following Neumann boundary controls:

$$\partial_\nu U = DH \quad \text{on } (0, +\infty) \times \Gamma_1, \tag{1.57}$$

where ∂_ν denotes the outward normal derivative.

Part 5 (from Chaps. 19 to 25) and Part 6 (from Chaps. 26 to 31) are devoted to the corresponding consideration for the coupled system of wave equations with the following coupled Robin boundary controls:

$$\partial_\nu U + BU = DH \quad \text{on } (0, +\infty) \times \Gamma_1. \tag{1.58}$$

where $B = (b_{ij})$ is a **boundary coupling matrix** of order N with constant elements.

Although the consideration in Part 3 and Part 5 (resp. Part 4 and Part 6) seems to be similar to that in Part 1 (resp. Part 2), however, the solution to the corresponding problem with Neumann or coupled Robin boundary conditions has less regularity than that with Dirichlet boundary conditions, moreover, there is a second coupling matrix in the coupled Robin boundary conditions, technically speaking, we will encounter more difficulties and some different and new treatments are necessarily needed. We omitted the details here.

In the last chapter—Chap. 32, we will give some related literature as well as certain prospects for the further study of exact and approximate boundary synchronizations.

Chapter 2
Algebraic Preliminaries

This chapter contains some algebraic preliminaries, which are useful in the whole book.

In this chapter, we denote by A a matrix of order N, by D a full column-rank matrix of order $N \times M$ with $M \leqslant N$, and by C_p a full row-rank matrix of order $(N-p) \times N$ with $0 < p < N$. All these matrices are of constant entries.

2.1 Bi-orthonormality

Definition 2.1 A subspace V of $\mathbb{R}^N$ is a **supplement** of a subspace W of $\mathbb{R}^N$, if

$$W + V = \mathbb{R}^N, \quad W \cap V = \{0\}, \tag{2.1}$$

where the sum of subspaces W and V is defined as

$$W + V = \{w + v : \quad w \in W, \quad v \in V\}. \tag{2.2}$$

In this case, W is also a supplement of V.

As a direct consequence of the dimension equality:

$$\dim(W + V) = \dim(W) + \dim(V) - \dim(W \cap V), \tag{2.3}$$

we have the following

Proposition 2.2 *A subspace V of $\mathbb{R}^N$ is a supplement of a subspace W of $\mathbb{R}^N$ if and only if*

T. Li and B. Rao, *Boundary Synchronization for Hyperbolic Systems*,
Progress in Nonlinear Differential Equations and Their Applications 94,
https://doi.org/10.1007/978-3-030-32849-8_2

$$dim(W) + dim(V) = N, \quad W \cap V = \{0\}. \tag{2.4}$$

In particular, W and $V^\perp$ (or $W^\perp$ and V) have the same dimension.

Definition 2.3 (*cf.* [15, 83]) Two d-dimensional ($0 < d \leqslant N$) subspaces V and W of $\mathbb{R}^N$ are **bi-orthonormal** if there exist a basis $(\epsilon_1, \dots, \epsilon_d)$ of V and a basis $(\eta_1, \dots, \eta_d)$ of W, such that

$$(\epsilon_k, \eta_l) = \delta_{kl}, \quad 1 \leqslant k, l \leqslant d, \tag{2.5}$$

where δ_{kl} is the Kronecker symbol.

Proposition 2.4 *Let V and W be two subspaces in $\mathbb{R}^N$. Then the equality*

$$dim(V^\perp \cap W) = dim(V \cap W^\perp) \tag{2.6}$$

holds if and only if V and W have the same dimension.

Proof First, by the relationship

$$(V^\perp + W)^\perp = V \cap W^\perp, \tag{2.7}$$

we have

$$\dim(V^\perp + W) = N - \dim(V \cap W^\perp). \tag{2.8}$$

Next, noting (2.3), we have

$$\dim(V^\perp + W) = \dim(V^\perp) + \dim(W) - \dim(V^\perp \cap W). \tag{2.9}$$

Then, it follows that

$$\dim(V^\perp) + \dim(W) - \dim(V^\perp \cap W) = N - \dim(V \cap W^\perp), \tag{2.10}$$

namely,

$$\dim(W) - \dim(V) = \dim(V^\perp \cap W) - \dim(V \cap W^\perp). \tag{2.11}$$

The proof is thus complete. □

Proposition 2.5 *Let V and W be two nontrivial subspaces of $\mathbb{R}^N$. Then V and W are bi-orthonormal if and only if*

$$V^\perp \cap W = V \cap W^\perp = \{0\}. \tag{2.12}$$

Proof Assume that (2.12) holds. By Proposition 2.4, V and W have the same dimension d.

In order to show that V and W are bi-orthonormal, we will construct, under condition (2.12), a basis $(\epsilon_1, \dots, \epsilon_d)$ of V and a basis $(\eta_1, \dots, \eta_d)$ of W, such that (2.5) holds. For this purpose, let $(\epsilon_1, \ldots, \epsilon_d)$ be a basis of V. For each i with $1 \leqslant i \leqslant d$ we define a subspace V_i of V by

$$V_i = \text{Span}\{\epsilon_1, \ldots, \epsilon_{i-1}, \epsilon_{i+1}, \ldots, \epsilon_d\}. \tag{2.13}$$

Since it is easy to see from (2.3) that

$$\begin{aligned} \dim(W \cap V_i^{\perp}) &\geqslant \dim(W) + \dim(V_i^{\perp}) - N & (2.14)\\ &= d + (N - d + 1) - N = 1, & (2.15) \end{aligned}$$

there exists a nontrivial vector $\eta_i \in W \cap V_i^{\perp}$, such that $(\epsilon_i, \eta_i) \neq 0$. Otherwise, we have $\eta_i \in V^{\perp}$, then $\eta_i \in W \cap V^{\perp} = \{0\}$ because of (2.12), which leads to a contradiction. Thus, we can choose η_i such that $(\epsilon_i, \eta_i) = 1$. Moreover, since $\eta_i \in V_i^{\perp}$, obviously we have

$$(\epsilon_j, \eta_i) = 0, \quad j = 1, \ldots, i-1, i+1, \ldots, d. \tag{2.16}$$

In this way, we obtain a family of vectors $(\eta_1, \ldots, \eta_d)$ of W, which satisfies the relation (2.5), therefore the vectors $\eta_1, \ldots, \eta_d$ are linearly independent. Since W has the same dimension d as V, the family of these linearly independent vectors $(\eta_1, \ldots, \eta_d)$ is in fact a basis of W.

Conversely, assume that V and W are bi-orthonormal. Let $(\epsilon_1, \ldots, \epsilon_d)$ and $(\eta_1, \ldots, \eta_d)$ be the bases of V and W, respectively, such that (2.5) holds. Since V and W have the same dimension, by Proposition 2.4, it is sufficient to check the first condition of (2.12). For this purpose, let $x \in V^{\perp} \cap W$. Since $x \in W$, there exist some coefficients $\alpha_k (k = 1, \ldots, d)$ such that

$$x = \sum_{k=1}^{d} \alpha_k \eta_k. \tag{2.17}$$

Noting that $x \in V^{\perp}$, the bi-orthonormal relationship (2.5) implies that

$$0 = (x, \epsilon_l) = \sum_{k=1}^{d} \alpha_k (\eta_k, \epsilon_l) = \alpha_l, \quad 1 \leqslant l \leqslant d. \tag{2.18}$$

Thus $x = 0$, therefore $V^{\perp} \cap W = \{0\}$. □

Definition 2.6 A subspace V of $\mathbb{R}^N$ is invariant for A, if

$$AV \subseteq V. \tag{2.19}$$

Proposition 2.7 (*cf.* [25]) *A subspace V of $\mathbb{R}^N$ is invariant for A if and only if its orthogonal complement $V^{\perp}$ is invariant for A^T. In particular, for any given matrix C, since $\{Ker(C)\}^{\perp} = Im(C^T)$, the subspace $Ker(C)$ is invariant for A if and only if the* *subspace $Im(C^T)$ is invariant for A^T. Furthermore, V admits a supplement W such that both V and W are invariant for A if and only if their orthogonal complements $V^\perp$ and $W^\perp$ are invariant for A^T, and $W^\perp$ is a supplement of $V^\perp$.*

Proof Assume that the subspace V of $\mathbb{R}^N$ is invariant for A. By definition, we have

$$(Ax, y)_{\mathbb{R}^N} = (x, A^T y)_{\mathbb{R}^N} = 0, \quad \forall x \in V, \quad \forall y \in V^\perp. \tag{2.20}$$

This proves the first part of the Proposition. Now let V admit a supplement W such that both V and W are invariant for A. By the first part of this Proposition, both $V^\perp$ and $W^\perp$ are invariant for A^T. Moreover, the relation

$$V^\perp \cap W^\perp = \{V \oplus W\}^\perp = \{0\} \tag{2.21}$$

implies that $W^\perp$ is a supplement of $V^\perp$. This finishes the proof of the second part of the Proposition. □

Proposition 2.8 *Let W be a nontrivial invariant subspace of A^T. Then W admits a supplement $V^\perp$, which is also invariant for A^T if and only if V is bi-orthonormal to W and invariant for A.*

Proof Let V be a subspace which is invariant for A and bi-orthonormal to W. By Proposition 2.7, $V^\perp$ is an invariant subspace for A^T. By Definition 2.1, (2.12) holds. By Proposition 2.5, V and W have the same dimension, then it follows that

$$\dim(V^\perp) + \dim(W) = N - \dim(V) + \dim(W) = N. \tag{2.22}$$

By Proposition 2.2, $V^\perp$ is a supplement of W.

Conversely, assume that $V^\perp$ is a supplement of W, and invariant for A^T. By Proposition 2.7, $(V^\perp)^\perp = V$ is invariant for A. On the other hand, since $V^\perp$ is a supplement of W, by Proposition 2.2, we have

$$N = \dim(V^\perp) + \dim(W) = N - \dim(V) + \dim(W), \tag{2.23}$$

so V and W have the same dimension $d > 0$. Moreover, noting that $V^\perp \cap W = \{0\}$, by Proposition 2.1, (2.12) holds. Then by Proposition 2.5, V and W are bi-orthonormal. The proof is then complete. □

By duality, Proposition 2.8 can be formulated as

Proposition 2.9 *Let V be a nontrivial invariant subspace of A. Then V admits a supplement $W^\perp$, which is also invariant for A if and only if W is bi-orthonormal to V and invariant for A^T.*

Remark 2.10 The subspace $W^\perp$ is a supplement of V and is invariant for A, so the matrix A is diagonalizable by blocks according to the decomposition $V \oplus W^\perp$, where $\oplus$ stands for the direct sum of subspaces.

Proposition 2.11 *Let C and K be two matrices of order $M \times N$ and $N \times L$, respectively. Then, the equality*

$$rank(CK) = rank(K) \tag{2.24}$$

holds if and only if

$$Ker(C) \cap Im(K) = \{0\}. \tag{2.25}$$

Proof Define the linear map $\mathcal{C}$ by $\mathcal{C}x = Cx$ for all $x \in \text{Im}(K)$. Then we have

$$\text{Im}(\mathcal{C}) = \text{Im}(CK), \quad \text{Ker}(\mathcal{C}) = \text{Ker}(C) \cap \text{Im}(K). \tag{2.26}$$

From the rank-nullity theorem:

$$\dim \text{Im}(\mathcal{C}) + \dim \text{Ker}(\mathcal{C}) = \text{rank}(K), \tag{2.27}$$

namely,

$$\text{rank }(CK) + \dim\,(\text{Ker}(C) \cap \text{Im}(K)) = \text{rank}(K), \tag{2.28}$$

we achieve the proof of Proposition 2.11. □

2.2 Kalman's Criterion

Proposition 2.12 *Let $d \geqslant 0$ be an integer. We have the following assertions:*

(i) The rank condition

$$rank(D, AD, \ldots, A^{N-1}D) \geqslant N - d \tag{2.29}$$

holds if and only if the dimension of any given invariant subspace of A^T, contained in $Ker(D^T)$, does not exceed d.

(ii) The rank condition

$$rank(D, AD, \ldots, A^{N-1}D) = N - d \tag{2.30}$$

holds if and only if the largest dimension of invariant subspaces of A^T, contained in $Ker(D^T)$, is equal to d.

Proof (i) Let

$$K = (D, AD, \ldots, A^{N-1}D) \tag{2.31}$$

and V be an invariant subspace of A^T, contained in $\text{Ker}(D^T)$. Clearly, we have

$$V \subseteq \text{Ker}(K^T). \tag{2.32}$$

Assume that (2.29) holds, then

$$\dim(V) \leqslant \dim \operatorname{Ker}(K^T) = N - \operatorname{rank}(K^T) \leqslant N - (N-d) = d. \tag{2.33}$$

Conversely, assume that (2.29) does not hold, then

$$\dim \operatorname{Ker}(K^T) = N - \operatorname{rank}(K^T) > N - (N-d) = d. \tag{2.34}$$

Hence, there exist $(d+1)$ linearly independent vectors $w_1, \ldots, w_d, w_{d+1} \in \operatorname{Ker}(K^T)$. In particular, we have

$$\operatorname{Span} \bigcup_{1 \leqslant k \leqslant d+1} \{w_k, A^T w_k, \ldots, (A^T)^{N-1} w_k\} \subseteq \operatorname{Ker}(D^T). \tag{2.35}$$

By Cayley–Hamilton's Theorem, the subspace

$$\operatorname{Span} \bigcup_{1 \leqslant k \leqslant d+1} \{w_k, A^T w_k, \ldots, (A^T)^{N-1} w_k\} \tag{2.36}$$

is invariant for A^T and its dimension is greater than or equal to $(d+1)$.

(ii) Equation (2.30) can be written as

$$\operatorname{rank}(D, AD, \ldots, A^{N-1}D) \geqslant N - d \tag{2.37}$$

and

$$\operatorname{rank}(D, AD, \ldots, A^{N-1}D) \leqslant N - d. \tag{2.38}$$

The rank condition (2.37) means that $\dim(V) \leqslant d$ for any given invariant subspace V of A^T, contained in $\operatorname{Ker}(D^T)$. While, by (i), the rank condition (2.38) implies the existence of a subspace V_0, which is invariant for A^T and is contained in $\operatorname{Ker}(D^T)$, such that $\dim(V_0) \geqslant d$. It proves (ii). The proof is complete. □

As a direct consequence, we get easily the following well-known Hautus test (cf. [17]).

Corollary 2.13 *Kalman's rank condition*

$$rank(D, AD, \ldots, A^{N-1}D) = N \tag{2.39}$$

is equivalent to Hautus test

$$rank(D, A - \lambda I) = N, \quad \forall \lambda \in \mathbb{C}. \tag{2.40}$$

Proof Noting that

$$\operatorname{Ker}(D, A - \lambda I)^T = \operatorname{Ker}(D^T) \cap \operatorname{Ker}(A^T - \lambda I), \tag{2.41}$$

Equation (2.40) is equivalent to

$$\mathrm{Ker}(D^T) \cap \mathrm{Ker}(A^T - \lambda I) = \{0\}, \tag{2.42}$$

which means that none of eigenvectors of A^T is contained in $\mathrm{Ker}(D^T)$. Therefore, $\mathrm{Ker}(D^T)$ does not contain any invariant subspace of A^T, which is just what Proposition 2.12 indicates in the case $d = 0$. □

2.3 Condition of C_p-Compatibility

Definition 2.14 The matrix A satisfies the condition of C_p-compatibility if there exists a unique matrix $\overline{A}_p$ of order $(N - p)$, such that

$$C_p A = \overline{A}_p C_p. \tag{2.43}$$

The matrix $\overline{A}_p$ will be called the reduced matrix of A by C_p.

Proposition 2.15 *The matrix A satisfies the condition of C_p-compatibility if and only if the kernel of C_p is an invariant subspace of A:*

$$A Ker(C_p) \subseteq Ker(C_p). \tag{2.44}$$

Moreover, the reduced matrix $\overline{A}_p$ is given by

$$\overline{A}_p = C_p A C_p^+, \tag{2.45}$$

where

$$C_p^+ = C_p^T (C_p C_p^T)^{-1} \tag{2.46}$$

is the Moore–Penrose inverse of C_p.

Proof Assume that (2.43) holds. Then

$$\mathrm{Ker}(C_p) \subseteq \mathrm{Ker}(\overline{A}_p C_p) = \mathrm{Ker}(C_p A). \tag{2.47}$$

Since $C_p A x = 0$ for any given $x \in \mathrm{Ker}(C_p)$, we get $Ax \in \mathrm{Ker}(C_p)$ for any given $x \in \mathrm{Ker}(C_p)$, namely, (2.44) holds.

Conversely, assume that (2.44) holds true. Then by Proposition 2.7, we have

$$A^T \mathrm{Im}(C_p^T) \subseteq \mathrm{Im}(C_p^T). \tag{2.48}$$

Then, there exists a matrix $\overline{A}_p$ of order $(N - p)$, such that

$$A^T C_p^T = C_p^T \overline{A}_p^T, \tag{2.49}$$

namely, (2.43) holds.

Moreover, let $(e_1, \ldots, e_p)$ be a basis of $\text{Ker}(C_p)$. Noting (2.44), we get

$$(C_p A - \overline{A}_p C_p)(e_1, \ldots, e_p, C_p^T) = (0, \ldots, 0, (C_p A - \overline{A}_p C_p) C_p^T). \tag{2.50}$$

Since the $N \times N$ matrix $(e_1, \ldots, e_p, C_p^T)$ is invertible, (2.43) is equivalent to

$$(C_p A - \overline{A}_p C_p) C_p^T = 0. \tag{2.51}$$

Since the matrix $C_p C_p^T$ is invertible, it follows that

$$\overline{A}_p = C_p A C_p^T (C_p C_p^T)^{-1}, \tag{2.52}$$

which gives just (2.45)–(2.46). The proof is complete. □

Proposition 2.16 *Assume that the matrix A satisfies the condition of C_p-compatibility. Then for given matrix D of order $N \times M$, we have*

$$\begin{aligned} &rank(C_p D, \overline{A}_p C_p D, \ldots, \overline{A}_p^{N-p-1} C_p D) \\ = &rank(C_p D, C_p A D, \ldots, C_p A^{N-1} D). \end{aligned} \tag{2.53}$$

Proof By Cayley–Hamilton's Theorem, we have

$$\begin{aligned} &\text{rank}(C_p D, \overline{A}_p C_p D, \ldots, \overline{A}_p^{N-p-1} C_p D) \\ = &\text{rank}(C_p D, \overline{A}_p C_p D, \ldots, \overline{A}_p^{N-1} C_p D). \end{aligned} \tag{2.54}$$

Then, noting $C_p A^l = \overline{A}_p^l C_p$ for any given integer $l \geqslant 0$, we get

$$\begin{aligned} &(C_p D, \overline{A}_p C_p D, \ldots, \overline{A}_p^{N-1} C_p D) \\ = &(C_p D, C_p A D, \ldots, C_p A^{N-1} D). \end{aligned} \tag{2.55}$$

The proof is complete. □

Definition 2.17 A subspace V is called A**-marked**, if it is invariant for A and there exists a Jordan basis of A in V, which can be extended (by adding new vectors) to a Jordan basis of A in $\mathbb{C}^N$. V is **strongly** A**-marked**, if it is invariant for A and every Jordan basis of A in V can be extended to a Jordan basis of A in $\mathbb{C}^N$.

Obviously, any subspace composed of eigenvectors of A is strongly A-marked. Moreover, if each eigenvalue λ of A is either semi-simple (λ has the same algebraic and geometrical multiplicity), or dim Ker$(A - \lambda I) = 1$, then every invariant subspace of A is strongly A-marked.

The existence of non-marked invariant subspaces is sometimes overlooked in linear algebra. We refer [9] for a complete discussion on this topic.

Definition 2.18 Assume that the matrix A satisfies the condition of C_p-compatibility (2.43). An $(N - q) \times N (0 \leqslant q < p)$ full row-rank matrix C_q^* is called the **induced extension matrix of** C_p, related to the matrix A, if

(a) Ker(C_q^*) is contained in Ker(C_p^T),

(b) Ker(C_q^*) is invariant for A and admits a supplement which is also invariant for A,

(c) Ker(C_q^*) is the largest one satisfying the previous two conditions.

By duality, the above requirements are equivalent to the following ones:

(i) Im(C_q^{*T}) contains Im(C_p^T),

(ii) Im(C_q^{*T}) is invariant for A^T and admits a supplement which is also invariant for A^T,

(iii) Im(C_q^{*T}) is the least one satisfying the previous two conditions.

This leads to the following explicit construction for the induced extension matrix C_q^*.

Let

$$\mathcal{E}_0^{(j)} = 0, \quad A^T \mathcal{E}_i^{(j)} = \lambda_j \mathcal{E}_i^{(j)} + \mathcal{E}_{i-1}^{(j)}, \quad 1 \leqslant i \leqslant m_j, \quad 1 \leqslant j \leqslant r \tag{2.56}$$

denote the Jordan chain of A^T *contained* in Im(C_p^T). Assume that the subspace Im$({C_p}^T)$ is A^T-marked. Then for any j with $1 \leqslant j \leqslant r$, there exist new root vectors $\mathcal{E}_{m_j+1}^{(j)}, \ldots, \mathcal{E}_{m_j^*}^{(j)}$ which extend the Jordan chain in Im$({C_p}^T)$ to a Jordan chain in $\mathbb{C}^N$:

$$\mathcal{E}_0^{(j)} = 0, \quad A^T \mathcal{E}_i^{(j)} = \lambda_j \mathcal{E}_i^{(j)} + \mathcal{E}_{i-1}^{(j)}, \quad 1 \leqslant i \leqslant m_j^*, \quad 1 \leqslant j \leqslant r. \tag{2.57}$$

Then, we define the $(N - q) \times N$ full row-rank matrix C_q^* by

$$C_q^{*T} = (\mathcal{E}_1^{(1)}, \ldots, \mathcal{E}_{m_1^*}^{(1)}; \ldots\ldots; \mathcal{E}_1^{(r)}, \ldots, \mathcal{E}_{m_r^*}^{(r)}), \tag{2.58}$$

where

$$q = N - \sum_{j=1}^{r} m_j^*. \tag{2.59}$$

We justify the above construction by

Proposition 2.19 *Assume that the matrix A satisfies the condition of C_p-compatibility (2.43) and that the subspace Im(C_p^T) is A^T-marked. Then the matrix C_q^* defined by* (2.58) *satisfies the requirements (i)–(iii).*

Proof Clearly, $\mathrm{Im}(C_q^{*T})$ contains $\mathrm{Im}(C_p^T)$. On the other hand, by Jordan's theorem, the space $\mathbb{C}^N$ can be decomposed into a direct sum of all the Jordan subspaces of A^T, therefore, $\mathrm{Im}(C_q^{*T})$, being direct sum of some Jordan subspaces, is invariant and admits a supplement which is invariant for A^T. Thus, we only have to check that $\mathrm{Im}(C_q^{*T})$ is the least one. If $q = p$, then $C_q^* = C_p$ is obviously the least one. Otherwise if $q < p$, there exists at least a root vector $\mathcal{E}_{m_j^*}^{(j)}$, which does not belong to $\mathrm{Im}({C_p}^T)$. Without loss of generality, assume that the root vector $\mathcal{E}_{m_1^*}^{(1)} \notin \mathrm{Im}(C_p^T)$. We remove it from $\mathrm{Im}(C_q^{*T})$, and let

$$\widehat{C}_q^{*T} = (\mathcal{E}_1^{(1)}, \ldots, \mathcal{E}_{m_1^*-1}^{(1)}; \ldots, \mathcal{E}_1^{(r)}, \ldots, \mathcal{E}_{m_r^*}^{(r)}). \tag{2.60}$$

We claim that $\mathrm{Im}(\widehat{C}_q^{*T})$ does not admit any supplement, which is invariant for A^T.

Otherwise, assume that $\mathrm{Im}(\widehat{C}_q^{*T})$ admits a supplement $\widehat{W}$, which is invariant for A^T. Then, there exist $\widehat{x} \in \mathrm{Im}(\widehat{C}_q^{*T})$ and $\widehat{y} \in \widehat{W}$, such that

$$\mathcal{E}_{m_1^*}^{(1)} = \widehat{x} + \widehat{y}. \tag{2.61}$$

Noting that

$$A^T \mathcal{E}_{m_1^*}^{(1)} = \lambda_1 \mathcal{E}_{m_1^*}^{(1)} + \mathcal{E}_{m_1^*-1}^{(1)}, \tag{2.62}$$

we have

$$A^T \widehat{x} + A^T \widehat{y} = \lambda_1 (\widehat{x} + \widehat{y}) + \mathcal{E}_{m_1^*-1}^{(1)}. \tag{2.63}$$

It then follows that

$$(A^T - \lambda_1)\widehat{x} - \mathcal{E}_{m_1^*-1}^{(1)} + (A^T - \lambda_1)\widehat{y} = 0. \tag{2.64}$$

On the other hand, noting that

$$(A^T - \lambda_1)\widehat{x} - \mathcal{E}_{m_1^*-1}^{(1)} \in \mathrm{Im}(\widehat{C}_q^{*T}) \quad \text{and} \quad (A^T - \lambda_1)\widehat{y} \in \widehat{W}, \tag{2.65}$$

it follows from (2.64) that

$$A^T \widehat{x} = \lambda_1 \widehat{x} + \mathcal{E}_{m_1^*-1}^{(1)} \quad \text{and} \quad A^T \widehat{y} = \lambda_1 \widehat{y}, \tag{2.66}$$

therefore,

$$A^T (\widehat{x} - \mathcal{E}_{m_1^*}^{(1)}) = \lambda_1 (\widehat{x} - \mathcal{E}_{m_1^*}^{(1)}). \tag{2.67}$$

Since $(\widehat{x} - \mathcal{E}_{m_1^*}^{(1)}) \in \mathrm{Span}\{\mathcal{E}_1^{(1)}, \ldots, \mathcal{E}_{m_1^*}^{(1)}\}$ and $\mathcal{E}_1^{(1)}$ is the only eigenvector of A^T in $\mathrm{Span}\{\mathcal{E}_1^{(1)}, \ldots, \mathcal{E}_{m_1^*}^{(1)}\}$, there exists $a \in \mathbb{R}$, such that $\widehat{x} - \mathcal{E}_{m_1^*}^{(1)} = a\mathcal{E}_1^{(1)}$, namely, $\widehat{x} - a\mathcal{E}_1^{(1)} = \mathcal{E}_{m_1^*}^{(1)}$. Since $\widehat{x}$ and $\mathcal{E}_1^{(1)}$ belong to $\mathrm{Im}(\widehat{C}_q^{*T})$, so does the vector $\widehat{x} - a\mathcal{E}_1^{(1)}$. But because of (2.60), $\mathcal{E}_{m_1^*}^{(1)} \notin \mathrm{Im}(\widehat{C}_q^{*T})$, we get thus a contradiction. □

Proposition 2.20 *Assume that the matrix A satisfies the condition of C_p-compatibility (2.43) and that $Ker(C_p)$ is A-marked. Then, there exists a subspace $Span\{e_1, \dots, e_q\}$ such that*

(a) $Ker(C_p)$ is contained in $Span\{e_1, \dots, e_q\}$.

(b) A^T admits an invariant subspace $Span\{E_1, \dots, E_q\}$ which is bi-orthonormal to $Span\{e_1, \dots, e_q\}$.

(c) $Span\{e_1, \dots, e_q\}$ is the least one satisfying the previous two conditions.

Proof By Proposition 2.9, it suffices to show that $\text{Ker}(C_p)$ has a least extension $\text{Span}\{e_1, \dots, e_q\}$, which is invariant for A and admits a supplement which is also invariant for A.

Let $\text{Ker}(C_p) = \text{Span}\{e_1, \dots, e_p\}$. Defining $D_p^T = (e_1, \dots, e_p)$, we have

$$\text{Im}(D_p^T) = \text{Ker}(C_p) \quad \text{and} \quad \text{Ker}(D_p) = \text{Im}(C_p^T). \tag{2.68}$$

By Proposition 2.7, the condition of C_p-compatibility (2.43) implies that $A^T \text{Im}(C_p^T) \subseteq \text{Im}(C_p^T)$, namely, $A^T \text{Ker}(D_p) \subseteq \text{Ker}(D_p)$. Therefore, A^T satisfies the condition of D_p-compatibility and $\text{Im}(D_p^T) = \text{Ker}(C_p)$ is A-marked. Then, by Proposition 2.19, D_p has a least extension D_q^* such that $\text{Im}(D_q^{*T})$ is invariant for A and admits a supplement which is also invariant for A. In other words, noting (2.68), $\text{Ker}(C_p)$ has a least extension $\text{Span}\{e_1, \dots, e_q\}$, which admits a supplement such that both two subspaces are invariant for A. The proof is then complete. □

Proposition 2.21 *Assume that the matrix A satisfies the condition of C_p-compatibility (2.43). Let $\{x_l^{(k)}\}_{1 \leqslant k \leqslant d, 1 \leqslant l \leqslant r_k}$ be a system of root vectors of the matrix A, corresponding to the eigenvalues $\lambda_k (1 \leqslant k \leqslant d)$, such that for each $k (1 \leqslant k \leqslant d)$ we have*

$$Ax_l^{(k)} = \lambda_k x_l^{(k)} + x_{l+1}^{(k)}, \quad 1 \leqslant l \leqslant r_k. \tag{2.69}$$

Define the following projected vectors by

$$\overline{x}_l^{(k)} = C_p x_l^{(k)}, \quad 1 \leqslant k \leqslant \overline{d}, \quad 1 \leqslant l \leqslant \overline{r}_k, \tag{2.70}$$

where $\overline{d}(1 \leqslant \overline{d} \leqslant d)$ and $\overline{r}_k(1 \leqslant \overline{r}_k \leqslant r_k)$ are given by (2.71) below, then the system $\{\overline{x}_l^{(k)}\}_{1 \leqslant k \leqslant \overline{d}, 1 \leqslant l \leqslant \overline{r}_k}$ forms a system of root vectors of the reduced matrix $\overline{A}_p$ given by (2.43). In particular, if A is similar to a real symmetric matrix, so is $\overline{A}_p$.

Proof Since $\text{Ker}(C_p)$ is an invariant subspace of A, without loss of generality, we may assume that there exist some integers $\overline{d}(1 \leqslant \overline{d} \leqslant d)$ and $\overline{r}_k(1 \leqslant \overline{r}_k \leqslant r_k)$, such that $\{x_l^{(k)}\}_{1 \leqslant l \leqslant \overline{r}_k, 1 \leqslant k \leqslant \overline{d}}$ forms a root system for the restriction of A to the invariant subspace $\text{Ker}(C_p)$. Then,

$$\text{Ker}(C_p) = \text{Span}\{x_l^{(k)} : 1 \leqslant k \leqslant \overline{d}, \ 1 \leqslant l \leqslant \overline{r}_k\}. \tag{2.71}$$

In particular, we have

$$\sum_{k=1}^{\bar{d}} (r_k - \bar{r}_k) = N - p. \tag{2.72}$$

Noting that $C_p^T (C_p C_p^T)^{-1} C_p$ is a projection from $\mathbb{R}^N$ onto $\operatorname{Im}(C_p^T)$, we have

$$C_p^T (C_p C_p^T)^{-1} C_p x = x, \quad \forall x \in \operatorname{Im}(C_p^T). \tag{2.73}$$

On the other hand, by $\mathbb{R}^N = \operatorname{Im}(C_p^T) \oplus \operatorname{Ker}(C_p)$ we can write

$$x_l^{(k)} = \widehat{x}_l^{(k)} + \widetilde{x}_l^{(k)} \quad \text{with} \quad \widehat{x}_l^{(k)} \in \operatorname{Im}(C_p^T), \quad \widetilde{x}_l^{(k)} \in \operatorname{Ker}(C_p). \tag{2.74}$$

It follows from (2.70) that

$$\overline{x}_l^{(k)} = C_p \widehat{x}_l^{(k)}, \quad 1 \leqslant k \leqslant \overline{d}, \quad 1 \leqslant l \leqslant \overline{r}_k. \tag{2.75}$$

Then, noting (2.45)–(2.46) and (2.73), we have

$$\overline{A}_p \overline{x}_l^{(k)} = C_p A C_p^T (C_p C_p^T)^{-1} C_p \widehat{x}_l^{(k)} = C_p A \widehat{x}_l^{(k)}. \tag{2.76}$$

Since $\operatorname{Ker}(C_p)$ is invariant for A, $A\widetilde{x}_l^{(k)} \in \operatorname{Ker}(C_p)$, then $C_p A \widetilde{x}_l^{(k)} = 0$. It follows that

$$\overline{A}_p \overline{x}_l^{(k)} = C_p A (\widehat{x}_l^{(k)} + \widetilde{x}_l^{(k)}) = C_p A x_l^{(k)}. \tag{2.77}$$

Then, using (2.69), it is easy to see that

$$\overline{A}_p \overline{x}_l^{(k)} = C_p (\lambda_k x_l^{(k)} + x_{l+1}^{(k)}) = \lambda_k \overline{x}_l^{(k)} + \overline{x}_{l+1}^{(k)}. \tag{2.78}$$

Therefore, $\overline{x}_1^{(k)}, \overline{x}_2^{(k)}, \ldots, \overline{x}_{\overline{r}_k}^{(k)}$ is a Jordan chain with length $\overline{r}_k$ of the reduced matrix $\overline{A}_p$, corresponding to the eigenvalue λ_k.

Since dim $\operatorname{Ker}(C_p) = p$, the projected system $\{\overline{x}_l^{(k)}\}_{1 \leqslant k \leqslant \overline{d}, 1 \leqslant l \leqslant \overline{r}_k}$ is of rank $(N - p)$. On the other hand, because of (2.72), the system $\{\overline{x}_l^{(k)}\}_{1 \leqslant l \leqslant \overline{r}_k, 1 \leqslant k \leqslant \overline{d}}$ contains $(N - p)$ vectors, therefore, forms a system of root vectors of the reduced matrix $\overline{A}_p$. The proof is complete. □

Part I
Synchronization for a Coupled System of Wave Equations with Dirichlet Boundary Controls: Exact Boundary Synchronization

We consider the following coupled system of wave equations with Dirichlet boundary controls:

$$\begin{cases} U'' - \Delta U + AU = 0 & in \quad (0, +\infty) \times \Omega, \\ U = 0 & on \quad (0, +\infty) \times \Gamma_0, \\ U = DH & on \quad (0, +\infty) \times \Gamma_1 \end{cases} \tag{I}$$

with the initial condition

$$t = 0: \quad U = \widehat{U}_0, \ U' = \widehat{U}_1 \quad in \quad \Omega, \tag{I0}$$

where $\Omega \subset \mathbb{R}^n$ is a bounded domain with smooth boundary $\Gamma = \Gamma_1 \cup \Gamma_0$ such that $\overline{\Gamma}_1 \cap \overline{\Gamma}_0 = \emptyset$ and $\text{mes}(\Gamma_1) > 0$; "$'$" stands for the time derivative; $\Delta = \sum_{k=1}^{n} \frac{\partial^2}{\partial x_k^2}$ is the Laplacian operator; $U = \left(u^{(1)}, \cdots, u^{(N)}\right)^T$ and $H = \left(h^{(1)}, \cdots, h^{(M)}\right)^T (M \leqslant N)$ denote the state variables and the boundary controls, respectively; the coupling matrix $A = (a_{ij})$ is of order N and D as the boundary control matrix is a full column-rank matrix of order $N \times M$, both with constant elements.

In this part, the exact boundary synchronization and the exact boundary Synchronization by groups for system (I) will be presented and discussed, while, correspondingly, the approximate boundary synchronization and the approximate boundary synchronization by groups for system (I) will be introduced and considered in the next part (Part 2).

Chapter 3
Exact Boundary Controllability and Non-exact Boundary Controllability

Since the exact boundary synchronization on a finite time interval is closely linked with the exact boundary null controllability, we first consider the exact boundary null controllability and the non-exact boundary null controllability for system (I) of wave equations with Dirichlet boundary controls in this chapter.

3.1 Exact Boundary Controllability

Since the exact boundary synchronization on a finite time interval is closely linked with the exact boundary null controllability, we first consider the exact boundary null controllability and the non-exact boundary null controllability in this chapter.

Let $\Omega \subset \mathbb{R}^n$ be a bounded domain with smooth boundary $\Gamma = \Gamma_1 \cup \Gamma_0$ with $\text{mes}(\Gamma_1) > 0$, such that $\overline{\Gamma}_1 \cap \overline{\Gamma}_0 = \emptyset$. Furthermore, we assume that there exists $x_0 \in \mathbb{R}^n$ such that, setting $m = x - x_0$, we have the following multiplier geometrical condition (cf. [7, 26, 61, 62]):

$$(m, \nu) > 0, \quad \forall x \in \Gamma_1; \qquad (m, \nu) \leqslant 0, \quad \forall x \in \Gamma_0, \tag{3.1}$$

where ν is the unit outward normal vector and $(\cdot, \cdot)$ denotes the inner product in $\mathbb{R}^n$.

Let

$$W = (w^{(1)}, \cdots, w^{(M)})^T, \quad \overline{H} = (\overline{h}^{(1)}, \cdots, \overline{h}^{(M)})^T. \tag{3.2}$$

Consider the following coupled system of wave equations:

$$\begin{cases} W'' - \Delta W + \overline{A} W = 0 \text{ in } (0, +\infty) \times \Omega, \\ W = 0 \qquad\qquad\qquad\quad \text{on } (0, +\infty) \times \Gamma_0, \\ W = \overline{H} \qquad\qquad\qquad\quad \text{on } (0, +\infty) \times \Gamma_1 \end{cases} \tag{3.3}$$

T. Li and B. Rao, *Boundary Synchronization for Hyperbolic Systems*, Progress in Nonlinear Differential Equations and Their Applications 94,
https://doi.org/10.1007/978-3-030-32849-8_3

with the initial condition

$$t = 0: \quad W = \widehat{W}_0, \; W' = \widehat{W}_1 \quad \text{in } \Omega, \tag{3.4}$$

where the coupling matrix $\overline{A} = (\overline{a}_{ij})$ of order M has real constant elements.

Definition 3.1 System (3.3) is **exactly null controllable** if there exists a positive constant $T > 0$, such that for any given initial data

$$(\widehat{W}_0, \widehat{W}_1) \in (L^2(\Omega))^M \times (H^{-1}(\Omega))^M, \tag{3.5}$$

one can find a boundary control $\overline{H} \in L^2(0, T; (L^2(\Gamma_1))^M)$, such that problem (3.3)–(3.4) admits a unique weak solution $W = W(t, x)$ with

$$(W, W') \in C^0([0, T]; (L^2(\Omega) \times H^{-1}(\Omega))^M), \tag{3.6}$$

which satisfies the final null condition

$$t = T: \quad W = 0, \quad W' = 0 \quad \text{in } \Omega. \tag{3.7}$$

In other words, one can find a control $\overline{H} \in L^2_{loc}(0, +\infty; (L^2(\Gamma_1))^M)$ with compact support in $[0, T]$, such that the solution $W = W(t, x)$ to the corresponding problem (3.3)–(3.4) satisfies the following condition:

$$t \geqslant T: \quad W \equiv 0 \quad \text{in } \Omega. \tag{3.8}$$

Remark 3.2 More generally, system (3.3) will be called to be exactly controllable if the final null condition (3.7) is replaced by the following inhomogeneous final condition

$$t = T: \quad W = \widetilde{W}_0, \quad W' = \widetilde{W}_1 \quad \text{in } \Omega. \tag{3.9}$$

However, it is well known that for a linear time-invertible system, the exact boundary null controllability with (3.7) is equivalent to the exact boundary controllability with (3.9) (cf. [61, 62, 73]). So, we will not distinguish these two notions of exact controllability later.

For a single wave equation, by means of the Hilbert Uniqueness Method (HUM) proposed by J.-L. Lions in [61], the exact boundary null controllability has been studied in the literature (cf. [26, 62]), but a few were done for general coupled systems of wave equations. If the coupling matrix $\overline{A}$ is symmetric and positive definite, then the exact boundary null controllability of system (3.3) can be transformed to the case of a single wave equation (cf. [61, 62]). However, in order to study the exact boundary synchronization, we have to establish the exact boundary null controllability for the coupled system (3.3) of wave equations with any given coupling matrix $\overline{A}$. In this chapter, we will use a result on the observability of compactly perturbed systems in

[69] to get the observability of the corresponding adjoint problem, and then the exact boundary null controllability follows from the standard HUM.

Now let

$$\Phi = (\phi^{(1)}, \cdots, \phi^{(M)})^T. \tag{3.10}$$

Consider the corresponding adjoint system:

$$\begin{cases} \Phi'' - \Delta\Phi + \overline{A}^T \Phi = 0 \text{ in } (0,T)\times\Omega, \\ \Phi = 0 \qquad\qquad\qquad\quad \text{on } (0,T)\times\Gamma, \end{cases} \tag{3.11}$$

$\overline{A}^T$ being the transpose of $\overline{A}$, with the initial condition

$$t = 0: \quad \Phi = \widehat{\Phi}_0, \ \Phi' = \widehat{\Phi}_1 \quad \text{in } \Omega. \tag{3.12}$$

It is well known (cf. [60, 64, 70]) that the adjoint problem (3.11)–(3.12) is well-posed in the space $\mathcal{V}\times\mathcal{H}$:

$$\mathcal{V} = \left(H_0^1(\Omega)\right)^M, \quad \mathcal{H} = \left(L^2(\Omega)\right)^M. \tag{3.13}$$

Moreover, we will prove the following direct and inverse inequalities.

Theorem 3.3 *Let $T > 0$ be suitably large. Then there exist positive constants c and c' such that for any given initial data$(\widehat{\Phi}_0, \widehat{\Phi}_1) \in \mathcal{V}\times\mathcal{H}$, the solution Φ to problem (3.11)–(3.12) satisfies the following inequalities:*

$$c\int_0^T\int_{\Gamma_1} |\partial_\nu\Phi|^2 d\Gamma dt \leqslant \|\widehat{\Phi}_0\|^2_{\mathcal{V}} + \|\widehat{\Phi}_1\|^2_{\mathcal{H}} \leqslant c'\int_0^T\int_{\Gamma_1} |\partial_\nu\Phi|^2 d\Gamma dt, \tag{3.14}$$

where ∂_ν stands for the outward normal derivative on the boundary.

Before proving Theorem 3.3, we first give a unique continuation result as follows.

Proposition 3.4 *Let B be a matrix of order M, and let $\Phi \in H^2(\Omega)$ be a solution to the following problem:*

$$\begin{cases} \Delta\Phi = B\Phi \ \text{in } \Omega, \\ \Phi = 0 \qquad \text{on } \Gamma. \end{cases} \tag{3.15}$$

Assume furthermore that

$$\partial_\nu\Phi = 0 \quad \text{on } \Gamma_1. \tag{3.16}$$

Then we have $\Phi \equiv 0$.

Proof Let

$$\widetilde{\Phi} = P\Phi \tag{3.17}$$

and

$$\widetilde{B} = PBP^{-1} = \begin{pmatrix} \widetilde{b}_{11} & 0 & \cdots & 0 \\ \widetilde{b}_{21} & \widetilde{b}_{22} & \cdots & 0 \\ & & \cdots & \\ \widetilde{b}_{M1} & \widetilde{b}_{M2} & \cdots & \widetilde{b}_{MM} \end{pmatrix}, \tag{3.18}$$

where $\widetilde{B}$ is a lower triangular matrix of complex entries. Then (3.15)–(3.16) can be reduced to

$$\begin{cases} \Delta\widetilde{\phi}^{(k)} = \sum_{p=1}^{k} \widetilde{b}_{kp}\widetilde{\phi}^{(p)} & \text{in } \Omega, \\ \widetilde{\phi}^{(k)} = 0 & \text{on } \Gamma, \\ \partial_\nu\widetilde{\phi}^{(k)} = 0 & \text{on } \Gamma_1 \end{cases} \tag{3.19}$$

for $k = 1, \cdots, M$.

In particular, for $k = 1$ we have

$$\begin{cases} \Delta\widetilde{\phi}^{(1)} = \widetilde{b}_{11}\widetilde{\phi}^{(1)} & \text{in } \Omega, \\ \widetilde{\phi}^{(1)} = 0 & \text{on } \Gamma, \\ \partial_\nu\widetilde{\phi}^{(1)} = 0 & \text{on } \Gamma_1. \end{cases} \tag{3.20}$$

Thanks to Carleman's unique continuation (cf. [14]), we get

$$\widetilde{\phi}^{(1)} \equiv 0. \tag{3.21}$$

Inserting (3.21) into the second set of (3.19) leads to

$$\begin{cases} \Delta\widetilde{\phi}^{(2)} = \widetilde{b}_{22}\widetilde{\phi}^{(2)} & \text{in } \Omega, \\ \widetilde{\phi}^{(2)} = 0 & \text{on } \Gamma, \\ \partial_\nu\widetilde{\phi}^{(2)} = 0 & \text{on } \Gamma_1 \end{cases} \tag{3.22}$$

and we can repeat the same procedure. Thus, by a simple induction, we get successively that

$$\widetilde{\phi}^{(k)} \equiv 0, \quad k = 1, \cdots, M. \tag{3.23}$$

This yields that

$$\widetilde{\Phi} \equiv 0 \Longrightarrow \Phi \equiv 0. \tag{3.24}$$

The proof is complete. □

Proof of Theorem 3.3. We rewrite system (3.11) as

$$\begin{pmatrix} \Phi \\ \Phi' \end{pmatrix}' = \mathcal{A}\begin{pmatrix} \Phi \\ \Phi' \end{pmatrix} + \mathcal{B}\begin{pmatrix} \Phi \\ \Phi' \end{pmatrix}, \tag{3.25}$$

where

$$\mathcal{A} = \begin{pmatrix} 0 & I_M \\ \Delta & 0 \end{pmatrix}; \quad \mathcal{B} = \begin{pmatrix} 0 & 0 \\ -\overline{A}^T & 0 \end{pmatrix}, \tag{3.26}$$

and I_M is the unit matrix of order M. It is easy to see that $\mathcal{A}$ is a skew-adjoint operator with compact resolvent in $\mathcal{V} \times \mathcal{H}$, and $\mathcal{B}$ is a compact operator in $\mathcal{V} \times \mathcal{H}$. Therefore, they can generate, respectively, C^0 groups $S_{\mathcal{A}}(t)$ and $S_{\mathcal{A}+\mathcal{B}}(t)$ in the energy space $\mathcal{V} \times \mathcal{H}$.

Following a perturbation result in [69], in order to prove the observability inequalities (3.14) for a system of this kind, it is sufficient to check the following assertions:

(i) The direct and inverse inequalities

$$c\int_0^T \int_{\Gamma_1} |\partial_\nu \widetilde{\Phi}|^2 d\Gamma dt \leqslant \|\widehat{\Phi}_0\|_{\mathcal{V}}^2 + \|\widehat{\Phi}_1\|_{\mathcal{H}}^2 \leqslant c' \int_0^T \int_{\Gamma_1} |\partial_\nu \widetilde{\Phi}|^2 d\Gamma dt \tag{3.27}$$

hold for the solution $\widetilde{\Phi} = S_{\mathcal{A}}(t)(\widehat{\Phi}_0, \widehat{\Phi}_1)$ to the decoupled problem (3.11)–(3.12) with $\overline{A} = 0$.

(ii) The system of root vectors of $\mathcal{A} + \mathcal{B}$ forms a Riesz basis of subspaces in $\mathcal{V} \times \mathcal{H}$. That is to say, there exists a family of subspaces $\mathcal{V}_i \times \mathcal{H}_i$ $(i \geqslant 1)$ composed of root vectors of $\mathcal{A} + \mathcal{B}$, such that for any given $x \in \mathcal{V} \times \mathcal{H}$, there exists a unique $x_i \in \mathcal{V}_i \times \mathcal{H}_i$ for each $i \geqslant 1$, such that

$$x = \sum_{i=1}^{+\infty} x_i, \quad c_1\|x\|^2 \leqslant \sum_{i=1}^{+\infty} \|x_i\|^2 \leqslant c_2\|x\|^2, \tag{3.28}$$

where c_1, c_2 are positive constants.

(iii) If $(\Phi, \Psi) \in \mathcal{V} \times \mathcal{H}$ and $\lambda \in \mathbb{C}$, such that

$$(\mathcal{A} + \mathcal{B})(\Phi, \Psi) = \lambda(\Phi, \Psi) \quad \text{and} \quad \partial_\nu \Phi = 0 \text{ on } \Gamma_1, \tag{3.29}$$

then $(\Phi, \Psi) \equiv 0$.

For simplification of notation, we will still denote by $\mathcal{V} \times \mathcal{H}$ the complex Hilbert space corresponding to $\mathcal{V} \times \mathcal{H}$.

Since the assertion (i) is well known under the multiplier geometrical condition (3.1), we only have to verify (ii) and (iii).

Verification of (ii). Let $\mu_i^2 > 0$ be an eigenvalue corresponding to an eigenvector ϕ_i of $-\Delta$ with homogeneous Dirichlet boundary condition:

$$\begin{cases} -\Delta\phi_i = \mu_i^2\phi_i & \text{in } \Omega, \\ \phi_i = 0 & \text{on } \Gamma. \end{cases} \tag{3.30}$$

Let

$$\mathcal{V}_i \times \mathcal{H}_i = \{(\alpha\phi_i, \beta\phi_i) : \quad \alpha, \beta \in \mathbb{C}^M\}. \tag{3.31}$$

Obviously, the subspaces $\mathcal{V}_i \times \mathcal{H}_i (i = 1, 2, \cdots)$ are mutually orthogonal and

$$\mathcal{V} \times \mathcal{H} = \bigoplus_{i \geqslant 1} \mathcal{V}_i \times \mathcal{H}_i, \tag{3.32}$$

where $\oplus$ stands for the direct sum of subspaces. In particular, for any given $x \in \mathcal{V} \times \mathcal{H}$, there exist $x_i \in \mathcal{V}_i \times \mathcal{H}_i$ for $i \geqslant 1$, such that

$$x = \sum_{i=1}^{+\infty} x_i, \quad \|x\|^2 = \sum_{i=1}^{+\infty} \|x_i\|^2. \tag{3.33}$$

On the other hand, for any given $i \geqslant 1$, $\mathcal{V}_i \times \mathcal{H}_i$ is an invariant subspace of $\mathcal{A} + \mathcal{B}$ and of finite dimension. Then, the restriction of $\mathcal{A} + \mathcal{B}$ to the subspace $\mathcal{V}_i \times \mathcal{H}_i$ is a linear bounded operator, therefore its root vectors constitute a basis in the finite-dimensional complex space $\mathcal{V}_i \times \mathcal{H}_i$. This together with (3.32)–(3.33) implies that the system of root vectors of $\mathcal{A} + \mathcal{B}$ form a Riesz basis of subspaces in $\mathcal{V} \times \mathcal{H}$.

Verification of (iii). Let $(\Phi, \Psi) \in \mathcal{V} \times \mathcal{H}$ and $\lambda \in \mathbb{C}$, such that (3.29) holds. We have

$$\Psi = \lambda \Phi \quad \text{and} \quad \Delta \Phi - \overline{A}^T \Phi = \lambda \Psi, \tag{3.34}$$

namely,

$$\begin{cases} \Delta \Phi = (\lambda^2 I + \overline{A}^T)\Phi & \text{in } \Omega, \\ \Phi = 0 & \text{on } \Gamma. \end{cases} \tag{3.35}$$

It follows from the classic elliptic theory that $\Phi \in H^2(\Omega)$. Moreover, we have

$$\partial_\nu \Phi = 0 \quad \text{on } \Gamma_1. \tag{3.36}$$

Thus, applying Proposition 3.4 to (3.35)–(3.36), we get $\Phi \equiv 0$ then $\Psi \equiv 0$. The proof is complete. □

By a standard application of HUM, from Theorem 3.3 we get the following

Theorem 3.5 *The coupled system (3.3) composed of M wave equations is exactly null controllable in the space $(L^2(\Omega))^M \times (H^{-1}(\Omega))^M$ by means of M boundary controls.*

Remark 3.6 In Theorem 3.5, we do not need any assumption on the coupling matrix $\overline{A}$.

Remark 3.7 Similar results on the exact boundary null controllability for a coupled system of 1-D wave equations in the framework of classical solutions can be found in [19, 32].

Let us denote by $\mathcal{U}_{ad}$ the admissible set of all boundary controls $\overline{H}$, which realize the exact boundary null controllability of system (3.3). Since system (3.3) is exactly null controllable, $\mathcal{U}_{ad}$ is not empty. Moreover, we have the following

Theorem 3.8 *Assume that system (3.3) is exactly null controllable in the space $(L^2(\Omega))^M \times (H^{-1}(\Omega))^M$. Then for $\epsilon > 0$ small enough, the values of $\overline{H} \in \mathcal{U}_{ad}$ on $(T - \epsilon, T) \times \Gamma_1$ can be arbitrarily chosen.*

Proof First recall that there exists a positive constant $T_0 > 0$ independent of the initial data, such that for all $T > T_0$ system (3.3) is exactly null controllable at the time T.

Next let $\epsilon > 0$ be such that $T - \epsilon > T_0$ and let

$$\widehat{H}_\epsilon \in L^2(T-\epsilon, T; (L^2(\Gamma_1))^M) \tag{3.37}$$

be arbitrarily given. We solve the backward problem for system (3.3) to get a solution $\widehat{W}_\epsilon$ on the time interval $[T-\epsilon, T]$ with the boundary function $\overline{H} = \widehat{H}_\epsilon$ and the final data

$$t = T: \quad \widehat{W}_\epsilon = \widehat{W}'_\epsilon = 0. \tag{3.38}$$

Since $T - \epsilon > T_0$, system (3.3) is still exactly controllable on the interval $[0, T-\epsilon]$, then we can find a boundary control

$$\widetilde{H}_\epsilon \in L^2(0, T-\epsilon; (L^2(\Gamma_1))^M), \tag{3.39}$$

such that the corresponding solution $\widetilde{W}_\epsilon$ satisfies the initial condition:

$$t = 0: \quad \widetilde{W}_\epsilon = W_0, \ \widetilde{W}'_\epsilon = W_1 \tag{3.40}$$

and the final condition:

$$t = T-\epsilon: \quad \widetilde{W}_\epsilon = \widehat{W}_\epsilon, \ \widetilde{W}'_\epsilon = \widehat{W}'_\epsilon. \tag{3.41}$$

Thus, setting

$$\overline{H} = \begin{cases} \widehat{H}_\epsilon, & t \in (T-\epsilon, T), \\ \widetilde{H}_\epsilon, & t \in (0, T-\epsilon) \end{cases} \tag{3.42}$$

and

$$W = \begin{cases} \widehat{W}_\epsilon, & t \in (T-\epsilon, T), \\ \widetilde{W}_\epsilon, & t \in (0, T-\epsilon), \end{cases} \tag{3.43}$$

we check easily that W is a weak solution to problem (3.3)–(3.4) and the boundary control $\overline{H}$ realizes the exact boundary null controllability. The proof is then complete. □

3.2 Non-exact Boundary Controllability

We have showed the exact boundary null controllability of the coupled system (3.3) composed of M wave equations by means of M boundary controls. In this paragraph, we will show that if the number of the boundary controls is fewer than M, then we cannot realize the exact boundary null controllability for the coupled system (3.3) composed of M wave equations for all initial data $(\widehat{W}_0, \widehat{W}_1)$ in the space $(L^2(\Omega))^M \times (H^{-1}(\Omega))^M$. For this purpose, we further investigate the exact boundary null controllability.

Now we are interested in finding a control $\overline{H}_0 \in \mathcal{U}_{ad}$, which has the least norm among all the others:

$$\|\overline{H}_0\|_{L^2(0,T;(L^2(\Gamma_1))^M)} = \inf_{\overline{H} \in \mathcal{U}_{ad}} \|\overline{H}\|_{L^2(0,T;(L^2(\Gamma_1))^M)}. \tag{3.44}$$

For any given $\widehat{H} \in L^2(0,T;(L^2(\Gamma_1))^M)$, we solve the backward problem

$$\begin{cases} V'' - \Delta V + \overline{A} V = 0 & \text{in } (0,T) \times \Omega, \\ V = 0 & \text{on } (0,T) \times \Gamma_0, \\ V = \widehat{H} & \text{on } (0,T) \times \Gamma_1, \\ t = T: \quad V = V' = 0 \text{ in } \Omega \end{cases} \tag{3.45}$$

and define the linear map

$$\mathcal{R}: \quad \widehat{H} \to (V(0), V'(0)) \tag{3.46}$$

which is continuous from $L^2(0,T;(L^2(\Gamma_1))^M)$ into $(L^2(\Omega))^M \times (H^{-1}(\Omega))^M$ because of the well-posedness. Let $\mathcal{N}$ be the kernel of $\mathcal{R}$. Problem (3.44) becomes

$$\inf_{q \in \mathcal{N}} \|\overline{H} - q\|_{L^2(0,T;(L^2(\Gamma_1))^M)}. \tag{3.47}$$

Let $\mathcal{P}$ denote the orthogonal projection from $L^2(0,T;(L^2(\Gamma_1))^M)$ to $\mathcal{N}$, which is a closed subspace. Then the boundary control $\overline{H}_0 = (\mathcal{I} - \mathcal{P})\overline{H}$ has the least norm. Moreover, we have the following

Theorem 3.9 *Assume that system (3.3) is exactly null controllable in $(L^2(\Omega))^M \times (H^{-1}(\Omega))^M$. Then there exists a positive constant $c > 0$, such that the control $\overline{H}_0$ with the least norm given by (3.44) satisfies the following estimate:*

$$\|\overline{H}_0\|_{L^2(0,T;(L^2(\Gamma_1))^M)} \leqslant c\|(\widehat{W}_0, \widehat{W}_1)\|_{(L^2(\Omega))^M \times (H^{-1}(\Omega))^M} \tag{3.48}$$

for any given $(\widehat{W}_0, \widehat{W}_1) \in (L^2(\Omega))^M \times (H^{-1}(\Omega))^M$.

Proof Consider the linear map $\mathcal{R} \circ (\mathcal{I} - \mathcal{P})$ from the quotient space $L^2(0,T;(L^2(\Gamma_1))^M)/\mathcal{N}$ into $(L^2(\Omega))^M \times (H^{-1}(\Omega))^M$.

First if $\mathcal{R} \circ (\mathcal{I} - \mathcal{P})\overline{H} = 0$, then $(\mathcal{I} - \mathcal{P})\overline{H} \in \mathcal{N}$, thus $\overline{H} = \mathcal{P}\overline{H} \in \mathcal{N}$. Therefore $\mathcal{R} \circ (\mathcal{I} - \mathcal{P})$ is an injective.

On the other hand, the exact boundary null controllability of system (3.3) implies that $\mathcal{R} \circ (\mathcal{I} - \mathcal{P})$ is a surjection, therefore a bijection which is continuous from $L^2(0,T;(L^2(\Gamma_1))^M)/\mathcal{N}$ into $(L^2(\Omega))^M \times (H^{-1}(\Omega))^M$. By Banach's theorem of closed graph, the inverse of $\mathcal{R} \circ (\mathcal{I} - \mathcal{P})$ is also bounded from $(L^2(\Omega))^M \times$

$(H^{-1}(\Omega))^M$ into $L^2(0,T;(L^2(\Gamma_1))^M)/\mathcal{N}$. This yields the inequality (3.48). The proof is complete. □

In the case of partial lack of boundary controls, we have the following negative result.

Theorem 3.10 *Assume that the number of boundary controls is fewer than M. Then, no matter how large $T>0$ is, the coupled system (3.3) composed of M wave equations is not exactly null controllable for all initial data $(\widehat{W}_0, \widehat{W}_1) \in (L^2(\Omega))^M \times (H^{-1}(\Omega))^M$.*

Proof Without loss of generality, we assume that $\overline{h}^{(1)} \equiv 0$. Let $\theta \in \mathcal{D}(\Omega)$. We choose the special initial data as

$$\widehat{W}_0 = (\theta, 0, \cdots, 0)^T, \quad \widehat{W}_1 = 0. \tag{3.49}$$

If system (3.3) is exactly null controllable, following Theorem 3.9, the boundary control H_0 with the least norm satisfies the following estimate:

$$\|\overline{H}_0\|_{L^2(0,T;(L^2(\Gamma_1))^M)} \leqslant c\|\theta\|_{L^2(\Omega)}. \tag{3.50}$$

Because of the well-posedness (cf. [60, 64, 70]), there exists a constant $c'>0$, such that

$$\begin{aligned}&\|W\|_{L^2(0,T;(L^2(\Omega))^M)} \\ \leqslant & c'\big(\|(\widehat{W}_0, \widehat{W}_1)\|_{(L^2(\Omega))^M\times(H^{-1}(\Omega))^M} + \|\overline{H}_0\|_{L^2(0,T;(L^2(\Gamma_1))^M)}\big).\end{aligned} \tag{3.51}$$

Then it follows that

$$\|W\|_{L^2(0,T;(L^2(\Omega))^M)} \leqslant c'(1+c)\|\theta\|_{L^2(\Omega)}. \tag{3.52}$$

Now consider the first set in problem (3.3)–(3.4) with $\overline{h}^{(1)} \equiv 0$ into the following backward problem:

$$\begin{cases} w_{tt}^{(1)} - \Delta w^{(1)} = -\sum_{j=1}^M \overline{a}_{1j} w^{(j)} & \text{in } (0,T)\times\Omega, \\ w^{(1)} = 0 & \text{on } (0,T)\times\Gamma, \\ t=T: \quad w^{(1)} = 0, \ \partial_t w^{(1)} = 0 & \text{in } \Omega \end{cases} \tag{3.53}$$

with the initial data

$$t=0: \quad w^{(1)} = \theta, \ \partial_t w^{(1)} = 0 \quad \text{in } \Omega. \tag{3.54}$$

Once again by the well-posedness for the backward problem (3.53), there exists a constant $c''>0$, such that

$$\|\theta\|_{H_0^1(\Omega)} \leqslant c''\|W\|_{L^2(0,T;(L^2(\Omega))^M)}. \tag{3.55}$$

This, together with (3.52), gives a contradiction:

$$\|\theta\|_{H_0^1(\Omega)} \leqslant c''c'(1+c)\|\theta\|_{L^2(\Omega)}, \quad \forall \theta \in \mathcal{D}(\Omega). \tag{3.56}$$

The proof is complete. □

In order to make the control problem more flexible, we introduce a matrix $\overline{D}$ of order M with constant elements, and consider the following mixed problem for a coupled system of wave equations:

$$\begin{cases} W'' - \Delta W + \overline{A}W = 0 \text{ in } (0,T)\times\Omega, \\ W = 0 \qquad\qquad\qquad \text{on } (0,T)\times\Gamma_0, \\ W = \overline{D}\,\overline{H} \qquad\qquad \text{on } (0,T)\times\Gamma_1 \end{cases} \tag{3.57}$$

with the initial condition

$$t = 0: \quad W = \widehat{W}_0, \; W' = \widehat{W}_1 \quad \text{in } \Omega. \tag{3.58}$$

Combining Theorems 3.5 and 3.10, we have the following result.

Theorem 3.11 *Under the multiplier geometrical condition (3.1), the coupled system (3.57) composed of M wave equations is exactly null controllable for any given initial data* $(\widehat{W}_0, \widehat{W}_1) \in (L^2(\Omega))^M \times (H^{-1}(\Omega))^M$ *if and only if the matrix* $\overline{D}$ *is of rank M.*

Chapter 4
Exact Boundary Synchronization and Non-exact Boundary Synchronization

In the case of partial lack of boundary controls, we consider the exact boundary synchronization and the non-exact boundary synchronization in this chapter for system (I) with Dirichlet boundary controls.

4.1 Definition

Let

$$U = (u^{(1)}, \cdots, u^{(N)})^T \text{and} H = (h^{(1)}, \cdots, h^{(M)})^T \tag{4.1}$$

with $M \leqslant N$. Consider the coupled system (I) with the initial condition (I0).

According to Theorem 3.11, system (I) is exactly null controllable if and only if rank$(D) = N$, namely, $M = N$ and D is invertible. In the case of partial lack of boundary controls, we now give the following

Definition 4.1 System (I) is **exactly synchronizable**, if there exists a positive constant $T > 0$, such that for any given initial data $(\widehat{U}_0, \widehat{U}_1) \in \left(L^2(\Omega)\right)^N \times \left(H^{-1}(\Omega)\right)^N$, there exists a suitable boundary control $H \in L^2_{loc}(0, +\infty; (L^2(\Gamma_1))^M)$ with compact support in $[0, T]$, such that the corresponding solution $U = U(t, x)$ to problem (I) and (I0) satisfies the following final condition

$$t \geqslant T: \quad u^{(1)} \equiv u^{(2)} \equiv \cdots \equiv u^{(N)} := u, \tag{4.2}$$

where $u = u(t, x)$, being unknown a priori, is called the **exactly synchronizable state**.

In the above definition, through the boundary control acts on the time interval $[0, T]$, the synchronization is required not only at the time $t = T$, but also for all

T. Li and B. Rao, *Boundary Synchronization for Hyperbolic Systems*,
Progress in Nonlinear Differential Equations and Their Applications 94,
https://doi.org/10.1007/978-3-030-32849-8_4

$t \geqslant T$, namely, after all boundary controls are eliminated. Thus, this kind of synchronization is not a short-lived one, but exists once and for all, as is needed in applications.

In fact, assuming that (4.2) is realized only at some moment $T > 0$ if we set hereafter $H \equiv 0$ for $t > T$, then, generally speaking, the corresponding solution to problem (I) and (I0) does not satisfy automatically the synchronization condition (4.2) for $t > T$. This is different from the exact boundary null controllability, where the solution vanishes with $H \equiv 0$ for $t \geqslant T$. To illustrate it, let us consider the following system:

$$\begin{cases} u'' - \Delta u = 0 \text{ in } (0, +\infty) \times \Omega, \\ v'' - \Delta v = u \text{ in } (0, +\infty) \times \Omega, \\ u = 0 \qquad\quad \text{on } (0, +\infty) \times \Gamma, \\ v = h \qquad\quad \text{on } (0, +\infty) \times \Gamma. \end{cases} \tag{4.3}$$

Since the first equation is separated from the second one, for any given initial data $(\widehat{u}_0, \widehat{u}_1)$ we can first find a solution u. Once u is determined, we can find (cf. [85]) a boundary control h such that the solution v to the second equation satisfies the final conditions coming from the requirement of synchronization:

$$t = T: \quad v = u, \quad v' = u'. \tag{4.4}$$

If we set $h \equiv 0$ for $t > T$, generally speaking, we cannot get $v \equiv u$ for $t \geqslant T$. So, in order to keep the synchronization for $t \geqslant T$, we have to maintain the boundary control h in action for $t \geqslant T$. However, for the sake of applications, what we want is to get the exact boundary synchronization by some boundary controls with compact support.

Remark 4.2 If system (I) is exactly null controllable, then we have certainly the exact boundary synchronization. This trivial situation should be excluded. Therefore, in Definition 4.1 we should restrict ourselves to the case that the number of boundary controls is fewer than N, namely, $M = \text{rank}(D) < N$, so that system (I) is not exactly null controllable.

4.2 Condition of Compatibility

Theorem 4.3 *Assume that system (I) is exactly synchronizable but not exactly null controllable. Then the coupling matrix $A = (a_{ij})$ should satisfy the following condition of compatibility* (**row-sum condition**)*:*

$$\sum_{p=1}^{N} a_{kp} := a, \quad k = 1, \cdots, N, \tag{4.5}$$

where a is a constant independent of $k = 1, \cdots, N$.

Remark 4.4 By Theorem 3.11, the rank condition

$$\text{rank}(D) < N \tag{4.6}$$

implies the non-exact boundary null controllability.

Proof of Theorem 4.3. By synchronization (4.2), there exist a constant $T > 0$ and a scalar function u such that

$$t \geqslant T: \quad u'' - \Delta u + \Big(\sum_{p=1}^{N} a_{kp}\Big)u = 0 \quad \text{in } \Omega, \quad k = 1, \cdots, N. \tag{4.7}$$

In particular, we have

$$t \geqslant T: \quad \Big(\sum_{p=1}^{N} a_{kp}\Big)u = \Big(\sum_{p=1}^{N} a_{lp}\Big)u \quad \text{in } \Omega, \quad k, l = 1, \cdots, N. \tag{4.8}$$

On the other hand, since system (I) is not exactly null controllable, there exists at least an initial data $(\widehat{U}_0, \widehat{U}_1)$ for which the corresponding synchronizable state u does not identically vanish for $t \geqslant T$, whatever the boundary controls H would be chosen. This yields the condition of compatibility (4.5). The proof is complete. □

Remark 4.5 In the study of synchronization for systems of ordinary differential equations, the row-sum condition (4.5) is imposed on the system according to physical meanings as a reasonable sufficient condition (in the most cases one takes $a = 0$ there). However, as we have seen before, for the synchronization of systems of partial differential equations on a finite time interval, the row-sum condition (4.5) is actually a necessary condition, which makes the theory of synchronization more complete for systems of partial differential equations.

Let the **synchronization matrix** C_1 be defined by

$$C_1 = \begin{pmatrix} 1 & -1 & 0 & \cdots & 0 \\ 0 & 1 & -1 & \cdots & 0 \\ \vdots & \vdots & \ddots & \ddots & \vdots \\ 0 & 0 & \cdots & 1 & -1 \end{pmatrix}_{(N-1)\times N}. \tag{4.9}$$

C_1 is an $(N-1) \times N$ full row-rank matrix, and the **exact boundary synchronization** (4.2) can be equivalently written as

$$t \geqslant T: \quad C_1 U \equiv 0. \tag{4.10}$$

Let

$$e_1 = (1, \cdots, 1)^T. \tag{4.11}$$

Then, the condition of compatibility (4.5) is equivalent to the fact that the vector e_1 is an eigenvector of A, corresponding to the eigenvalue a:

$$Ae_1 = ae_1. \tag{4.12}$$

On the other hand, since $\text{Ker}(C_1)=\text{Span}\{e_1\}$, condition (4.12) means that $\text{Ker}(C_1)$ is a one-dimensional invariant subspace of A:

$$A\text{Ker}(C_1) \subseteq \text{Ker}(C_1). \tag{4.13}$$

Moreover, by Proposition 2.15 (in which we take $p = 1$), the condition (4.13) is also equivalent to the existence of a unique matrix $\overline{A}_1$ of order $(N - 1)$, such that

$$C_1 A = \overline{A}_1 C_1. \tag{4.14}$$

Such a matrix $\overline{A}_1 = (\overline{a}_{ij})$ is called the **reduced matrix of A by C_1**, which can be given explicitly by

$$\overline{A}_1 = C_1 A C_1^+, \tag{4.15}$$

where C_1^+ denotes the Moore–Penrose generalized inverse of C_1:

$$C_1^+ = C_1^T (C_1 C_1^T)^{-1}, \tag{4.16}$$

in which C_1^T is the transpose of C_1. More precisely, the elements of $\overline{A}_1$ are given by

$$\overline{a}_{ij} = \sum_{p=1}^{j} (a_{ip} - a_{i+1,p}) = \sum_{p=j+1}^{N} (a_{i+1,p} - a_{ip}) \quad i, j = 1, \cdots, N-1. \tag{4.17}$$

Remark 4.6 The synchronization matrix C_1 given by (4.9) is not the unique choice. In fact, noting that $\text{Ker}(C_1) = \text{Span}\{e_1\}$ with $e_1 = (1, \cdots, 1)^T$, any given $(N - 1) \times N$ full row-rank matrix C_1 with the zero row sum, or equivalently, such that $\text{Ker}(C_1)$ is an invariant subspace of A (namely, (4.13) holds), can be always taken as a synchronization matrix. However, for fixing the idea, in what follows we always use the synchronization matrix C_1 given by (4.9).

4.3 Exact Boundary Synchronization and Non-exact Boundary Synchronization

We first give a rank condition for the boundary control matrix D, which is necessary for the exact boundary synchronization of system (I) in the space $(L^2(\Omega))^N \times (H^{-1}(\Omega))^N$.

Theorem 4.7 *Assume that system (I) is exactly synchronizable. Then we necessarily have*

$$rank(C_1 D) = N - 1. \tag{4.18}$$

Proof If $\text{Ker}(D^T) \cap \text{Im}(C_1^T) = \{0\}$, then by Proposition 2.11, we have

$$\text{rank}(C_1 D) = \text{rank}(D^T C_1^T) = \text{rank}(C_1^T) = N - 1. \tag{4.19}$$

Otherwise, there exists a unit vector $E \in \text{Ker}(D^T) \cap \text{Im}(C_1^T)$. Assume that system (I) is exactly synchronizable, then for any given $\theta \in \mathcal{D}(\Omega)$, there exists a boundary control H such that the solution U to system (I) with the following initial condition

$$t = 0: \quad U = E\theta, \ U' = 0 \ \text{ in } \Omega \tag{4.20}$$

satisfies the final condition (4.10). Then, applying E to problem (I) and (I0) and setting $\phi = (E, U)$, it is easy to see that

$$\begin{cases} \phi'' - \Delta\phi = -(E, AU) & \text{in } (0, +\infty) \times \Omega, \\ \phi = 0 & \text{on } (0, +\infty) \times \Gamma, \\ t = T: \quad \phi \equiv 0 & \text{in } \Omega. \end{cases} \tag{4.21}$$

Moreover, by a similar procedure as in the proof of Theorem 3.9, the control H can be chosen such that

$$\|H\|_{L^2(0,T;(L^2(\Gamma_1))^N)} \leqslant c\|\theta\|_{L^2(\Omega)}, \tag{4.22}$$

where $c > 0$ is a positive constant. By the well-posedness of problem (I) and (I0), there exists a constant $c' > 0$, such that

$$\|U\|_{L^2(0,T;(L^2(\Omega))^N)} \leqslant c'\|\theta\|_{L^2(\Omega)}. \tag{4.23}$$

Then, by the well-posedness for the backward problem given by (4.21), there exists a constant $c'' > 0$, such that

$$\|\theta\|_{H_0^1(\Omega)} \leqslant c''\|U\|_{L^2(0,T;(L^2(\Omega))^N)} \leqslant c''c'\|\theta\|_{L^2(\Omega)}, \quad \forall \theta \in \mathcal{D}(\Omega). \tag{4.24}$$

We get thus a contradiction which achieves the proof. □

As an immediate consequence of Theorem 4.7, we have the following

Corollary 4.8 *If*

$$rank(C_1 D) < N - 1, \tag{4.25}$$

in particular if

$$rank(D) < N - 1, \tag{4.26}$$

then, no matter how large $T > 0$ is, system (I) cannot be exactly synchronizable at the time T.

Let

$$W_1 = (w^{(1)}, \cdots, w^{(N-1)})^T. \tag{4.27}$$

Then, under the condition of compatibility (4.5) and noting (4.14), the original problem (I) and (I0) for the variable U can be reduced to the following self-closed problem for the variable $W = C_1 U$:

$$\begin{cases} W_1'' - \Delta W_1 + \overline{A}_1 W_1 = 0 & \text{in } (0, +\infty) \times \Omega, \\ W_1 = 0 & \text{on } (0, +\infty) \times \Gamma_0, \\ W_1 = C_1 D H & \text{on } (0, +\infty) \times \Gamma_1, \end{cases} \tag{4.28}$$

associated with the initial data

$$t = 0: \quad W_1 = C_1 \widehat{U}_0, \; W_1' = C_1 \widehat{U}_1 \quad \text{in } \Omega, \tag{4.29}$$

where $\overline{A}_1$ is given by (4.15).

Proposition 4.9 *Under the condition of compatibility (4.5), the exact boundary synchronization of the original system (I) is equivalent to the exact boundary null controllability of the* **reduced system** *(4.28).*

Proof Clearly, the linear map

$$(\widehat{U}_0, \widehat{U}_1) \to (C_1 \widehat{U}_0, C_1 \widehat{U}_1) \tag{4.30}$$

is surjective from the space $(L^2(\Omega))^N \times (H^{-1}(\Omega))^N$ onto the space $(L^2(\Omega))^{N-1} \times (H^{-1}(\Omega))^{N-1}$. Then the exact boundary synchronization of system (I) implies the exact boundary null controllability of the reduced system (4.28). On the other hand, the exact boundary synchronization of system (I) obviously follows from the exact boundary null controllability of system (4.28). The proof is complete. □

By Theorem 3.11 and Proposition 4.9, we immediately get the following

Theorem 4.10 *Under the multiplier geometrical condition (3.1) and the condition of compatibility (4.5), system (I) is exactly synchronizable in the space $(L^2(\Omega))^N \times (H^{-1}(\Omega))^N$, provided that the rank condition (4.18) is satisfied.*

Chapter 5
Exactly Synchronizable States

When system (I) possesses the exact boundary synchronization, the corresponding exactly synchronizable states will be studied in this chapter.

5.1 Attainable Set of Exactly Synchronizable States

In the case that system (I) possesses the exact boundary synchronization at the time $T > 0$, it is easy to see that for $t \geqslant T$, the exactly synchronizable state $u = u(t, x)$ defined by (4.2) satisfies the following wave equation with homogenous Dirichlet boundary condition:

$$\begin{cases} u'' - \Delta u + au = 0 & \text{in } (T, +\infty) \times \Omega, \\ u = 0 & \text{on } (T, +\infty) \times \Gamma, \end{cases} \tag{5.1}$$

where a is given by (4.5). Hence, the evolution of the exactly synchronizable state $u = u(t, x)$ with respect to t is completely determined by the values of (u, u_t) at the time $t = T$:

$$t = T: \quad u = \widehat{u}_0, \ u' = \widehat{u}_1 \quad \text{in } \Omega. \tag{5.2}$$

Theorem 5.1 *Assume that the coupling matrix A satisfies the condition of compatibility (4.5). Then the attainable set of the values (u, u') at the time $t = T$ of the exactly synchronizable state $u = u(t, x)$ is actually the whole space $L^2(\Omega) \times H^{-1}(\Omega)$ as the initial data $(\widehat{U}_0, \widehat{U}_1)$ vary in the space $\left(L^2(\Omega)\right)^N \times \left(H^{-1}(\Omega)\right)^N$.*

Proof For any given $(\widehat{u}_0, \widehat{u}_1) \in L^2(\Omega) \times H^{-1}(\Omega)$, by solving the following backward problem

$$\begin{cases} u'' - \Delta u + au = 0 & \text{in } (0, T) \times \Omega, \\ u = 0 & \text{on } (0, T) \times \Gamma \end{cases} \tag{5.3}$$

T. Li and B. Rao, *Boundary Synchronization for Hyperbolic Systems*, Progress in Nonlinear Differential Equations and Their Applications 94,
https://doi.org/10.1007/978-3-030-32849-8_5

with the final condition

$$t = T: \quad u = \widehat{u}_0, \; u' = \widehat{u}_1 \quad \text{in } \Omega, \tag{5.4}$$

we get the corresponding solution $u = u(t, x)$. Then, under the condition of compatibility (4.5), the function

$$U(t, x) = u(t, x)e_1, \tag{5.5}$$

in which $e_1 = (1, \cdots, 1)^T$, is the solution to problem (I) and (I0) with the null control $H \equiv 0$ and the initial condition

$$t = 0: \quad U = u(0, x)e_1, \; U' = u'(0, x)e_1. \tag{5.6}$$

Therefore, by solving problem (I) and (I0) with the null boundary control and the initial condition (5.6), we can reach any given exactly synchronizable state $(\widehat{u}_0, \widehat{u}_1)$ at the time $t = T$. This fact shows that any given state $(\widehat{u}_0, \widehat{u}_1) \in L^2(\Omega) \times H^{-1}(\Omega)$ can be expected to be a exactly synchronizable state. Consequently, the set of the values $(u(T), u'(T))$ of the exactly synchronizable state $u = (t, x)$ at the time T is actually the whole space $L^2(\Omega) \times H^{-1}(\Omega)$ as the initial data $(\widehat{U}_0, \widehat{U}_1)$ vary in the space $\left(L^2(\Omega)\right)^N \times \left(H^{-1}(\Omega)\right)^N$. The proof is complete. □

5.2 Determination of Exactly Synchronizable States

We now try to determine the exactly synchronizable state of system (I) for each given initial data $(\widehat{U}_0, \widehat{U}_1)$.

As shown in Sect. 4.2, the vector $e_1 = (1, \cdots, 1)^T$ is an eigenvector of A, corresponding to the real eigenvalue a given by (4.5).

Let $\epsilon_1, \cdots, \epsilon_r$ (resp. $\mathcal{E}_1, \cdots, \mathcal{E}_r$) be a Jordan chain of length r of A (resp. of A^T), such that

$$\begin{cases} A\epsilon_l = a\epsilon_l + \epsilon_{l+1}, & 1 \leqslant l \leqslant r, \\ A^T\mathcal{E}_k = a\mathcal{E}_k + \mathcal{E}_{k-1}, & 1 \leqslant k \leqslant r, \\ (\mathcal{E}_k, \epsilon_l) = \delta_{kl}, & 1 \leqslant k, l \leqslant r, \end{cases} \tag{5.7}$$

where

$$\epsilon_r = (1, \cdots, 1)^T, \quad \epsilon_{r+1} = 0, \quad \mathcal{E}_0 = 0. \tag{5.8}$$

Clearly $\epsilon_r = e_1$ is an eigenvector of A, respectively, $\mathcal{E}_1 = E_1$ is an eigenvector of A^T associated with the same eigenvalue a.

Consider the projection P on the subspace $\mathrm{Span}\{\epsilon_1, \cdots, \epsilon_r\}$ as follows:

$$P = \sum_{k=1}^{r} \epsilon_k \otimes \mathcal{E}_k, \tag{5.9}$$

where $\otimes$ stands for the tensor product such that

$$(\epsilon_k \otimes \mathcal{E}_k)U = (\mathcal{E}_k, U)\epsilon_k, \quad \forall U \in \mathbb{R}^N. \tag{5.10}$$

P can be represented by a matrix of order N. We can then decompose

$$\mathbb{R}^N = \mathrm{Im}(P) \oplus \mathrm{Ker}(P), \tag{5.11}$$

where $\oplus$ stands for the direct sum of subspaces. Moreover, we have

$$\mathrm{Im}(P) = \mathrm{Span}\{\epsilon_1, \cdots, \epsilon_r\}, \tag{5.12}$$

$$\mathrm{Ker}(P) = \big(\mathrm{Span}\{\mathcal{E}_1, \cdots, \mathcal{E}_r\}\big)^{\perp} \tag{5.13}$$

and

$$PA = AP. \tag{5.14}$$

Now let $U = U(t, x)$ be the solution to problem (I) and (I0). We define

$$\begin{cases} U_c := (I - P)U, \\ U_s := PU. \end{cases} \tag{5.15}$$

If system (I) is exactly synchronizable, we have

$$t \geqslant T: \quad U = u\epsilon_r, \tag{5.16}$$

where $u = u(t, x)$ is the exactly synchronizable state and $\epsilon_r = (1, \cdots, 1)^T$. Then, noting (5.15)–(5.16), we have

$$t \geqslant T: \quad \begin{cases} U_c = u(I - P)\epsilon_r = 0, \\ U_s = uP\epsilon_r = u\epsilon_r. \end{cases} \tag{5.17}$$

Thus U_c and U_s will be called the **controllable part** and the **synchronizable part** of U, respectively.

Noting (5.14) and applying the projection P on problem (I) and (I0), we get immediately

Proposition 5.2 *The controllable part U_c is the solution to the following problem:*

$$\begin{cases} U_c'' - \Delta U_c + AU_c = 0 & \text{in } (0, +\infty) \times \Omega, \\ U_c = 0 & \text{on } (0, +\infty) \times \Gamma_0, \\ U_c = (I - P)DH & \text{on } (0, +\infty) \times \Gamma_1, \\ t = 0: \quad U_c = (I - P)\widehat{U}_0, \ U_c' = (I - P)\widehat{U}_1 \text{ in } \Omega, \end{cases} \tag{5.18}$$

while the synchronizable part U_s is the solution to the following problem:

$$\begin{cases} U_s'' - \Delta U_s + AU_s = 0 & \text{in } (0, +\infty) \times \Omega, \\ U_s = 0 & \text{on } (0, +\infty) \times \Gamma_0, \\ U_s = PDH & \text{on } (0, +\infty) \times \Gamma_1, \\ t = 0: \quad U_s = P\widehat{U}_0, \ U_s' = P\widehat{U}_1 \text{ in } \Omega. \end{cases} \tag{5.19}$$

Remark 5.3 In fact, the boundary control H realizes the exact boundary null controllability for U_c with the initial data $((I - P)\widehat{U}_0, (I - P)\widehat{U}_1) \in \text{Ker}(P) \times \text{Ker}(P)$ on one hand, and the exact boundary synchronization for U_s with the initial data $(P\widehat{U}_0, P\widehat{U}_1) \in \text{Im}(P) \times \text{Im}(P)$ on the other hand.

We define

$$\mathcal{D}_{N-1} = \{D \in \mathbb{M}^{N \times (N-1)} : \text{rank}(D) = \text{rank}(C_1 D) = N - 1\}. \tag{5.20}$$

Proposition 5.4 *Let a matrix D of order $N \times (N-1)$ be defined by*

$$Im(D) = \big(Span\{\mathcal{E}_r\}\big)^{\perp}. \tag{5.21}$$

We have $D \in \mathcal{D}_{N-1}$.

Proof Clearly, $\text{rank}(D) = N - 1$. On the other hand, since $(\epsilon_r, \mathcal{E}_r) = 1$, we have $\epsilon_r \notin \big(\text{Span}\{\mathcal{E}_r\}\big)^{\perp} = \text{Im}(D)$. Noting that $\text{Ker}(C_1) = \text{Span}\{\epsilon_r\}$, we have $\text{Im}(D) \cap \text{Ker}(C_1) = \{0\}$. Then, by Proposition 2.11, it follows that $\text{rank}(C_1 D) = \text{rank}(D) = N - 1$. The proof is complete. □

Theorem 5.5 *When $r = 1$, we can take a boundary control matrix $D \in \mathcal{D}_{N-1}$, such that the synchronizable part U_s is independent of boundary controls H. Inversely, if the synchronizable part U_s is independent of boundary controls H, then we necessarily have $r = 1$.*

Proof When $r = 1$, by Proposition 5.4, we can take a boundary control matrix $D \in \mathcal{D}_{N-1}$ as in (5.21). Noting (5.13), we have

$$\text{Ker}(P) = \big(\text{Span}\{\mathcal{E}_1\}\big)^{\perp} = \text{Im}(D). \tag{5.22}$$

Then $PD = 0$, therefore (5.19) becomes a problem with homogeneous Dirichlet boundary condition. Consequently, the solution U_s is independent of boundary controls H.

Inversely, let H_1 and H_2 be two boundary controls which realize simultaneously the exact boundary synchronization of system (I). If the corresponding solutions U_s to problem (5.19) are independent of the boundary controls H_1 and H_2, then we have

$$PD(H_1 - H_2) = 0 \quad \text{on } (0, T) \times \Gamma_1. \tag{5.23}$$

By Theorem 3.8 and Proposition 4.9, the values of $C_1 D(H_1 - H_2)$ on $(T - \epsilon, T) \times \Gamma_1$ can be arbitrarily chosen. Since $C_1 D$ is invertible, the values of $(H_1 - H_2)$ on $(T - \epsilon, T) \times \Gamma_1$ can be arbitrarily chosen. This yields that $PD = 0$. It follows that

$$\text{Im}(D) \subseteq \text{Ker}(P). \tag{5.24}$$

Noting (5.13), we have $\dim \text{Ker}(P) = N - r$ and $\dim \text{Im}(D) = N - 1$, then $r = 1$. The proof is complete. □

Corollary 5.6 *Assume that both Ker(C_1) and Im(C_1^T) are invariant subspaces of A. Then there exists a boundary control matrix $D \in \mathcal{D}_{N-1}$, such that system (I) is exactly synchronizable and the synchronizable part U_s is independent of boundary controls H.*

Proof Recall that Ker(C_1)=Span$\{e_1\}$ with $e_1 = (1, \cdots, 1)^T$. Since Im(C_1^T) is an invariant subspace of A, by Proposition 2.7, $\left(\text{Im}(C_1^T)\right)^\perp = \text{Ker}(C_1)$ is an invariant subspace of A^T. Thus, e_1 is also an eigenvector of A^T, corresponding to the same eigenvalue a given by (4.5). Then taking $E_1 = e_1/N$, we have $(E_1, e_1) = 1$ and then $r = 1$. Thus, by Theorem 5.5 we can chose a boundary control matrix $D \in \mathcal{D}_{N-1}$, such that the synchronizable part U_s of system (I) is independent of boundary controls H. The proof is complete. □

Remark 5.7 Under the condition $r = 1$, there exists an eigenvector E_1 of A^T, such that $(E_1, e_1) = 1$. Clearly, if A is symmetric or A^T satisfies also the condition of compatibility (4.5), then e_1 is also an eigenvector of A^T. Consequently, we can take $E_1 = e_1/N$ such that $(E_1, e_1) = 1$. However, this condition is not always satisfied for any given matrix A. For example, let

$$A = \begin{pmatrix} 2 & -1 \\ 1 & 0 \end{pmatrix}. \tag{5.25}$$

We have

$$a = 1, \quad e_1 = \begin{pmatrix} 1 \\ 1 \end{pmatrix}, \quad E_1 = \begin{pmatrix} 1 \\ -1 \end{pmatrix}, \tag{5.26}$$

then

$$(E_1, e_1) = 0. \tag{5.27}$$

In general, if (5.27) holds, then

$$E_1 \in \big(\text{Span}\{e_1\}\big)^{\perp} = \big(\text{Ker }(C_1)\big)^{\perp} = \text{Im}(C_1^T). \tag{5.28}$$

This means that (E_1, U) is just a linear combination of the components of the vector $C_1 U$, therefore it does not provide any new information for the synchronizable part U_s.

Remark 5.8 Let $(\widehat{c}_1, \cdots, \widehat{c}_{N-1})$ be a basis of $\big(Span\{E_1\}\big)^{\perp}$. Then

$$A(e_1, \widehat{c}_1, \cdots, \widehat{c}_{N-1}) = (e_1, \widehat{c}_1, \cdots, \widehat{c}_{N-1}) \begin{pmatrix} a & 0 \\ 0 & A_{22} \end{pmatrix}, \tag{5.29}$$

where A_{22} is a matrix of order $(N-1)$. Therefore, A is diagonalizable by blocks under the basis $(e_1, \widehat{c}_1, \cdots, \widehat{c}_{N-1})$.

We next discuss the general case $r \geqslant 1$. Let us denote

$$\phi_k = (\mathcal{E}_k, U), \quad 1 \leqslant k \leqslant r \tag{5.30}$$

and write

$$U_s = \sum_{k=1}^{r} (\mathcal{E}_k, U)\epsilon_k = \sum_{k=1}^{r} \phi_k \epsilon_k. \tag{5.31}$$

Then, $(\phi_1, \cdots, \phi_r)$ are the coordinates of U_s on the bi-orthonormal basis $\{\epsilon_1, \cdots, e_r\}$ and $\{\mathcal{E}_1, \cdots, \mathcal{E}_r\}$.

Theorem 5.9 *Let $\epsilon_1, \cdots, \epsilon_r$ (resp. $\mathcal{E}_1, \cdots, \mathcal{E}_r$) be a Jordan chain of A (resp. A^T) corresponding to the eigenvalue a and $\epsilon_r = (1, \cdots, 1)^T$. Then the synchronizable part $U_s = (\phi_1, \cdots, \phi_r)$ can be determined by the solution of the following problem $(1 \leqslant k \leqslant r)$:*

$$\begin{cases} \phi_k'' - \Delta\phi_k + a\phi_k + \phi_{k-1} = 0 & \text{in } (0, +\infty) \times \Omega, \\ \phi_k = 0 & \text{on } (0, +\infty) \times \Gamma_0, \\ \phi_k = h_k & \text{on } (0, +\infty) \times \Gamma_1, \\ t = 0: \quad \phi_k = (\mathcal{E}_k, \widehat{U}_0), \ \phi_k' = (\mathcal{E}_k, \widehat{U}_1) \text{ in } \Omega, \end{cases} \tag{5.32}$$

where

$$\phi_0 = 0 \quad \text{and} \quad h_k = (\mathcal{E}_k, DH). \tag{5.33}$$

Moreover, the exactly synchronizable state is given by $u = \phi_r$ for $t \geqslant T$.

Proof First, for $1 \leqslant k \leqslant r$, we have

$$(\mathcal{E}_k, U) = (\mathcal{E}_k, U_s) = \phi_k, \tag{5.34}$$

$$(\mathcal{E}_k, PDH) = \sum_{l=1}^{r} (\mathcal{E}_l, DH)(\mathcal{E}_k, \epsilon_l) = (\mathcal{E}_k, DH) \tag{5.35}$$

and

$$(\mathcal{E}_k, PU_0) = (\mathcal{E}_k, U_0), \quad (\mathcal{E}_k, PU_1) = (\mathcal{E}_k, U_1). \tag{5.36}$$

Taking the inner product of (5.19) with $\mathcal{E}_k$, we get (5.32)–(5.33).

On the other hand, noting (5.16), we have

$$t \geqslant T: \quad \phi_k = (\mathcal{E}_k, U) = (\mathcal{E}_k, u\epsilon_r) = u\delta_{kr} \tag{5.37}$$

for $1 \leqslant k \leqslant r$. Thus, the exactly synchronizable state u is given by

$$t \geqslant T: \quad u = u(t,x) = \phi_r(t,x). \tag{5.38}$$

The proof is complete. □

In the special case $r = 1$, by Theorems 5.5 and 5.9, we have

Corollary 5.10 *When $r = 1$, we can take $D \in \mathcal{D}_{N-1}$, such that $D^T \mathcal{E}_1 = 0$. Then the exactly synchronizable state u is determined by $u = \phi$ for $t \geqslant T$, where ϕ is the solution to the following problem with homogeneous Dirichlet boundary condition:*

$$\begin{cases} \phi'' - \Delta\phi + a\phi = 0 & \text{in } (0,+\infty) \times \Omega, \\ \phi = 0 & \text{on } (0,+\infty) \times \Gamma, \\ t = 0: \quad \phi = (\mathcal{E}_1, \widehat{U}_0), \ \phi' = (\mathcal{E}_1, \widehat{U}_1) \text{ in } \Omega. \end{cases} \tag{5.39}$$

Inversely if the synchronizable part $U_s = (\phi_1, \cdots, \phi_r)$ is independent of boundary controls H, then we have necessarily

$$r = 1 \quad \text{and} \quad D^T \mathcal{E}_1 = 0. \tag{5.40}$$

Consequently, the exactly synchronizable state u is given by $u = \phi$ for $t \geqslant T$, where ϕ is the solution to problem (5.39). In particular, if

$$(\mathcal{E}_1, \widehat{U}_0) = (\mathcal{E}_1, \widehat{U}_1) = 0, \tag{5.41}$$

then system (I) is exactly null controllable for such initial data $(\widehat{U}_0, \widehat{U}_1)$.

5.3 Approximation of Exactly Synchronizable States

The relation (5.37) shows that only the last component ϕ_r is synchronized, while the others are steered to zero. However, in order to get ϕ_r, we have to solve the whole problem (5.32)–(5.33) for $(\phi_1, \cdots, \phi_r)$. Therefore, except in the case $r = 1$, the exactly synchronizable state u depends on boundary controls which realize the exact boundary synchronization, and then, generically speaking, one cannot uniquely determine the exactly synchronizable state u. However, we have the following result.

Theorem 5.11 *Assume that system (I) is exactly synchronizable by means of a boundary control matrix $D \in \mathcal{D}_{N-1}$. Let ϕ be the solution to the following homogeneous problem:*

$$\begin{cases} \phi'' - \Delta\phi + a\phi = 0 & \textit{in } (0, +\infty) \times \Omega, \\ \phi = 0 & \textit{on } (0, +\infty) \times \Gamma, \\ t = 0: \quad \phi = (\mathcal{E}_r, \widehat{U}_0), \ \phi' = (\mathcal{E}_r, \widehat{U}_1) & \textit{in } \Omega. \end{cases} \tag{5.42}$$

Assume furthermore that

$$D^T \mathcal{E}_r = 0. \tag{5.43}$$

Then there exists a positive constant $c_T > 0$, depending on T, such that the exactly synchronizable state u satisfies the following estimate:

$$\begin{aligned} &\|(u, u'(T) - (\phi, \phi')(T)\|_{H_0^1(\Omega)\times L^2(\Omega)} \\ &\leqslant c_T \|C_1(\widehat{U}_0, \widehat{U}_1)\|_{(L^2(\Omega))^{N-1}\times(H^{-1}(\Omega))^{N-1}}. \end{aligned} \tag{5.44}$$

Proof By Proposition 5.4, we can take a boundary control matrix $D \in \mathcal{D}_{N-1}$, such that (5.43) is satisfied. Then, considering the rth equation in (5.32), we get the following problem with homogeneous Dirichlet boundary condition:

$$\begin{cases} \phi_r'' - \Delta\phi_r + a\phi_r = -\phi_{r-1} & \text{in } (0, +\infty) \times \Omega, \\ \phi_r = 0 & \text{on } (0, +\infty) \times \Gamma, \\ t = 0: \quad \phi_r = (\mathcal{E}_r, \widehat{U}_0), \ \phi_r' = (\mathcal{E}_r, \widehat{U}_1) & \text{in } \Omega. \end{cases} \tag{5.45}$$

From (5.37) we have

$$t \geqslant T: \quad \phi_r \equiv u, \quad \phi_{r-1} \equiv 0. \tag{5.46}$$

Noting that problems (5.42) and (5.45) have the same initial data and the same homogeneous Dirichlet boundary condition, by well-posedness, there exists a positive constant $c_1 > 0$, such that

$$\|(u,u')(T)-(\phi,\phi')(T)\|^2_{H^1_0(\Omega)\times L^2(\Omega)} \leqslant c_1\int_0^T \|\psi_{r-1}(s)\|^2_{L^2(\Omega)}ds. \tag{5.47}$$

Noting that the condition $(\mathcal{E}_{r-1},\epsilon_r)=0$ implies that

$$\mathcal{E}_{r-1}\in \big(\text{Span}\{\epsilon_r\}\big)^\perp = \big(\text{Ker}(C_1)\big)^\perp = \text{Im}(C_1^T), \tag{5.48}$$

$\mathcal{E}_{r-1}$ is a linear combination of the columns of C_1^T. Therefore, there exists a positive constant $c_2>0$, such that

$$\|\phi_{r-1}(s)\|^2_{L^2(\Omega)} = \|(\mathcal{E}_{r-1},U(s))\|^2_{L^2(\Omega)} \leqslant c_2\|C_1U(s)\|^2_{(L^2(\Omega))^{N-1}}. \tag{5.49}$$

Recall that $W=C_1U$, due to the exact boundary null controllability of the reduced system (4.28), there exists a positive constant $c_T>0$, such that

$$\int_0^T \|C_1U(s)\|^2_{(L^2(\Omega))^{N-1}}ds \leqslant c_T\|C_1(\widehat{U}_0,\widehat{U}_1)\|^2_{(L^2(\Omega))^{N-1}\times(H^{-1}(\Omega))^{N-1}}. \tag{5.50}$$

Finally, inserting (5.49)–(5.50) into (5.47), we get (5.44). The proof is complete. □

Remark 5.12 When $r>1$, since $(\mathcal{E}_1,\epsilon_r)=0$, the exactly synchronizable state u of system (I) cannot be determined independently of applied boundary controls H. Nevertheless, by Theorem 5.11, if $C_1(\widehat{U}_0,\widehat{U}_1)$ is suitably small, then u is closed to the solution to problem (5.42), the initial data of which is given by the weighted average of the original initial data $(\widehat{U}_0,\widehat{U}_1)$ with the weight $\mathcal{E}_r$ (a root vector of A^T).

Chapter 6
Exact Boundary Synchronization by Groups

The exact boundary synchronization by groups will be considered in this chapter for system (I) with further lack of Dirichlet boundary controls.

6.1 Definition

Let

$$U = \left(u^{(1)}, \cdots, u^{(N)}\right)^T \quad \text{and} \quad H = \left(h^{(1)}, \cdots, h^{(M)}\right)^T \tag{6.1}$$

with $M \leqslant N$. Consider the coupled system (I) of wave equations with Dirichlet boundary controls with the initial condition (I0).

It was shown in Chaps. 3 and 4 that system (I) is neither exactly null controllable nor exactly synchronizable with fewer boundary controls ($M < N$ or $M < N - 1$, respectively). In order to consider the situation that the number of boundary controls is further reduced, we will investigate the exact boundary synchronization by groups for system (I).

Let $p \geqslant 1$ be an integer and let

$$0 = n_0 < n_1 < n_2 < \cdots < n_p = N \tag{6.2}$$

be integers such that $n_r - n_{r-1} \geqslant 2$ for all $1 \leqslant r \leqslant p$.

We rearrange the components of U into p groups

$$(u^{(1)}, \cdots, u^{(n_1)}), (u^{(n_1+1)}, \cdots, u^{(n_2)}), \cdots, (u^{(n_{p-1}+1)}, \cdots, u^{(n_p)}). \tag{6.3}$$

Definition 6.1 System (I) is **exactly synchronizable by p-groups** at the time $T > 0$ if, for any given initial data $(\widehat{U}_0, \widehat{U}_1) \in (L^2(\Omega))^N \times (H^{-1}(\Omega))^N$, there exist a boundary control $H \in L^2_{loc}(0, +\infty; (L^2(\Gamma_1))^M)$ with compact support in $[0, T]$, such that

T. Li and B. Rao, *Boundary Synchronization for Hyperbolic Systems*, Progress in Nonlinear Differential Equations and Their Applications 94,
https://doi.org/10.1007/978-3-030-32849-8_6

the corresponding solution $U = U(t,x)$ to the mixed initial-boundary value problem (I) and (I0) satisfies the following final conditions:

$$t \geqslant T: \quad \begin{cases} u^{(1)} \equiv \cdots \equiv u^{(n_1)} := u_1, \\ u^{(n_1+1)} \equiv \cdots \equiv u^{(n_2)} := u_2, \\ \cdots \\ u^{(n_{p-1}+1)} \equiv \cdots \equiv u^{(n_p)} := u_p, \end{cases} \tag{6.4}$$

where $u = (u_1, \cdots, u_p)^T$ is called the **exactly synchronizable state by p-groups**, which is a priori unknown.

Remark 6.2 The functions $u_1, \cdots, u_p$ depend on the initial data and on applied boundary controls. In the whole monograph, when we claim that $u_1, \cdots, u_p$ are linearly independent, it means that there exists at least an initial data and an applied boundary control such that the corresponding $u_1, \cdots, u_p$ are linearly independent; while, when we claim that $u_1, \cdots, u_p$ are linearly dependent, it means that they are linearly dependent for any given initial data and for any given applied boundary control.

Let S_r (*cf.* Remark 4.6) be the following $(n_r - n_{r-1} - 1) \times (n_r - n_{r-1})$ matrix:

$$S_r = \begin{pmatrix} 1 & -1 & 0 & \cdots & 0 \\ 0 & 1 & -1 & \cdots & 0 \\ \vdots & \vdots & \ddots & \ddots & \vdots \\ 0 & 0 & \cdots & 1 & -1 \end{pmatrix} \tag{6.5}$$

and let C_p be the following $(N-p) \times N$ full row-rank **matrix of synchronization by p-groups**:

$$C_p = \begin{pmatrix} S_1 & & & \\ & S_2 & & \\ & & \ddots & \\ & & & S_p \end{pmatrix}. \tag{6.6}$$

The **exact boundary synchronization by p-groups** (6.4) is equivalent to

$$t \geqslant T: \quad C_p U \equiv 0. \tag{6.7}$$

For $r = 1, \cdots, p$, setting

$$(e_r)_i = \begin{cases} 1, & n_{r-1} + 1 \leqslant i \leqslant n_r, \\ 0, & \text{otherwise}, \end{cases} \tag{6.8}$$

it is clear that

$$\text{Ker}(C_p) = \text{Span}\{e_1, e_2, \cdots, e_p\} \tag{6.9}$$

and the exact boundary synchronization by p-groups (6.4) becomes

$$t \geqslant T: \quad U = \sum_{r=1}^{p} u_r e_r. \tag{6.10}$$

We first observe that the exact boundary synchronization by p-groups is achieved not only at the time T by means of the action of boundary controls, but is also maintained forever. Since the control is removed from the time T, the coupling matrix A should satisfy some additional conditions in order to maintain the exact boundary synchronization by p-groups. As in Chap. 4, we should derive the corresponding condition of compatibility for the exact boundary synchronization by p-groups of system (I).

6.2 A Basic Lemma

In what follows, we will prove the necessity of the condition of compatibility by a repeating reduction procedure. For clarity, this section is entirely devoted to this basic result.

Lemma 6.3 *Suppose that C_p is a full row-rank matrix of order $(N-p) \times N$, such that*

$$t \geqslant T: \quad C_p U \equiv 0 \tag{6.11}$$

for any given solution U to the system of equations

$$U'' - \Delta U + AU = 0 \quad in \quad (0, +\infty) \times \Omega, \tag{6.12}$$

we have either

$$AKer(C_p) \subseteq Ker(C_p), \tag{6.13}$$

or C_p has a full row-rank extension $\widehat{C}_{p-1}$ of order $(N-p+1) \times N$, such that

$$t \geqslant T: \quad \widehat{C}_{p-1} U \equiv 0. \tag{6.14}$$

Proof et $e_r \in \mathbb{R}^N$ $(r = 1, \cdots, p)$ such that

$$\mathrm{Ker}(C_p) = \mathrm{Span}\{e_1, \cdots, e_p\}. \tag{6.15}$$

Noting that (6.11) implies (6.10) and applying the matrix C_p to system (6.12), we get

$$t \geqslant T: \quad \sum_{r=1}^{p} u_r C_p A e_r = 0. \tag{6.16}$$

If $C_pAe_r = 0$ $(r = 1, \cdots, p)$, then we get the inclusion (6.13). Otherwise, noting that $C_pAe_r (r = 1, \cdots, p)$ are constant vectors, there exist some real constant coefficients α_r $(r = 1, \cdots, p-1)$ such that

$$u_p = \sum_{r=1}^{p-1} \alpha_r u_r. \tag{6.17}$$

Then (6.10) becomes

$$t \geqslant T: \quad U = \sum_{r=1}^{p-1} u_r(e_r + \alpha_r e_p). \tag{6.18}$$

Setting

$$\widehat{e_r} = e_r + \alpha_r e_p, \quad r = 1, \cdots, p-1, \tag{6.19}$$

we get

$$t \geqslant T: \quad U = \sum_{r=1}^{p-1} u_r \widehat{e_r}. \tag{6.20}$$

We next construct an enlarged matrix $\widehat{C}_{p-1}$ such that (6.14) holds. For this purpose, let $\widehat{c}_{p+1}$ be a row vector defined by

$$\widehat{c}_{p+1}^T = \frac{e_p}{\|e_p\|^2} - \sum_{l=1}^{p-1} \frac{\alpha_l e_l}{\|e_l\|^2}. \tag{6.21}$$

Noting (6.19) and the orthogonality of the set $\{e_1, \cdots, e_p\}$, it is easy to check that

$$\begin{aligned} \widehat{c}_{p+1}\widehat{e_r} = & \frac{(e_p, e_r)}{\|e_p\|^2} - \sum_{l=1}^{p-1} \frac{\alpha_l (e_l, e_r)}{\|e_l\|^2} \\ & + \alpha_r \Big(\frac{(e_p, e_p)}{\|e_p\|^2} - \sum_{l=1}^{p-1} \frac{\alpha_l (e_l, e_p)}{\|e_l\|^2} \Big) = -\alpha_r + \alpha_r = 0. \end{aligned} \tag{6.22}$$

for all r with $1 \leqslant r \leqslant p-1$. Then we define the enlarged matrix $\widehat{C}_{p-1}$ by

$$\widehat{C}_{p-1} = \begin{pmatrix} C_p \\ \widehat{c}_{p+1} \end{pmatrix}. \tag{6.23}$$

Noting that $\widehat{e_r} \in \mathrm{Ker}(C_p)$ for $r = 1, \cdots, p-1$ and using (6.22), we have

$$\widehat{C}_{p-1}\widehat{e_r} = \begin{pmatrix} C_p \\ \widehat{c}_{p+1} \end{pmatrix} \widehat{e_r} = \begin{pmatrix} C_p\widehat{e_r} \\ \widehat{c}_{p+1}\widehat{e_r} \end{pmatrix} = \begin{pmatrix} 0 \\ 0 \end{pmatrix}, \quad r = 1, \cdots, p-1, \tag{6.24}$$

then it follows immediately from (6.20) that

$$t \geqslant T: \quad \widehat{C}_{p-1}U \equiv 0. \tag{6.25}$$

Finally, noting that

$$\widehat{c}_{p+1}^T \in \mathrm{Ker}(C_p) = \{\mathrm{Im}(C_p^T)\}^{\perp}, \tag{6.26}$$

we have $\widehat{c}_{p+1}^T \notin \mathrm{Im}(C_p^T)$, then $\mathrm{rank}(\widehat{C}_{p-1}) = N - p + 1$, namely, $\widehat{C}_{p-1}$ is a full row-rank matrix of order $(N - p + 1) \times N$. The proof is complete. □

6.3 Condition of C_p-Compatibility

We will show that the exact boundary synchronization by p-groups of system (I) requires at least $(N - p)$ boundary controls. In particular, if $\mathrm{rank}(D) = N - p$, then we get the following **condition of C_p-compatibility**:

$$A\mathrm{Ker}(C_p) \subseteq \mathrm{Ker}(C_p), \tag{6.27}$$

which is necessary for the exact boundary synchronization by p-groups of system (I). Conversely, we will show in the next section that under the condition of C_p-compatibility (6.27), there exists a boundary control matrix D with $M = N - p$, such that system (I) is exactly synchronizable by p-groups.

We first give the lower bound estimate on the rank of the boundary control matrix D, which is necessary for the exact boundary synchronization by p-groups.

Theorem 6.4 *Assume that system (I) is exactly synchronizable by p-groups. Then we necessarily have*

$$\mathit{rank}(C_pD) = N - p. \tag{6.28}$$

In particular, we have

$$\mathit{rank}(D) \geqslant N - p. \tag{6.29}$$

Proof From (6.7), we have $C_pU \equiv 0$ for $t \geqslant T$. If $A\mathrm{Ker}(C_p) \nsubseteq \mathrm{Ker}(C_p)$, then by Lemma 6.3, we can construct a full row-rank $(N - p + 1) \times N$ matrix $\widehat{C}_{p-1}$ such that $\widehat{C}_{p-1}U \equiv 0$ for $t \geqslant T$. If $A\mathrm{Ker}(\widehat{C}_{p-1}) \nsubseteq \mathrm{Ker}(\widehat{C}_{p-1})$, once again by Lemma 6.3, we can construct a full row-rank $(N - p + 2) \times N$ matrix $\widehat{C}_{p-2}$ such that $\widehat{C}_{p-2}U \equiv 0$ for $t \geqslant T$, and so forth. The procedure should stop at some step r with $0 \leqslant r \leqslant p$. So, we get an enlarged full row-rank $(N - p + r) \times N$ matrix $\widehat{C}_{p-r}$ (when $r = 0$, we take $\widehat{C}_p = C_p$, and this special situation means that the condition of compatibility (6.27) holds) such that

$$t \geqslant T: \quad \widehat{C}_{p-r}U \equiv 0 \tag{6.30}$$

and

$$A\mathrm{Ker}(\widehat{C}_{p-r}) \subseteq \mathrm{Ker}(\widehat{C}_{p-r}). \tag{6.31}$$

Then, by Proposition 2.15, there exists a unique matrix $\overline{A}_{p-r}$ of order $(N-p+r)$, such that

$$\widehat{C}_{p-r}A = \overline{A}_{p-r}\widehat{C}_{p-r}. \tag{6.32}$$

Applying $\widehat{C}_{p-r}$ to problem (I) and (I0) and setting $W = \widehat{C}_{p-r}U$, we get the following reduced system:

$$\begin{cases} W'' - \Delta W + \overline{A}_{p-r}W = 0 & \text{in } (0,+\infty)\times\Omega, \\ W = 0 & \text{on } (0,+\infty)\times\Gamma_0, \\ W = \widehat{C}_{p-r}DH & \text{on } (0,+\infty)\times\Gamma_1 \end{cases} \tag{6.33}$$

with the initial condition

$$t = 0: \quad W = \widehat{C}_{p-r}\widehat{U}_0, \; W' = \widehat{C}_{p-r}\widehat{U}_1 \quad \text{in } \Omega. \tag{6.34}$$

Moreover, (6.30) implies that

$$t \geqslant T: \quad W \equiv 0. \tag{6.35}$$

On the other hand, since $\widehat{C}_{p-r}$ is a $(N-p+r)\times N$ full row-rank matrix, the linear map

$$(\widehat{U}_0, \widehat{U}_1) \to (\widehat{C}_{p-r}\widehat{U}_0, \widehat{C}_{p-r}\widehat{U}_1) \tag{6.36}$$

is surjective from the space $(L^2(\Omega))^N \times (H^{-1}(\Omega))^N$ onto the space $(L^2(\Omega))^{N-p+r} \times (H^{-1}(\Omega))^{N-p+r}$. We thus get the exact boundary null controllability of the reduced system (6.33) in the space $(L^2(\Omega))^{N-p+r} \times (H^{-1}(\Omega))^{N-p+r}$. By Theorem 3.11, the rank of the corresponding boundary control matrix $\widehat{C}_{p-r}D$ satisfies

$$\mathrm{rank}(\widehat{C}_{p-r}D) = N - p + r. \tag{6.37}$$

Finally, noting that the $(N-p+r)\times N$ matrix $\widehat{C}_{p-r}D$ is of full row-rank, then the sub-matrix C_pD composed of the first $(N-p)$ rows of $\widehat{C}_{p-r}D$ is of full row-rank, we get thus (6.28). □

Theorem 6.5 *Assume that system (I) is exactly synchronizable by p-groups under the minimal rank condition*

$$M = rank(D) = N - p. \tag{6.38}$$

Then, we necessarily have the condition of C_p-compatibility (6.27).

Proof Noting (6.37) and (6.38), it follows easily from

$$\text{rank}(D) \geqslant \text{rank}(\widehat{C}_{p-r}D)$$

that $r = 0$. It means that it is not necessary to proceed the extension in the proof of Theorem 6.4, we have already the condition of C_p-compatibility (6.27). □

Remark 6.6 The condition of C_p-compatibility (6.27) is necessary for the exact boundary synchronization by p-groups of system (I) only under the minimal rank condition (6.38), namely, under the minimal number of boundary controls.

Remark 6.7 Noting (6.9), the condition of C_p-compatibility (6.27) can be equivalently written as

$$Ae_r = \sum_{s=1}^{p} \alpha_{sr} e_s, \quad 1 \leqslant r \leqslant p, \tag{6.39}$$

where α_{sr} are some constant coefficients. Moreover, because of the specific expression of e_r given in (6.8), the above expression (6.39) can be written in the form of **row-sum condition by blocks**:

$$\sum_{j=n_{s-1}+1}^{n_s} a_{ij} = \alpha_{rs} \tag{6.40}$$

for all $1 \leqslant r, s \leqslant p$ and $n_{r-1} + 1 \leqslant i \leqslant n_r$, which is a natural generalization of the row-sum condition (4.5) in the case $p = 1$.

Remark 6.8 By Proposition 2.15, the condition of C_p-compatibility (6.27) is equivalent to the existence of a unique matrix $\overline{A}_p$ of order $(N - p)$, such that

$$C_p A = \overline{A}_p C_p. \tag{6.41}$$

$\overline{A}_p$ is called the **reduced matrix of** A **by** C_p.

6.4 Exact Boundary Synchronization by p-Groups

Theorem 6.4 shows that the exact boundary synchronization by p-groups of system (I) requires at least $(N - p)$ boundary controls, namely, we have the following

Corollary 6.9 *If*

$$rank(C_p D) < N - p, \tag{6.42}$$

or, in particular, if

$$rank(D) < N - p, \tag{6.43}$$

then, no matter how large $T > 0$ is, system (I) cannot be exactly synchronizable by p-groups at the time T.

The following result shows that the converse also holds.

Theorem 6.10 *Assume that the coupling matrix A satisfies the condition of C_p-compatibility (6.27). Let $M = N - p$ and let the $N \times (N - p)$ boundary control matrix D satisfy the rank condition*

$$rank(C_p D) = N - p. \tag{6.44}$$

Then, under the multiplier geometrical condition (3.1), system (I) is exactly synchronizable by p-groups in the space $(L^2(\Omega))^N \times (H^{-1}(\Omega))^N$.

Proof By Remark 6.8, applying C_p to problem (I) and (I0) and setting $W_p = C_p U$, we get the following self-closed reduced system:

$$\begin{cases} W_p'' - \Delta W_p + \overline{A}_p W_p = 0 & \text{in } (0, +\infty) \times \Omega, \\ W_p = 0 & \text{on } (0, +\infty) \times \Gamma_0, \\ W_p = C_p D H & \text{on } (0, +\infty) \times \Gamma_1 \end{cases} \tag{6.45}$$

with

$$t = 0: \quad W_p = C_p \widehat{U}_0, \ W_p' = C_p \widehat{U}_1 \quad \text{in } \Omega. \tag{6.46}$$

Then, by Theorem 3.11 and noting the rank condition (6.44), the reduced system (6.45) is exactly null controllable at the time $T > 0$ by means of controls $H \in L^2_{loc}(0, +\infty; (L^2(\Gamma_1))^{N-p})$ with compact support in $[0, T]$. Thus, we have

$$t \geqslant T: \quad C_p U \equiv W_p \equiv 0.$$

The proof is then complete. □

Remark 6.11 The rank M of the boundary control matrix D presents the number of boundary controls applied to the original system (I), while, the rank of the matrix $C_p D$ presents the number of boundary controls effectively applied to the reduced system (6.45). Let us write

$$DH = H_0 + H_1 \text{ with } H_0 \in \text{Ker}(C_p) \quad \text{and} \quad H_1 \in \text{Im}(C_p^T). \tag{6.47}$$

The part H_0 will disappear in the reduced system (6.45), and is useless for the exact boundary synchronization by p-groups of the original system (I). The exact boundary null controllability of the reduced system (6.45), therefore, the exact boundary synchronization by p-groups of the original system (I) is in fact realized only by the part H_1. So, in order to minimize the number of boundary controls, we are interested in the matrices D such that

$$\mathrm{Im}(D) \cap \mathrm{Ker}(C_p) = \{0\}, \tag{6.48}$$

or, by Proposition 2.11, such that

$$\mathrm{rank}(C_p D) = \mathrm{rank}(D) = N - p. \tag{6.49}$$

Proposition 6.12 *Let $\mathcal{D}_{N-p}$ be the set of $N \times (N-p)$ matrices satisfying* (6.49), *namely,*

$$\mathcal{D}_{N-p} = \{D \in \mathbb{M}^{N\times(N-p)} : \quad rank(D) = rank(C_p D) = N - p\}. \tag{6.50}$$

Then, for any given boundary control matrix $D \in \mathcal{D}_{N-p}$, the subspaces $Ker(D^T)$ and $Ker(C_p)$ are bi-orthonormal. Moreover, we have

$$\mathcal{D}_{N-p} = \{C_p^T D_1 + (e_1, \cdots, e_p) D_0\}, \tag{6.51}$$

where D_1 is an invertible matrix of order $(N-p)$, D_0 is a matrix of order $p \times (N-p)$, and the vectors $e_1, \cdots, e_p$ are given by (6.8).

Proof Noting that $\{\mathrm{Ker}(D^T)\}^\perp = \mathrm{Im}(D)$, and $\mathrm{Ker}(D^T)$ and $\mathrm{Ker}(C_p)$ have the same dimension p, by Propositions 2.4 and 2.5, to prove that $\mathrm{Ker}(D^T)$ and $\mathrm{Ker}(C_p)$ are bi-orthonormal, it suffices to show that $\mathrm{Ker}(C_p) \cap \mathrm{Im}(D) = \{0\}$, which, by Proposition 2.11, is equivalent to the condition $\mathrm{rank}(C_p D) = \mathrm{rank}(D)$.

Now we prove (6.51). Let D be a matrix of order $N \times (N-p)$. Noting that $\mathrm{Im}(C_p^T) \oplus \mathrm{Ker}(C_p) = \mathbb{R}^N$, there exist a matrix D_0 of order $p \times (N-p)$ and a square matrix D_1 of order $(N-p)$, such that

$$D = C_p^T D_1 + (e_1, \cdots, e_p) D_0, \tag{6.52}$$

where $e_1, \cdots, e_p$ are given by (6.8). Moreover, $C_p D = C_p C_p^T D_1$ is of rank $(N-p)$, if and only if D_1 is invertible. On the other hand, let $x \in \mathbb{R}^{N-p}$, such that

$$Dx = C_p^T D_1 x + (e_1, \cdots, e_p) D_0 x = 0. \tag{6.53}$$

Since $\mathrm{Im}\,(C_p^T) \perp \mathrm{Ker}(C_p)$, it follows that

$$C_p^T D_1 x = (e_1, \cdots, e_p) D_0 x = 0. \tag{6.54}$$

Then, noting that $C_p^T D_1$ is of full column-rank, we have $x = 0$. Hence, D is of full column-rank $(N-p)$. This proves (6.51). □

Remark 6.13 Under the multiplier geometrical condition 3.1, for any given matrix $D \in \mathcal{D}_{N-p}$, the condition of C_p-compatibility (6.27) is necessary and sufficient for the exact boundary synchronization by p-groups of system (I) by Theorem 6.10 and Remark 6.6. In particular, if we take $D = C_p^T C_p \in \mathcal{D}_{N-p}$, we can really achieve the exact boundary synchronization by p-groups for system (I) by means of $(N - p)$ boundary controls.

Chapter 7
Exactly Synchronizable States by p-Groups

When system (I) possesses the exact boundary synchronization by p-groups, the corresponding exactly synchronizable states by p-groups will be studied in this chapter.

7.1 Introduction

Under the condition of C_p-compatibility (6.27), it is easy to see that for $t \geqslant T$, the **exactly synchronizable state by p-groups** $u = (u_1, \cdots u_p)^T$ satisfies the following coupled system of wave equations with homogenous Dirichlet boundary condition:

$$\begin{cases} u'' - \Delta u + \tilde{A} u = 0 \text{ in } (T, +\infty) \times \Omega, \\ u = 0 \qquad\qquad\qquad \text{on } (T, +\infty) \times \Gamma, \end{cases} \tag{7.1}$$

where $\tilde{A} = (\alpha_{rs})$ is given by (6.39). Hence, the evolution of the exactly synchronizable state by p-groups $u = (u_1, \cdots u_p)^T$ with respect to t is completely determined by the values of (u, u_t) at the time $t = T$:

$$t = T: \quad u = \widehat{u}_0, \ u' = \widehat{u}_1. \tag{7.2}$$

As in Theorem 5.1 for the case $p = 1$, we can show that the attainable set of all possible values of (u, u') at $t = T$ is the whole space $(L^2(\Omega))^p \times (H^{-1}(\Omega))^p$, when the initial data $(\widehat{U}_0, \widehat{U}_1)$ vary in the space $(L^2(\Omega))^N \times (H^{-1}(\Omega))^N$.

In this chapter, we will discuss the determination of exactly synchronizable states by p-groups $u = (u_1, \cdots u_p)^T$ for each given initial data $(\widehat{U}_0, \widehat{U}_1)$. Since there is an infinity of boundary controls which can realize the exact boundary synchronization by p-groups for system (I), exactly synchronizable states by p-groups $u = (u_1, \cdots u_p)^T$ naturally depend on applied boundary controls H. However, for some coupling matrices A, for example, the symmetric ones, the exactly synchronizable states by

T. Li and B. Rao, *Boundary Synchronization for Hyperbolic Systems*,
Progress in Nonlinear Differential Equations and Their Applications 94,
https://doi.org/10.1007/978-3-030-32849-8_7

p-groups $u = (u_1, \cdots u_p)^T$ could be independent of applied boundary controls H. In the general case, exactly synchronizable states by p-groups $u = (u_1, \cdots u_p)^T$ depend on applied boundary controls, however we can give an estimate on the difference between each exactly synchronizable state by p-groups $u = (u_1, \cdots u_p)^T$ and the solution to a problem which is independent of applied boundary controls (cf. Theorems 7.1 and 7.2 below).

7.2 Determination of Exactly Synchronizable States by p-Groups

Now we return to the determination of exactly synchronizable states by p-groups. The case $p = 1$ was considered in Sect. 5.2.

We first consider the case that A^T admits an invariant subspace $\text{Span}\{E_1, \cdots, E_p\}$, which is bi-orthonormal to $\text{Span}\{e_1, \cdots, e_p\}$, namely, we have

$$(e_i, E_j) = \delta_{ij}, \quad 1 \leqslant i, j \leqslant p, \tag{7.3}$$

where $e_1, \cdots, e_p$ are given by (6.8)–(6.9).

Theorem 7.1 *Assume that the matrix A satisfies the condition of C_p-compatibility (6.27). Assume furthermore that A^T admits an invariant subspace $\text{Span}\{E_1, \cdots, E_p\}$, which is bi-orthonormal to $\text{Ker}(C_p) = \text{Span}\{e_1, \cdots, e_p\}$. Then there exists a boundary control matrix $D \in \mathcal{D}_{N-p}$ (cf. (6.51)), such that the exactly synchronizable state by p-groups $u = (u_1, \cdots, u_p)^T$ is uniquely determined by*

$$t \geqslant T: \quad u = \phi, \tag{7.4}$$

where $\phi = (\phi_1, \cdots, \phi_p)^T$ is the solution to the following problem independent of applied boundary controls H: for $s = 1, 2, \cdots, p$,

$$\begin{cases} \phi_s'' - \Delta\phi_s + \sum_{r=1}^p \alpha_{sr}\phi_r = 0 & \text{in } (0, +\infty) \times \Omega, \\ \phi_s = 0 & \text{on } (0, +\infty) \times \Gamma, \\ t = 0: \quad \phi_s = (E_s, \widehat{U}_0), \ \phi_s' = (E_s, \widehat{U}_1) \text{ in } \Omega, \end{cases} \tag{7.5}$$

where α_{sr} are given by (6.39).

Proof Since the subspaces $\text{Span}\{E_1, \cdots, E_p\}$ and $\text{Span}\{e_1, \cdots, e_p\}$ are bi-orthonormal, then taking

$$D_1 = I_{N-p}, \quad D_0 = -E^T C_p^T \text{ with } E = (E_1, \cdots, E_p) \tag{7.6}$$

in (6.51), we obtain a boundary control matrix $D \in \mathcal{D}_{N-p}$, such that

$$E_s \in \text{Ker}(D^T), \quad s = 1, \cdots, p. \tag{7.7}$$

On the other hand, since Span$\{E_1, \cdots, E_p\}$ is invariant for A^T, and noting (6.39) and (7.3), it is easy to check that

$$A^T E_s = \sum_{r=1}^{p} \alpha_{sr} E_r, \quad s = 1, \cdots, p. \tag{7.8}$$

Then, taking the inner product of E_s with problem (I) and (I0), and setting $\phi_s = (E_s, U)$, we get problem (7.5). Finally, by the exact boundary synchronization by p-groups (6.10) and the relationship (7.3), we get

$$t \geqslant T: \quad \phi_s = (E_s, U) = \sum_{r=1}^{p} (E_s, e_r) u_r = u_s, \quad 1 \leqslant s \leqslant p. \tag{7.9}$$

The proof is then complete. □

Theorem 7.2 *Assume that the condition of C_p-compatibility (6.27) holds. Then for any given boundary control matrix $D \in \mathcal{D}_{N-p}$, there exists a positive constant c_T independent of initial data, but depending on T, such that each exactly synchronizable state by p-groups $u = (u_1, \cdots, u_p)^T$ satisfies the following estimate:*

$$\begin{aligned} &\|(u, u')(T) - (\phi, \phi')(T)\|_{(H_0^1(\Omega))^p \times (L^2(\Omega))^p} \\ \leqslant & c_T \|C_p(\widehat{U}_0, \widehat{U}_1)\|_{(L^2(\Omega))^{N-p} \times (H^{-1}(\Omega))^{N-p}}, \end{aligned} \tag{7.10}$$

where $\phi = (\phi_1, \cdots, \phi_p)^T$ is the solution to problem (7.5), in which Span$\{E_1, \cdots, E_p\}$ is bi-orthonormal to Span$\{e_1, \cdots, e_p\}$.

Proof By Proposition 6.12, the subspaces Ker(D^T) and Ker(C_p) are bi-orthonormal. Then we can chose $E_1, \cdots, E_r \in \text{Ker}(D^T)$, such that Span$\{E_1, \cdots, E_p\}$ and Span$\{e_1, \cdots, e_p\}$ are bi-orthonormal.

Moreover, noting (6.39) and (7.3), by a direct computation, we get

$$\begin{aligned} &(A^T E_s - \sum_{r=1}^{p} \alpha_{sr} E_r, e_k) \\ =&(E_s, Ae_k) - \sum_{r=1}^{p} \alpha_{sr} (E_r, e_k) \\ =&\sum_{l=1}^{p} \alpha_{lk} (E_s, e_l) - \alpha_{sk} = \alpha_{sk} - \alpha_{sk} = 0 \end{aligned} \tag{7.11}$$

for all $s, k = 1, \cdots, p$, and then

$$A^T E_s - \sum_{r=1}^{p} \alpha_{sr} E_r \in \{\mathrm{Ker}(C_p)\}^{\perp} = \mathrm{Im}(C_p^T), \quad s = 1, \cdots, p. \tag{7.12}$$

Thus, there exists a vector $R_s \in \mathbb{R}^{N-p}$, such that

$$A^T E_s - \sum_{r=1}^{q} \alpha_{sr} E_r = -C_p^T R_s. \tag{7.13}$$

Taking the inner product of E_s with problem (I) and (I0), and setting $\psi_s = (E_s, U)$ $(s = 1, \cdots, p)$, it is easy to see that

$$\begin{cases} \psi_s'' - \Delta\psi_s + \sum_{r=1}^{p} \alpha_{sr}\psi_r = (R_s, C_p U) & \text{in } (0, +\infty) \times \Omega, \\ \psi_s = 0 & \text{on } (0, +\infty) \times \Gamma, \\ t = 0: \quad \psi_s = (E_s, \widehat{U}_0), \ \psi_s' = (E_s, \widehat{U}_1) \text{ in } \Omega. \end{cases} \tag{7.14}$$

By the well-posedness of problems (7.5) and (7.14), there exists a constant $c > 0$ independent of initial data, such that

$$\begin{aligned} &\|(\psi, \psi')(T) - (\phi, \phi')(T)\|^2_{(H_0^1(\Omega))^p \times (L^2(\Omega))^p} \\ \leqslant c &\int_0^T \|C_p U(s)\|^2_{(L^2(\Omega))^{N-p}} ds. \end{aligned} \tag{7.15}$$

Since $C_p U = W_p$, the exact boundary null controllability of the reduced system (6.45) shows that there exists another positive constant $c_T > 0$ independent of initial data, but depending on T, such that

$$\begin{aligned} &\int_0^T \|C_p U(s)\|^2_{(L^2(\Omega))^{N-p}} ds \\ \leqslant &c_T \|C_p(\widehat{U}_0, \widehat{U}_1)\|^2_{(L^2(\Omega))^{N-p} \times (H^{-1}(\Omega))^{N-p}}. \end{aligned} \tag{7.16}$$

Finally, noting that

$$t \geqslant T: \quad \psi_s = (E_s, U) = \sum_{r=1}^{p} (E_s, e_r) u_r = u_s, \quad s = 1, \cdots, p, \tag{7.17}$$

and inserting (7.16)–(7.17) into (7.15), we get (7.10). The proof is complete. □

Remark 7.3 Since $\{e_1, \cdots, e_p\}$ is an orthonormal system, we can take $E_s = e_s/\|e_s\|^2$ $(s = 1, \cdots, p)$ as a specific choice. Accordingly, the boundary control matrix D has the tendency to drive the initial data to the average value.

Remark 7.4 Let $\text{Ker}(C_p)$ be invariant for A and bi-orthonormal to $W = \text{Span}\{E_1, E_2, \cdots, E_p\}$. By Proposition 2.9, $W^\perp$ is a supplement of $\text{Ker}(C_p)$ and invariant for the matrix A if and only if W is invariant for the matrix A^T. Accordingly, the coupling matrix A can be diagonalized by blocks under the decomposition $\mathbb{R}^N = Ker(C_p) \oplus W^\perp$.

In particular, if A^T satisfies the condition of C_p-compatibility (6.27), then $\text{Ker}(C_p)$ is also invariant for A^T. By Proposition 2.7, $\text{Im}(C_p^T)$ is an invariant subspace of A. Then, we can write

$$A(e_1, \cdots, e_p, C_p^T) = (e_1, \cdots, e_p, C_p^T) \begin{pmatrix} \widetilde{A} & 0 \\ 0 & \widehat{A} \end{pmatrix}, \tag{7.18}$$

where $\widetilde{A} = (\alpha_{rs})$ is given by (6.40) and $\widehat{A}$ is given by

$$\widehat{A} = (C_p C_p^T)^{-1} C_p A C_p^T. \tag{7.19}$$

In that case, Theorem 7.1 shows that the exactly synchronizable state by p-groups is independent of applied boundary controls. Otherwise, it depends on applied boundary controls; however, as shown in Theorem 7.2, we can give an estimate on the difference between each exactly synchronizable state by p-groups and the solution to a problem independent of applied boundary controls.

7.3 Determination of Exactly Synchronizable States by p-Groups (Continued)

We now consider the determination of exactly synchronizable states by p-groups in the case that $\text{Ker}(C_p) = \text{Span}\{e_1, \cdots, e_p\}$ is an invariant subspace of A, but A^T does not admit any invariant subspace which is bi-orthonormal to $\text{Ker}(C_p)$. In this case, we will extend the subspace $\text{Span}\{e_1, \cdots, e_p\}$ to an invariant subspace $\text{Span}\{e_1, \cdots, e_q\}$ of A with $q \geqslant p$, so that A^T admits an invariant subspace $\text{Span}\{E_1, \cdots, E_q\}$ that is bi-orthonormal to $\text{Span}\{e_1, \cdots, e_q\}$.

When $\text{Ker}(C_p)$ is A-marked, the procedure for obtaining the subspace $\text{Span}\{e_1, \cdots, e_q\}$ is given in Proposition 2.20.

When $\text{Ker}(C_p)$ is not A-marked, let $\lambda_j (1 \leqslant j \leqslant d)$ denote the eigenvalues of the restriction of A to $\text{Ker}(C_p)$. We define

$$\text{Span}\{e_1, \cdots, e_q\} = \bigoplus_{j=1}^{d} \text{Ker}(A - \lambda_j I)^{m_j} \tag{7.20}$$

and

$$\mathrm{Span}\{E_1, \cdots, E_q\} = \bigoplus_{j=1}^{d} \mathrm{Ker}(A^T - \lambda_j I)^{m_j}, \tag{7.21}$$

where m_j is an integer such that $\mathrm{Ker}(A^T - \lambda_j I)^{m_j} = \mathrm{Ker}(A^T - \lambda_j I)^{m_j+1}$ and

$$q = \sum_{j=1}^{d} \mathrm{Dim}\ \mathrm{Ker}(A^T - \lambda_j I)^{m_j}. \tag{7.22}$$

Clearly, the subspaces given by (7.20) and (7.21) satisfy well the first two conditions in Proposition 2.20. However, since $\mathrm{Ker}(C_p)$ is not A-marked, the subspace $\mathrm{Span}\{e_1, \cdots, e_q\}$ constructed by this way is not a priori the least one.

In the above two cases, we can define the projection P on the subspace $\mathrm{Span}\{e_1, \cdots, e_q\}$ as follows:

$$P = \sum_{r=1}^{q} e_r \otimes E_r, \tag{7.23}$$

where the tensor product $\otimes$ is defined by

$$(e_r \otimes E_r)U = (E_r, U)e_r, \quad \forall U \in \mathbb{R}^N. \tag{7.24}$$

Then we have

$$\mathrm{Im}(P) = \mathrm{Span}\{e_1, \cdots, e_q\}, \tag{7.25}$$

$$\mathrm{Ker}(P) = \big(\mathrm{Span}\{E_1, \cdots, E_q\}\big)^{\perp} \tag{7.26}$$

and

$$PA = AP. \tag{7.27}$$

Let $U = U(t, x)$ be the solution to problem (I) and (I0). We may define the **synchronizable part** U_s and the **controllable part** U_c by

$$U_s = PU, \quad U_c = (I - P)U, \tag{7.28}$$

respectively. In fact, if system (I) is exactly synchronizable by p-groups, we have

$$U \in \mathrm{Span}\{e_1, \cdots, e_p\} \subseteq \mathrm{Span}\{e_1, \cdots, e_q\} = \mathrm{Im}(P) \tag{7.29}$$

for all $t \geqslant T$, so that

$$t \geqslant T: \quad U_s = PU \equiv U, \quad U_c \equiv 0. \tag{7.30}$$

Moreover, noting (7.27), we have

$$\begin{cases} U_s'' - \Delta U_s + AU_s = 0 & \text{in } (0, +\infty) \times \Omega, \\ U_s = 0 & \text{on } (0, +\infty) \times \Gamma_0, \\ U_s = PDH & \text{on } (0, +\infty) \times \Gamma_1, \\ t = 0: \quad U_s = P\widehat{U}_0, \ U_s' = P\widehat{U}_1 \text{ in } \Omega. \end{cases} \tag{7.31}$$

The condition $p = q$ means exactly that the matrix A satisfies the condition of C_p-compatibility (6.27), and A^T admits an invariant subspace, which is bi-orthonormal to $\text{Ker}(C_p)$. In this case, by Theorem 7.1, there exists a boundary control matrix $D \in \mathcal{D}_{N-p}$, such that the exactly synchronizable state by p-groups $u = (u_1, \cdots, u_p)^T$ is independent of applied boundary controls H. Conversely, we have the following

Theorem 7.5 *Assume that the matrix A satisfies the condition of C_p-compatibility (6.27). Assume that system (I) is exactly synchronizable by p-groups. If the synchronizable part U_s is independent of applied boundary controls H, then A^T admits an invariant subspace which is bi-orthonormal to Ker(C_p).*

Proof Let H_1 and H_2 be two boundary controls, which realize simultaneously the exact boundary synchronization by p-groups for system (I). If the corresponding solution U_s to problem (7.31) is independent of applied boundary controls H_1 and H_2, then we have

$$PD(H_1 - H_2) = 0 \quad \text{on } (0, T) \times \Gamma_1. \tag{7.32}$$

By Theorem 3.8, it is easy to see that the values of $C_pD(H_1 - H_2)$ on $(T - \epsilon, T) \times \Gamma_1$ can be arbitrarily chosen for $\epsilon > 0$ small enough. Since C_pD is invertible, the values of $(H_1 - H_2)$ on $(T - \epsilon, T) \times \Gamma_1$ can be arbitrarily chosen, too. This yields that $PD = 0$ so that

$$\text{Im}(D) \subseteq \text{Ker}(P). \tag{7.33}$$

Noting (7.26), we have

$$\dim \text{Ker}(P) = N - q, \tag{7.34}$$

however, by Theorem 6.4, we have

$$\dim \text{Im}(D) \geqslant \dim \text{Im}(C_pD) = \text{rank}(C_pD) = N - p. \tag{7.35}$$

Then, it follows from (7.33) that $p = q$. Then $\text{Span}\{E_1, \cdots, E_p\}$ will be the desired subspace. The proof is complete. □

7.4 Precise Consideration on the Exact Boundary Synchronization by 2-Groups

The case $p = 1$ was considered in Chap. 5. The condition $p = q = 1$ is equivalent to the existence of an eigenvector E_1 of A^T, such that $(E_1, e_1) = 1$. In this section, we give the precise consideration for the case $p = 2$.

Let $m \geqslant 2$ be an integer such that $N - m \geqslant 2$. We rewrite the exact boundary synchronization by 2-groups as

$$t \geqslant T: \quad u^{(1)} = \cdots u^{(m)} := u, \quad u^{(m+1)} = \cdots = u^{(N)} := v. \tag{7.36}$$

Let

$$C_2 = \begin{pmatrix} S_m & \\ & S_{N-m} \end{pmatrix} \tag{7.37}$$

be the matrix of synchronization by 2-groups and let

$$e_1 = (\overbrace{1, \cdots, 1}^{m}, \overbrace{0, \cdots, 0}^{N-m})^T, \quad e_2 = (\overbrace{0, \cdots, 0}^{m}, \overbrace{1, \cdots, 1}^{N-m})^T. \tag{7.38}$$

Clearly, we have

$$\text{Ker}(C_2) = \text{Span}\{e_1, e_2\} \tag{7.39}$$

and the synchronization condition by 2-groups (6.10) means

$$t \geqslant T: \quad U = ue_1 + ve_2. \tag{7.40}$$

(i) Assume that A admits two eigenvectors ϵ_r and $\tilde{\epsilon}_s$, associated, respectively, with the eigenvalues λ and μ, contained in the invariant subspace $\text{Ker}(C_2)$. Let $\epsilon_1, \cdots, \epsilon_r$, respectively, $\tilde{\epsilon}_1, \cdots, \tilde{\epsilon}_s$ denote the corresponding Jordan chains of A:

$$\begin{cases} A\epsilon_k = \lambda\epsilon_k + \epsilon_{k+1}, \ 1 \leqslant k \leqslant r, \ \epsilon_{r+1} = 0, \\ A\tilde{\epsilon}_i = \mu\tilde{\epsilon}_i + \tilde{\epsilon}_{i+1}, \ \ 1 \leqslant i \leqslant s, \ \tilde{\epsilon}_{s+1} = 0. \end{cases} \tag{7.41}$$

Accordingly, let $\mathcal{E}_1, \cdots, \mathcal{E}_r$ and $\widetilde{\mathcal{E}}_1 \cdots, \widetilde{\mathcal{E}}_s$ denote the corresponding Jordan chains of A^T:

$$\begin{cases} A^T\mathcal{E}_k = \lambda\mathcal{E}_k + \mathcal{E}_{k-1}, \ 1 \leqslant k \leqslant r, \ \mathcal{E}_0 = 0, \\ A^T\widetilde{\mathcal{E}}_i = \mu\widetilde{\mathcal{E}}_i + \widetilde{\mathcal{E}}_{i-1}, \ \ 1 \leqslant i \leqslant s, \ \widetilde{\mathcal{E}}_0 = 0. \end{cases} \tag{7.42}$$

Moreover, for any $1 \leqslant k, l \leqslant r$ and $1 \leqslant i, j \leqslant s$, we have

$$\begin{cases} (\epsilon_k, \mathcal{E}_l) = \delta_{kl}, \quad (\tilde{\epsilon}_i, \widetilde{\mathcal{E}}_j) = \delta_{ij}, \\ (\epsilon_k, \widetilde{\mathcal{E}}_i) = 0, \quad\ (\tilde{\epsilon}_j, \mathcal{E}_l) = 0. \end{cases} \tag{7.43}$$

Taking the inner product of problem (I) and (I0) with $\mathcal{E}_k$ and $\widetilde{\mathcal{E}}_i$, and setting $\phi_k = (\mathcal{E}_k, U)$ and $\tilde{\phi}_i = (\widetilde{\mathcal{E}}_i, U)$, we get the subsystem

$$\begin{cases} \phi_k'' - \Delta\phi_k + \lambda\phi_k + \phi_{k-1} = 0 & \text{in } (0, +\infty) \times \Omega, \\ \phi_k = 0 & \text{on } (0, +\infty) \times \Gamma_0, \\ \phi_k = (\mathcal{E}_k, DH) & \text{on } (0, +\infty) \times \Gamma_1, \\ t = 0: \quad \psi_k = (\mathcal{E}_k, \widehat{U}_0), \ \phi_k' = (\mathcal{E}_k, \widehat{U}_1) & \text{in } \Omega \end{cases} \tag{7.44}$$

for $k = 1, \cdots, r$, and the subsystem

$$\begin{cases} \tilde{\phi}_i'' - \Delta\tilde{\phi}_i + \mu\tilde{\phi}_i + \tilde{\phi}_{i-1} = 0 & \text{in } (0, +\infty) \times \Omega, \\ \tilde{\phi}_i = 0 & \text{on } (0, +\infty) \times \Gamma_0, \\ \tilde{\phi}_i = (\widetilde{\mathcal{E}}_i, DH) & \text{on } (0, +\infty) \times \Gamma_1, \\ t = 0: \quad \tilde{\phi}_i = (\widetilde{\mathcal{E}}_i, \widehat{U}_0), \ \tilde{\phi}_i' = (\widetilde{\mathcal{E}}_i, \widehat{U}_1) & \text{in } \Omega \end{cases} \tag{7.45}$$

for $i = 1, \cdots, s$, respectively.

Once the solutions $(\phi_1, \cdots, \phi_r)$ and $(\tilde{\phi}_1, \cdots, \tilde{\phi}_s)$ are determined, we look for the corresponding exactly synchronizable state by 2-groups $(u, v)^T$.

Noting that $\epsilon_r, \tilde{\epsilon}_s \in \text{Span}\{e_1, e_2\}$, we can write

$$e_1 = \alpha\epsilon_r + \beta\tilde{\epsilon}_s, \quad e_2 = \gamma\epsilon_r + \delta\tilde{\epsilon}_s \tag{7.46}$$

with $\alpha\delta - \beta\gamma \neq 0$. Then the exact boundary synchronization by 2-groups (7.40) gives

$$t \geqslant T: \quad U = (\alpha u + \gamma v)\epsilon_r + (\beta u + \delta v)\tilde{\epsilon}_s. \tag{7.47}$$

Noting (7.43), it follows that

$$\phi_r = \alpha u + \gamma v, \quad \tilde{\phi}_s = \beta u + \delta v. \tag{7.48}$$

By solving this linear system, we get the exactly synchronizable state by 2-groups (u, v).

In particular, if $r = s = 1$, we can choose a boundary control matrix $D \in \mathcal{D}_{N-2}$, such that $D^T\mathcal{E}_1 = D^T\widetilde{\mathcal{E}}_1 = 0$. Then the exactly synchronizable state by 2-groups $(u, v)^T$ can be uniquely determined independently of applied boundary controls. Otherwise, we have to solve the whole systems (7.44) and (7.45) to get $\phi_1, \cdots, \phi_r$ and $\tilde{\phi}_1, \cdots, \tilde{\phi}_s$, even though we only need ϕ_r and $\tilde{\phi}_s$.

(ii) Assume that A admits only one eigenvector ϵ_r in the invariant subspace $\text{Ker}(C_2)$, and $\text{Ker}(C_2)$ is A-marked. Let $\epsilon_1, \cdots, \epsilon_r$ denote a Jordan chain of A. Respectively, let $\mathcal{E}_1, \cdots, \mathcal{E}_r$ denote the corresponding Jordan chain of A^T. We get again the subsystem (7.44). Since $\text{Ker}(C_2)$ is A-marked, then $\epsilon_{r-1}, \epsilon_r \in \text{Span}\{e_1, e_2\}$ so that we can write

$$e_1 = \alpha\epsilon_r + \beta\epsilon_{r-1}, \quad e_2 = \gamma\epsilon_r + \delta\epsilon_{r-1} \tag{7.49}$$

with $\alpha\delta - \beta\gamma \neq 0$. The exact boundary synchronization by 2-groups (7.40) becomes

$$t \geqslant T: \quad U = (\alpha u + \gamma v)\epsilon_r + (\beta u + \delta v)\epsilon_{r-1}. \tag{7.50}$$

Noting the first formula in (7.43), it follows that

$$\phi_r = \alpha u + \gamma v, \quad \phi_{r-1} = \beta u + \delta v, \tag{7.51}$$

which gives the exactly synchronizable state by 2-groups (u, v).

In particular, if $r = 2$, we can choose a boundary control matrix $D \in \mathcal{D}_{N-2}$, such that $D^T \mathcal{E}_1 = D^T \mathcal{E}_2 = 0$. Then the exactly synchronizable state by 2-groups $(u, v)^T$ can be uniquely determined independently of applied boundary controls. Otherwise, we have to solve the whole systems (7.44) to get $\phi_1, \cdots, \phi_r$, even though we only need ϕ_{r-1} and ϕ_r.

Remark 7.6 The above consideration can be used to the general case $p \geqslant 1$ without any essential difficulties. Later in Sect. 15.4, we will give the details in the case $p = 3$ for Neumann boundary controls.

Part II
Synchronization for a Coupled System of Wave Equations with Dirichlet Boundary Controls: Approximate Boundary Synchronization

From the results given in Part 1, we know that, roughly speaking, if the domain satisfies the multiplier geometrical condition, and the coupling matrix A satisfies the corresponding condition of compatibility in the case of various kinds of synchronizations, then the exact boundary null controllability, the exact boundary synchronization and the exact boundary synchronization by groups can be realized for system (I) with Dirichlet boundary controls, respectively, provided that the control time $T > 0$ *is large enough and there are enough boundary controls. However, when the domain does not satisfy the multiplier geometrical condition or there is a lack of boundary controls, "what weakened controllability or synchronization can be obtained" becomes a very interesting and practically important problem.*

In this part, in order to answer this question, we will introduce the concept of approximate boundary null controllability, approximate boundary synchronization and approximate boundary synchronization by groups, and establish the corresponding theory for a coupled system of wave equations with Dirichlet boundary controls. Moreover, we will show that Kalman's criterion of various kinds will play an important role in the discussion.

Chapter 8
Approximate Boundary Null Controllability

In this chapter we will define the approximate boundary null controllability for system (I) and the D-observability for the adjoint problem, and show that these two concepts are equivalent to each other. Moreover, the corresponding Kalman's criterion is introduced and studied.

8.1 Definition

Let

$$U = (u^{(1)}, \cdots, u^{(N)})^T \quad \text{and} \quad H = (h^{(1)}, \cdots, h^{(M)})^T \tag{8.1}$$

with $M \leqslant N$. Consider the coupled system (I) with the initial condition (I0).

Let

$$\mathcal{H}_0 = L^2(\Omega), \quad \mathcal{H}_1 = H_0^1(\Omega), \quad \mathcal{L} = L^2_{loc}(0, +\infty; L^2(\Gamma_1)). \tag{8.2}$$

The dual of $\mathcal{H}_1$ is denoted by $\mathcal{H}_{-1} = H^{-1}(\Omega)$.

By Theorem 3.11, as $M < N$, system (I) is not exactly null controllable in the space $(\mathcal{H}_0)^N \times (\mathcal{H}_{-1})^N$. So we look for some weakened controllability, for example, the **approximate boundary null controllability** as follows.

Definition 8.1 System (I) is approximately null controllable at the time $T > 0$ if for any given initial data $(\widehat{U}_0, \widehat{U}_1) \in (\mathcal{H}_0)^N \times (\mathcal{H}_{-1})^N$, there exists a sequence $\{H_n\}$ of boundary controls in $\mathcal{L}^M$ with compact support in $[0, T]$, such that the corresponding sequence $\{U_n\}$ of solutions to problem (I) and (I0) satisfies the following condition:

$$(U_n, U_n') \to (0, 0) \quad \text{in } C^0_{loc}([T, +\infty); (\mathcal{H}_0 \times \mathcal{H}_{-1})^N) \text{ as } n \to +\infty. \tag{8.3}$$

T. Li and B. Rao, *Boundary Synchronization for Hyperbolic Systems*, Progress in Nonlinear Differential Equations and Their Applications 94,
https://doi.org/10.1007/978-3-030-32849-8_8

Remark 8.2 Since H_n has a compact support in $[0, T]$, the corresponding solution U_n satisfies the homogeneous Dirichlet boundary condition on Γ for $t \geqslant T$. Hence, there exist positive constants c and ω such that

$$\begin{aligned} &\|(U_n(t), U_n'(t))\|_{(\mathcal{H}_0)^N \times (\mathcal{H}_{-1})^N} \\ \leqslant & ce^{\omega(t-T)} \|(U_n(T), U_n'(T))\|_{(\mathcal{H}_0)^N \times (\mathcal{H}_{-1})^N} \end{aligned} \tag{8.4}$$

for all $t \geqslant T$ and for all $n \geqslant 0$. Then the convergence

$$(U_n(T),\ U_n'(T)) \to (0, 0) \quad \text{in } (\mathcal{H}_0)^N \times (\mathcal{H}_{-1})^N \text{ as } n \to +\infty \tag{8.5}$$

implies the convergence (8.3).

Remark 8.3 In Definition 8.1, the convergence (8.3) of the sequence $\{U_n\}$ of solutions does not imply the convergence of the sequence $\{H_n\}$ of boundary controls. We don't even know if the sequence $\{U_n, U_n'\}$ is bounded in $C^0([0, T]; (\mathcal{H}_0 \times \mathcal{H}_{-1})^N)$. However, since $\{H_n\}$ has a compact support in $[0, T]$, the sequence $\{U_n, U_n'\}$ converges uniformly to $(0, 0)$ in $C^0_{loc}([T, +\infty); (\mathcal{H}_0 \times \mathcal{H}_{-1})^N)$ as $n \to +\infty$.

8.2 *D*-Observability for the Adjoint Problem

Let

$$\Phi = (\phi^{(1)}, \cdots, \phi^{(N)})^T. \tag{8.6}$$

Consider the adjoint problem

$$\begin{cases} \Phi'' - \Delta\Phi + A^T\Phi = 0 & \text{in } (0, +\infty) \times \Omega, \\ \Phi = 0 & \text{on } (0, +\infty) \times \Gamma, \\ t = 0: \quad \Phi = \widehat{\Phi}_0,\ \Phi' = \widehat{\Phi}_1 \text{ in } \Omega. \end{cases} \tag{8.7}$$

Definition 8.4 The adjoint problem (8.7) is ***D*-observable** on the interval $[0, T]$ if the observation

$$D^T \partial_\nu \Phi \equiv 0 \quad \text{on } [0, T] \times \Gamma_1 \tag{8.8}$$

implies that $(\widehat{\Phi}_0, \widehat{\Phi}_1) \equiv 0$, then $\Phi \equiv 0$.

Let $\mathcal{C}$ be the set of all the initial states $(V(0), V'(0))$ given by the following backward problem:

$$\begin{cases} V'' - \Delta V + AV = 0 & \text{in } (0, T) \times \Omega, \\ V = 0 & \text{on } (0, T) \times \Gamma_0, \\ V = DH & \text{on } (0, T) \times \Gamma_1, \\ t = T: \quad V = V' = 0 \text{ in } \Omega \end{cases} \tag{8.9}$$

as the boundary control H varies in $\mathcal{L}^M$ with compact support in $[0, T]$.

Lemma 8.5 *System (I) is approximately null controllable in* $(\mathcal{H}_0)^N \times (\mathcal{H}_{-1})^N$ *if and only if*

$$\overline{\mathcal{C}} = (\mathcal{H}_0)^N \times (\mathcal{H}_{-1})^N. \tag{8.10}$$

Proof Assume that (8.10) holds. Then for any given $(\widehat{U}_0, \widehat{U}_1) \in (\mathcal{H}_0)^N \times (\mathcal{H}_{-1})^N$ there exists a sequence $\{H_n\}$ of boundary controls in $\mathcal{L}^M$, such that the corresponding sequence $\{V_n\}$ of solutions to problem (8.9) satisfies

$$(V_n(0), V_n'(0)) \to (\widehat{U}_0, \widehat{U}_1) \quad \text{in } (\mathcal{H}_0)^N \times (\mathcal{H}_{-1})^N \text{ as } n \to +\infty. \tag{8.11}$$

Let

$$\mathcal{R}: \quad (\widehat{U}_0, \widehat{U}_1, H) \to (U, U') \tag{8.12}$$

be the resolution of problem (I) and (I0). $\mathcal{R}$ being linear, we have

$$\begin{aligned} &\mathcal{R}(\widehat{U}_0, \widehat{U}_1, H_n) \\ =&\mathcal{R}(\widehat{U}_0 - V_n(0), \widehat{U}_1 - V_n'(0), 0) + \mathcal{R}(V_n(0), V_n'(0), H_n). \end{aligned} \tag{8.13}$$

By the definition of V_n, we have

$$\mathcal{R}(V_n(0), V_n'(0), H_n)(T) = 0, \tag{8.14}$$

then

$$\mathcal{R}(\widehat{U}_0, \widehat{U}_1, H_n)(T) = \mathcal{R}(\widehat{U}_0 - V_n(0), \widehat{U}_1 - V_n'(0), 0)(T). \tag{8.15}$$

By means of the well-posedness of problem (I) and (I0), there exists a positive constant c such that

$$\begin{aligned} &\|\mathcal{R}(\widehat{U}_0, \widehat{U}_1, H_n)(T)\|_{(\mathcal{H}_0)^N \times (\mathcal{H}_{-1})^N} \\ \leqslant &c\|(\widehat{U}_0 - V_n(0), \widehat{U}_1 - V_n'(0))\|_{(\mathcal{H}_0)^N \times (\mathcal{H}_{-1})^N}. \end{aligned} \tag{8.16}$$

Then, it follows from (8.11) that

$$\|\mathcal{R}(\widehat{U}_0, \widehat{U}_1, H_n)(T)\|_{(\mathcal{H}_0)^N \times (\mathcal{H}_{-1})^N} \to 0 \tag{8.17}$$

as $n \to +\infty$. This gives the approximate boundary null controllability of system (I).

Inversely, assume that system (I) is approximately null controllable. Then for any given $(\widehat{U}_0, \widehat{U}_1) \in (\mathcal{H}_0)^N \times (\mathcal{H}_{-1})^N$, there exists a sequence $\{H_n\}$ of boundary controls in $\mathcal{L}^M$ with compact support in $[0, T]$, such that the corresponding solution U_n of problem (I) and (I0) satisfies

$$
\begin{aligned}
&(U_n(T), U_n'(T)) \\
=&\mathcal{R}(\widehat{U}_0, \widehat{U}_1, H_n)(T) \to (0,0) \quad \text{in } (\mathcal{H}_0)^N \times (\mathcal{H}_{-1})^N
\end{aligned} \tag{8.18}
$$

as $n \to +\infty$. Taking the boundary control $H = H_n$, we solve the backward problem (8.9) and denote by V_n the corresponding solution. By linearity, we have

$$
\begin{aligned}
&\mathcal{R}(\widehat{U}_0, \widehat{U}_1, H_n) - \mathcal{R}(V_n(0), V_n'(0), H_n) \\
=&\mathcal{R}(\widehat{U}_0 - V_n(0), \widehat{U}_1 - V_n'(0), 0).
\end{aligned} \tag{8.19}
$$

Once again, by means of the well-posedness of problem (I) and (I0) regarding as a backward problem, and noting (8.18), we get

$$
\begin{aligned}
&\|\mathcal{R}(\widehat{U}_0 - V_n(0), \widehat{U}_1 - V_n'(0), 0)(0)\|_{(\mathcal{H}_0)^N \times (\mathcal{H}_{-1})^N} \\
\leqslant &c\|(U_n(T) - V_n(T), U_n'(T) - V_n'(T))\|_{(\mathcal{H}_0)^N \times (\mathcal{H}_{-1})^N} \\
=&c\|(U_n(T), U_n'(T)\|_{(\mathcal{H}_0)^N \times (\mathcal{H}_{-1})^N} \to 0
\end{aligned} \tag{8.20}
$$

as $n \to +\infty$, which together with (8.19) implies that

$$
\begin{aligned}
&\|(\widehat{U}_0, \widehat{U}_1) - (V_n(0), V_n'(0))\|_{(\mathcal{H}_0)^N \times (\mathcal{H}_{-1})^N} \\
=&\|\mathcal{R}(\widehat{U}_0, \widehat{U}_1, H_n)(0) - \mathcal{R}(V_n(0), V_n'(0), H_n)(0)\|_{(\mathcal{H}_0)^N \times (\mathcal{H}_{-1})^N} \\
\leqslant &c\|\mathcal{R}(\widehat{U}_0 - V_n(0), \widehat{U}_1 - V_n'(0), 0)(0)\|_{(\mathcal{H}_0)^N \times (\mathcal{H}_{-1})^N} \to 0
\end{aligned} \tag{8.21}
$$

as $n \to +\infty$. This shows $\overline{\mathcal{C}} = (\mathcal{H}_0)^N \times (\mathcal{H}_{-1})^N$. The proof is complete. □

Theorem 8.6 *System (I) is approximately null controllable at the time $T > 0$ if and only if the adjoint problem (8.7) is D-observable on the interval* $[0, T]$.

Proof Assume that system (I) is not approximately null controllable. Then, by Lemma 8.5, there exists a nontrivial element $(-\widehat{\Phi}_1, \widehat{\Phi}_0) \in \mathcal{C}^{\perp}$. Here, the orthogonality is defined in the sense of duality, therefore $(-\widehat{\Phi}_1, \widehat{\Phi}_0) \in (\mathcal{H}_0)^N \times (\mathcal{H}_1)^N$. Taking $(\widehat{\Phi}_0, \widehat{\Phi}_1)$ as the initial data, we solve the adjoint problem (8.7) to get a solution Φ. Next, multiplying the backward problem (8.9) by Φ and integrating by parts, we get

$$
\int_{\Omega}(V(0), \widehat{\Phi}_1)dx - \int_{\Omega}(V'(0), \widehat{\Phi}_0)dx = \int_0^T \int_{\Gamma_1}(DH, \partial_\nu \Phi)d\Gamma dt. \tag{8.22}
$$

Noting that $(-\widehat{\Phi}_1, \widehat{\Phi}_0) \in \mathcal{C}^{\perp}$, it follows that

$$
\int_0^T \int_{\Gamma_1}(DH, \partial_\nu \Phi)d\Gamma dt = 0 \tag{8.23}
$$

for all $H \in \mathcal{L}^M$. This gives the observation (8.8) with $\Phi \not\equiv 0$, therefore, contradicts the D-observation of the adjoint problem (8.7).

Inversely, assume that the adjoint problem (8.7) is not D-observable. Then there exists a nontrivial initial data $(\widehat{\Phi}_0, \widehat{\Phi}_1) \in (\mathcal{H}_1)^N \times (\mathcal{H}_0)^N$, such that the corresponding solution Φ to the adjoint problem (8.7) satisfies the observation (8.8). Now for any given $(\widehat{U}_0, \widehat{U}_1) \in \overline{\mathcal{C}}$, by the definition of $\overline{\mathcal{C}}$, there exists a sequence $\{H_n\}$ in $\mathcal{L}^M$ with compact support in $[0, T]$, such that the corresponding solution V_n to the backward problem (8.9) satisfies

$$(V_n(0), V_n'(0)) \to (\widehat{U}_0, \widehat{U}_1) \quad \text{in } (\mathcal{H}_0)^N \times (\mathcal{H}_{-1})^N \text{ as } n \to +\infty. \tag{8.24}$$

Noting (8.8), the relation (8.22) becomes

$$\int_\Omega (V_n(0), \widehat{\Phi}_1)dx - \int_\Omega (V_n'(0), \widehat{\Phi}_0)dx = 0. \tag{8.25}$$

Passing to the limit as $n \to +\infty$, we get

$$\langle (\widehat{U}_0, \widehat{U}_1), (-\widehat{\Phi}_1, \widehat{\Phi}_0) \rangle_{(\mathcal{H}_0)^N \times (\mathcal{H}_{-1})^N; (\mathcal{H}_0)^N \times (\mathcal{H}_1)^N} = 0 \tag{8.26}$$

for all $(\widehat{U}_0, \widehat{U}_1) \in \overline{\mathcal{C}}$. In particular, we get $(-\widehat{\Phi}_1, \widehat{\Phi}_0) \in \overline{\mathcal{C}}^{\perp}$. Therefore $\overline{\mathcal{C}} \neq (\mathcal{H}_0)^N \times (\mathcal{H}_{-1})^N$. □

Corollary 8.7 *If $M = N$, then system (I) is always approximately null controllable.*

Proof Since $M = N$, D is invertible, then the observation (8.8) gives that

$$\partial_\nu \Phi \equiv 0 \quad \text{on } [0, T] \times \Gamma_1. \tag{8.27}$$

By means of Holmgren's uniqueness theorem (cf. Theorem 8.2 in [62]), we deduce the D-observability of the adjoint problem (8.7). Then by means of Theorem 8.6, we get the approximate boundary null controllability of system (I). The proof is complete. □

Remark 8.8 Corollary 8.7 is a unique continuation result, which is not sufficient for the exact boundary null controllability. By Theorem 3.11, under the multiplier geometrical condition (3.1), system (I) is exactly null controllable by means of N boundary controls, however, without the multiplier geometrical condition (3.1), generally speaking, we cannot conclude the exact boundary null controllability even though we have applied N boundary controls for system (I) of N wave equations. We will discuss later the approximate boundary null controllability with fewer than N boundary controls.

8.3 Kalman's Criterion. Total (Direct and Indirect) Controls

Theorem 8.9 *Assume that the adjoint problem (8.7) is D-observable. Then, we have necessarily the following* **Kalman's criterion***:*

$$rank(D, AD, \cdots, A^{N-1}D) = N. \tag{8.28}$$

Proof By (ii) of Proposition 2.12, it is easy to see that we only need to prove that Ker (D^T) does not contain a nontrivial subspace V which is invariant for A^T.

Let ϕ_n be the solution to the following eigenvalue problem with $\mu_n > 0$:

$$\begin{cases} -\Delta\phi_n = \mu_n^2\phi_n & \text{in } \Omega, \\ \phi_n = 0 & \text{on } \Gamma. \end{cases} \tag{8.29}$$

Assume that A^T possesses a nontrivial invariant subspace $V \subseteq Ker(D^T)$. For any fixed integer $n > 0$, we define

$$W = \{\phi_n w : \quad w \in V\}. \tag{8.30}$$

Clearly, W is a finite-dimensional invariant subspace of $-\Delta + A^T$. Therefore, we can solve the adjoint problem (8.7) in W and the corresponding solutions are given by $\Phi = \phi_n w(t)$, where $w(t) \in V$ satisfies

$$\begin{cases} w'' + (\mu_n^2 I + A^T)w = 0, & 0 < t < \infty, \\ t = 0: \quad w = \widehat{w}_0 \in V, \ w' = \widehat{w}_1 \in V. \end{cases} \tag{8.31}$$

Since $w(t) \in V$ for all $t \geqslant 0$, it follows that

$$D^T \partial_\nu \Phi = \partial_\nu \phi_n D^T w(t) \equiv 0 \quad \text{on } [0, T] \times \Gamma_1. \tag{8.32}$$

Clearly, $\Phi \not\equiv 0$, then we get a contradiction. The proof is complete. □

Thus, by Theorem 8.6, if system (I) is approximately null controllable, then we have necessarily the Kalman's criterion (8.28).

For the exact boundary null controllability, by Theorem 3.11 the number $M =$ rank(D), namely, the number of boundary controls, should be equal to N, the number of state variables. However, as we will see in what follows, the approximate boundary null controllability of system (I) could be realized if the number $M =$ rank(D) is very small, even if $M =$ rank$(D) = 1$. Nevertheless, Theorems 8.6 and 8.9 show that if system (I) is approximately null controllable, then the enlarged matrix $(D, AD, \cdots, A^{N-1}D)$, composed of the coupling matrix A and the boundary control matrix D, should be of full row-rank. That is to say, even if the rank of D might be small, but because of the existence and influence of the coupling matrix

A, in order to realize the approximate boundary null controllability, the rank of the enlarged matrix $(D, AD, \cdots, A^{N-1}D)$ should be still equal to N, the number of state variables. From this point of view, we may say that the rank M of D is the **number of "direct" boundary controls** acting on Γ_1, and rank$(D, AD, \cdots, A^{N-1}D)$ denotes the **"total" number of direct and indirect controls**, while the number of "indirect" controls is given by the difference: rank$(D, AD, \cdots, A^{N-1}D)$ – rank(D), which is equal to $(N - M)$ in the case of approximate boundary null controllability. It is different from the exact boundary null controllability, in which only the number rank(D) of direct boundary controls is concerned and $M = \text{rank}(D)$ should be equal to N, that for the approximate boundary null controllability, we should consider not only the number of direct boundary controls, but also the number of indirect controls, namely, the number of the total (direct and indirect) controls.

Remark 8.10 It is well known that Kalman's criterion (8.28) is necessary and sufficient for the exact controllability of systems of ODEs (cf. [23, 73]). But, the situation is more complicated for hyperbolic distributed parameter systems. Because of the finite speed of wave propagation, it is natural to consider Kalman's criterion only in the case that $T > 0$ is sufficiently large. However, the following theorem precisely shows that even on the infinite observation interval $[0, +\infty)$, the sufficiency of Kalman's criterion may fail. So, some algebraic assumptions on the coupling matrix A should be required to guarantee the sufficiency of Kalman's criterion.

Theorem 8.11 *Let μ_n^2 and ϕ_n be defined by (8.29). Assume that the set*

$$\Lambda = \{(m,n): \quad \mu_m \neq \mu_n, \quad \partial_\nu \phi_m = \partial_\nu \phi_n \text{ on } \Gamma_1\} \tag{8.33}$$

is not empty. For any given $(m, n) \in \Lambda$, let

$$\epsilon = \frac{\mu_m^2 - \mu_n^2}{2}. \tag{8.34}$$

Then the adjoint system

$$\begin{cases} \phi'' - \Delta\phi + \epsilon\psi = 0 \text{ in } (0, +\infty) \times \Omega, \\ \psi'' - \Delta\psi + \epsilon\phi = 0 \text{ in } (0, +\infty) \times \Omega, \\ \phi = \psi = 0 \qquad\quad \text{on } (0, +\infty) \times \Gamma \end{cases} \tag{8.35}$$

admits a nontrivial solution (ϕ, ψ) such that

$$\partial_\nu \phi \equiv 0 \quad \text{on } [0, +\infty) \times \Gamma_1, \tag{8.36}$$

hence the corresponding adjoint problem (8.7) is not D-observable with $D = (1, 0)^T$.

Proof Let

$$\phi_\lambda = (\phi_n - \phi_m), \quad \psi_\lambda = (\phi_n + \phi_m), \quad \lambda^2 = \frac{\mu_m^2 + \mu_n^2}{2}. \tag{8.37}$$

We check easily that $(\phi_\lambda, \psi_\lambda)$ satisfies the following eigensystem:

$$\begin{cases} \lambda^2\phi_\lambda + \Delta\phi_\lambda - \epsilon\psi_\lambda = 0 \text{ in } \Omega, \\ \lambda^2\psi_\lambda + \Delta\psi_\lambda - \epsilon\phi_\lambda = 0 \text{ in } \Omega, \\ \phi_\lambda = \psi_\lambda = 0 \quad\quad\quad \text{on } \Gamma. \end{cases} \tag{8.38}$$

Moreover, noting the definition (8.33) of Λ, we have

$$\partial_\nu\phi_\lambda \equiv 0 \quad \text{on } \Gamma_1. \tag{8.39}$$

Now, let

$$\phi = e^{i\lambda t}\phi_\lambda, \quad \psi = e^{i\lambda t}\psi_\lambda. \tag{8.40}$$

It is easy to see that (ϕ, ψ) is a nontrivial solution to system (8.35) and satisfies the observation (8.36).

In order to complete the proof, we examine the following situations (there are many others!), in which the set Λ is indeed not empty.

(i) $\Omega = (0, \pi)$ with $\Gamma_1 = \{0\}$. In this case, we have

$$\mu_n = n, \quad \phi_n = \frac{1}{n}\sin nx \text{ and } \phi_n'(0) = 1. \tag{8.41}$$

So, $(m, n) \in \Lambda$ for all $m \neq n$.

(ii) $\Omega = (0, \pi) \times (0, \pi)$ with $\Gamma_1 = \{0\} \times [0, \pi]$. Setting

$$\mu_{m,n} = \sqrt{m^2 + n^2}, \quad \phi_{m,n} = \frac{1}{m}\sin mx \sin ny, \tag{8.42}$$

we have

$$\frac{\partial}{\partial x}\phi_{m,n}(0, y) = \frac{\partial}{\partial x}\phi_{m',n}(0, y) = \sin ny, \quad 0 \leqslant y \leqslant \pi. \tag{8.43}$$

So $\big(\{m, n\}, \{m', n\}\big) \in \Lambda$ for all $m \neq m'$ and $n \geqslant 1$.

The proof is thus complete. □

Remark 8.12 Note that

$$D = \begin{pmatrix} 1 \\ 0 \end{pmatrix}, \quad A = \begin{pmatrix} 0 & \epsilon \\ \epsilon & 0 \end{pmatrix}, \quad (D, AD) = \begin{pmatrix} 1 & 0 \\ 0 & \epsilon \end{pmatrix} \tag{8.44}$$

correspond to the adjoint system (8.35) with the observation (8.36). Since the matrices A and D satisfy well the corresponding Kalman's criterion (8.28), Theorem 8.11 shows that even in the infinite interval of observation, Kalman's criterion is not sufficient for the D-observability of the adjoint problem (8.35) at least in the two cases mentioned above. Therefore, in order to get the D-observability, some additional algebraic assumptions on A should be imposed.

8.4 Sufficiency of Kalman's Criterion for $T > 0$ Large Enough for the Nilpotent System

A matrix A of order N is called to be nilpotent if there exists an integer k with $1 \leqslant k \leqslant N$, such that $A^k = 0$. Then it is easy to see that A is nilpotent if and only if all the eigenvalues of A are equal to zero. Thus, under a suitable basis B, a **nilpotent matrix** A can be written in a diagonal form of Jordan blocks:

$$B^{-1}AB = \begin{pmatrix} J_p & & \\ & J_q & \\ & & \ddots \end{pmatrix}, \tag{8.45}$$

where J_p is the following Jordan block of order p:

$$J_p = \begin{pmatrix} 0 & 1 & & & \\ & 0 & 1 & & \\ & & \ddots & \ddots & \\ & & & 0 & 1 \\ & & & & 0 \end{pmatrix}. \tag{8.46}$$

When A is a sole Jordan block, so-called the cascade matrix, it was shown in [2] that the observation on the last component of adjoint variable of the adjoint problem (8.7) is sufficient for the corresponding D-observability. In this section, we will generalize this result to the nilpotent system.

We first consider a very special case.

Proposition 8.13 *Let $A = aI$, where a is a real number. If Kalman's criterion (8.28) holds, then the adjoint problem (8.7) is D-observable, provided that $T > 0$ is large enough.*

Proof In this case, Kalman's criterion (8.28) implies that $M = N$. Consequently, the observation (8.8) implies that

$$\partial_\nu \Phi \equiv 0 \quad \text{on } [0, T] \times \Gamma_1. \tag{8.47}$$

Thus, by the classical Holmgren's uniqueness theorem, we get $\Phi \equiv 0$, provided that $T > 0$ is large enough. In this situation, we don't need any multiplier geometrical condition on the domain Ω. The proof is complete. □

Lemma 8.14 *Assume that there exists an invertible matrix P such that $PA = AP$. Then the adjoint problem (8.7) is D-observable if and only if it is PD-observable.*

Proof Let $\widetilde{\Phi} = P^{-T}\Phi$. Noting $PA = AP$, the new variable $\widetilde{\Phi}$ satisfies the same system as in problem (8.7). On the other hand, since

$$D^T \partial_\nu \Phi = (PD)^T \partial_\nu \widetilde{\Phi} \quad \text{on } \Gamma_1, \tag{8.48}$$

the D-observability on Φ is equivalent to the PD-observability on $\widetilde{\Phi}$. The proof is complete. □

Proposition 8.15 *Let P be an invertible matrix. Define*

$$\tilde{A} = PAP^{-1} \quad \text{and} \quad \widetilde{D} = PD.$$

Then the matrices A and D satisfy Kalman's criterion (8.28) if and only if the matrices $\tilde{A}$ and $\widetilde{D}$ do so.

Proof It is sufficient to note that

$$[\widetilde{D}, \tilde{A}\widetilde{D}, \cdots, \tilde{A}^{N-1}\widetilde{D} = P[D, AD, \cdots, A^{N-1}D]$$

and that P is invertible. □

Theorem 8.16 *Assume that $\Omega \subset \mathbb{R}^n$ satisfies the multiplier geometrical condition (3.1). Let $T > 0$ be large enough. Assume that the coupling matrix A is nilpotent. Then Kalman's criterion (8.28) is sufficient for the D-observability of the adjoint system (8.7).*

Proof (i) Case that A is a Jordan block (the cascade matrix):

$$A = \begin{pmatrix} 0 & 1 & & & \\ & 0 & 1 & & \\ & & \cdot & \cdot & \\ & & & 0 & 1 \\ & & & & 0 \end{pmatrix} =: J_N. \tag{8.49}$$

Noting that $E = (0, \cdots, 0, 1)^T$ is the only eigenvector of A^T, by Proposition 2.12 (ii), A and D satisfy Kalman's criterion (8.28) if and only if

$$D^T E \neq 0, \tag{8.50}$$

namely, if and only if the last row of D is not a zero vector. We denote by $d = (d_1, d_2, \cdots, d_N)^T$ a column of D with $d_N \neq 0$ and let

$$P = \begin{pmatrix} d_N & d_{N-1} & \cdot & d_1 \\ 0 & d_N & \cdot & d_2 \\ \cdot & \cdot & \cdot & \cdot \\ 0 & 0 & d_N & d_{N-1} \\ 0 & 0 & 0 & d_N \end{pmatrix}. \tag{8.51}$$

Obviously, P is invertible. Noting that

$$P = d_N I + d_{N-1} J_N + \cdots + d_1 J_N^{N-1}, \tag{8.52}$$

it is easy to see that $PA = AP$. On the other hand, under the multiplier geometrical condition (3.1) on Ω, it was shown in [2] that the adjoint system (8.7) with the coupling matrix (8.49) is D_0-observable with

$$D_0 = (0, \cdots, 0, 1)^T. \tag{8.53}$$

Thus, by Lemma 8.14, the same system is PD_0-observable, then, noting that

$$PD_0 = (d_1, \cdots, d_N)^T \tag{8.54}$$

is a submatrix of D, problem (8.7) must be D-observable.

(ii) Case that A is composed of two Jordan blocks of the same size:

$$A = \begin{pmatrix} J_p & 0 \\ 0 & J_p \end{pmatrix}, \tag{8.55}$$

where J_p is the Jordan block of order p.

First, let ϵ_i be defined by

$$\epsilon_i = (0, \cdots, 0, \overset{(i)}{1}, 0, \cdots, 0)^T \tag{8.56}$$

for $i = 1, \cdots, 2p$. Consider the special boundary control matrix

$$D_0 = (\epsilon_p, \epsilon_{2p}). \tag{8.57}$$

Noting that in this situation the adjoint system (8.7) and the observation (8.8) are entirely decoupled into two independent subsystems, each of them satisfies Kalman's criterion (8.28) with $N = p$. Then, by the consideration in the previous case, the adjoint system (8.7) is D_0-observable.

Now, let us consider the general boundary control matrix of order $2p \times M$:

$$D = \begin{pmatrix} a_1 & c_1 & \cdots\cdots \\ \vdots & \vdots & \\ a_p & c_p & \cdots\cdots \\ b_1 & d_1 & \cdots\cdots \\ \vdots & \vdots & \\ b_p & d_p & \cdots\cdots \end{pmatrix}. \tag{8.58}$$

Noting that ϵ_p and ϵ_{2p} are the only two eigenvectors of A^T, associated with the same eigenvalue zero, then for any given real numbers α and β with $\alpha^2 + \beta^2 \neq 0$, $\alpha\epsilon_p + \beta\epsilon_{2p}$ is also an eigenvector of A^T, by Proposition 2.12 (ii), Kalman's criterion (8.28) holds if and only if $D^T\epsilon_p$ and $D^T\epsilon_{2p}$, namely, the row vectors

$$(a_p, c_p, \cdots\cdots), \quad (b_p, d_p, \cdots\cdots), \tag{8.59}$$

are linearly independent. Without loss of generality, we may assume that

$$a_p d_p - b_p c_p \neq 0. \tag{8.60}$$

Let the matrix P of order $2p$ be defined by

$$P = \left(\begin{array}{cccc|cccc} a_p & a_{p-1} & \cdots & a_1 & b_p & b_{p-1} & \cdots & b_1 \\ 0 & a_p & \cdots & a_2 & 0 & b_p & \cdots & b_2 \\ \vdots & \vdots & \ddots & \vdots & \vdots & \vdots & \ddots & \vdots \\ 0 & 0 & \cdots & a_p & 0 & 0 & \cdots & b_p \\ \hline c_p & c_{p-1} & \cdots & c_1 & d_p & d_{p-1} & \cdots & d_1 \\ 0 & c_p & \cdots & c_2 & 0 & d_p & \cdots & d_2 \\ \vdots & \vdots & \ddots & \vdots & \vdots & \vdots & \ddots & \vdots \\ 0 & 0 & \cdots & c_p & 0 & 0 & \cdots & d_p \end{array}\right) = \begin{pmatrix} P_{11} & P_{12} \\ P_{21} & P_{22} \end{pmatrix}. \tag{8.61}$$

Since P_{11}, P_{12}, P_{21}, and P_{22} have a similar structure as (8.52) for P, we check easily that

$$PA = \begin{pmatrix} P_{11}J_p & P_{12}J_p \\ P_{21}J_p & P_{22}J_p \end{pmatrix} = \begin{pmatrix} J_pP_{11} & J_pP_{12} \\ J_pP_{21} & J_pP_{22} \end{pmatrix} = AP. \tag{8.62}$$

Moreover, P is invertible under condition (8.60). Since the adjoint system (8.7) is D_0-observable, by Lemma 8.14, it is PD_0-observable, then, D-observable since PD_0 is composed of the first two columns of D.

(iii) Case that A is composed of two Jordan blocks with different sizes:

$$A = \begin{pmatrix} J_p & 0 \\ 0 & J_q \end{pmatrix} \tag{8.63}$$

with $q < p$. In this case, the adjoint system (8.7) is composed of the first subsystem

$$i = 1, \cdots, p: \quad \begin{cases} \phi_i'' - \Delta\phi_i + \phi_{i-1} = 0 & \text{in } (0, +\infty) \times \Omega, \\ \phi_i = 0 & \text{on } (0, +\infty) \times \Gamma \end{cases} \tag{8.64}$$

with $\phi_0 = 0$, and of the second subsystem

$$j = p - q + 1, \cdots, p: \quad \begin{cases} \psi_j'' - \Delta\psi_j + \psi_{j-1} = 0 & \text{in } (0, +\infty) \times \Omega, \\ \psi_j = 0 & \text{on } (0, +\infty) \times \Gamma \end{cases} \tag{8.65}$$

with $\psi_{p-q} = 0$. These two subsystems (8.64) and (8.65) are coupled together by the D-observations:

$$\begin{cases} \sum_{i=1}^{p} a_i \partial_\nu \phi_i + \sum_{j=p-q+1}^{p} b_j \partial_\nu \psi_j = 0 & \text{on } (0,T) \times \Gamma_1, \\ \sum_{i=1}^{p} c_i \partial_\nu \phi_i + \sum_{j=p-q+1}^{p} d_j \partial_\nu \psi_j = 0 & \text{on } (0,T) \times \Gamma_1, \\ \cdots\cdots\cdots\cdots \\ \cdots\cdots\cdots\cdots \end{cases} \tag{8.66}$$

In order to transfer the problem to the case $p = q$, we expand the second subsystem (8.65) from $\{p-q+1, \cdots, p\}$ to $\{1, \cdots, p\}$:

$$j = 1, \cdots, p: \quad \begin{cases} \psi_j'' - \Delta \psi_j + \psi_{j-1} = 0 & \text{in } (0, +\infty) \times \Omega, \\ \psi_j = 0 & \text{on } (0, +\infty) \times \Gamma \end{cases} \tag{8.67}$$

with $\psi_0 = 0$, so that the two subsystems (8.64) and (8.67) have the same size.

Accordingly, the D-observations (8.66) can be extended to the $\widetilde{D}$-observations:

$$\begin{cases} \sum_{i=1}^{p} a_i \partial_\nu \phi_i + \sum_{j=1}^{p} b_j \partial_\nu \psi_j = 0 & \text{on } (0,T) \times \Gamma_1, \\ \sum_{i=1}^{p} c_i \partial_\nu \phi_i + \sum_{j=1}^{p} d_j \partial_\nu \psi_j = 0 & \text{on } (0,T) \times \Gamma_1, \\ \cdots\cdots\cdots\cdots \\ \cdots\cdots\cdots\cdots \end{cases} \tag{8.68}$$

with arbitrarily given b_j and d_j for $j = 1, \cdots, p-q$.

Let us write the matrix D of order $(p+q) \times M$ of the observations (8.66) as

$$D = \begin{pmatrix} a_1 & c_1 & \cdots\cdots \\ \vdots & \vdots & \\ a_p & c_p & \cdots\cdots \\ b_{p-q+1} & d_{p-q+1} & \cdots\cdots \\ \vdots & \vdots & \\ b_p & d_p & \cdots\cdots \end{pmatrix}. \tag{8.69}$$

Similarly to the case (ii), ϵ_p and ϵ_{p+q} are the only two eigenvectors of A^T, associated with the same eigenvalue zero, then for any given real numbers α and β with $\alpha^2 + \beta^2 \neq 0$, $\alpha\epsilon_p + \beta\epsilon_{p+q}$ is also an eigenvector of A^T. By Proposition 2.12 (ii), Kalman's criterion (8.28) holds if and only if $D^T\epsilon_p$ and $D^T\epsilon_{p+q}$, namely, the row vectors

$$(a_p, c_p, \cdots\cdots), \quad (b_p, d_p, \cdots\cdots), \tag{8.70}$$

are linearly independent. Without loss of generality, we may assume that (8.60) holds.

Similarly, we write the matrix $\widetilde{D}$ of order $2p \times M$ of the extended observations (8.68) as

$$\widetilde{D} = \begin{pmatrix} a_1 & c_1 & \cdots\cdots \\ \vdots & \vdots & \\ a_p & c_p & \cdots\cdots \\ b_1 & d_1 & \cdots\cdots \\ \vdots & \vdots & \\ b_{p-q} & d_{p-q} & \\ b_{p-q+1} & d_{p-q+1} & \cdots\cdots \\ \vdots & \vdots & \\ b_p & d_p & \cdots\cdots \end{pmatrix}. \tag{8.71}$$

The matrix $\widetilde{A}$ of order $2p$ of the extended adjoint system, composed of (8.64) and (8.67), is the same as given in (8.55). Then, $\widetilde{A}$ and $\widetilde{D}$ satisfy the corresponding Kalman's criterion if and only if the condition (8.60) holds. Then by the conclusion of (ii), the extended adjoint system composed of (8.64) and (8.67) is $\widetilde{D}$-observable in the space $(H_0^1(\Omega))^{2p} \times (L^2(\Omega))^{2p}$.

In particular, if we choose the specific initial data such that

$$t = 0: \quad \psi_1 = \cdots = \psi_{p-q} = 0 \quad \text{and} \quad \psi_1' = \cdots = \psi_{p-q}' = 0 \tag{8.72}$$

for the extended subsystem (8.67), then by well-posedness we have

$$\psi_1 \equiv \cdots \equiv \psi_{p-q} \equiv 0 \quad \text{in } (0, +\infty) \times \Omega. \tag{8.73}$$

Thus, the extended adjoint system composed of (8.64) and (8.67) with the $\widetilde{D}$-observations (8.68) reduces to the original adjoint system composed of (8.64) and (8.65) with the D-observations (8.66), hence we get that the original adjoint system is D-observable in the space $(H_0^1(\Omega))^{p+q} \times (L^2(\Omega))^{p+q}$.

The case that A is composed of several Jordan blocks can be carried on similarly.

Since any given nilpotent matrix can be decomposed in a diagonal form of Jordan blocks under a suitable basis, by Proposition 8.15, the previous conclusion is still valid for any given nilpotent matrix A. The proof is complete. □

Theorem 8.17 *Assume that $\Omega \subset \mathbb{R}^n$ satisfies the multiplier geometrical condition (3.1). Let $T > 0$ be large enough. Assume furthermore that the coupling matrix A admits one sole eigenvalue $\lambda \geqslant 0$. Then the adjoint problem (8.7) is D-observable if and only if the matrix D satisfies Kalman's criterion (8.28).*

Proof In fact, the operator $-\Delta + \lambda I$ is still self-adjoint and coercive in $L^2(\Omega)$ and $A - \lambda I$ is nilpotent. So, Theorem 8.16 remains true as $-\Delta$ is replaced by $-\Delta + \lambda I$. □

8.5 Sufficiency of Kalman's Criterion for $T > 0$ Large Enough for 2×2 Systems

Theorem 8.18 *Let $\Omega \subset \mathbb{R}^n$ satisfy the multiplier geometrical condition (3.1). Assume that the 2×2 matrix A has real eigenvalues $\lambda \geqslant 0$ and $\mu \geqslant 0$ such that*

$$|\lambda - \mu| \leqslant \epsilon_0, \tag{8.74}$$

where $\epsilon_0 > 0$ is small enough. Then the adjoint problem (8.7) is D-observable for $T > 0$ large enough if and only if the matrix D satisfies Kalman's criterion (8.28).

Proof By Theorem 8.9, we only need to prove the sufficiency.

If D is invertible, then the observation (8.8) implies that

$$\partial_\nu \Phi \equiv 0 \quad \text{on } [0, T] \times \Gamma_1 \tag{8.75}$$

and the classical Holmgren's uniqueness theorem, implies that $\Phi \equiv 0$, provided that $T > 0$ is large enough. Noting that in this case, we do not need the multiplier geometrical condition (3.1) on Ω. Thus, we only need to consider the case that D is of rank one. Without loss of generality, we may consider the following three cases.

(a) If $\lambda = \mu$ with two linearly independent eigenvectors, then any nontrivial vector is an eigenvector of A^T. By (ii) of Lemma 2.12 with $d = 0$, $\text{Ker}(D^T)$ is reduced to $\{0\}$, then the matrix D is invertible, which gives a contradiction.

(b) If $\lambda = \mu$ with one eigenvector, then

$$A \sim \begin{pmatrix} \lambda & 1 \\ 0 & \lambda \end{pmatrix}. \tag{8.76}$$

Since $\lambda \geqslant 0$, it suffices to apply Theorem 8.17.

(c) If $\lambda \neq \mu$, then

$$A \sim \begin{pmatrix} \frac{\lambda+\mu}{2} & 0 \\ 0 & \frac{\lambda+\mu}{2} \end{pmatrix} + \begin{pmatrix} 0 & \frac{\lambda-\mu}{2} \\ \frac{\lambda-\mu}{2} & 0 \end{pmatrix}. \tag{8.77}$$

Since $\lambda + \mu \geqslant 0$, the operator $-\Delta + \frac{\lambda+\mu}{2} I$ is still coercive in $H_0^1(\Omega)$, then it is sufficient to consider the adjoint problem (8.7), in which $-\Delta$ is replaced by $-\Delta + \frac{\lambda+\mu}{2} I$, with the matrix

$$A = \begin{pmatrix} 0 & \frac{\lambda-\mu}{2} \\ \frac{\lambda-\mu}{2} & 0 \end{pmatrix}. \tag{8.78}$$

Since Ω satisfies the multiplier geometrical condition (3.1), following a result in [1] (cf. also [65] in the case of different speeds of wave propagation), the corresponding adjoint problem (8.7) is D_0-observable with

$$D_0 = \begin{pmatrix} 1 \\ 0 \end{pmatrix}. \tag{8.79}$$

On the other hand, since D is of rank one, we have

$$D = \begin{pmatrix} a \\ b \end{pmatrix}, \quad A = \begin{pmatrix} 0 & \epsilon \\ \epsilon & 0 \end{pmatrix}, \quad (D, AD) = \begin{pmatrix} a & \epsilon b \\ b & \epsilon a \end{pmatrix} \tag{8.80}$$

with $\epsilon = \frac{\lambda - \mu}{2}$. So, Kalman's criterion (8.28) is satisfied if and only if $a^2 \neq b^2$. Then the matrix

$$P = \begin{pmatrix} a & b \\ b & a \end{pmatrix} \tag{8.81}$$

is invertible and commutes with A. Thus, by Lemma 8.14, the adjoint problem (8.7) is also PD_0-observable, but

$$PD_0 = \begin{pmatrix} a \\ b \end{pmatrix} = D. \tag{8.82}$$

The proof is complete. □

Remark 8.19 Theorem 8.11 shows that the condition "$\epsilon_0 > 0$ is small enough" in Theorem 8.18 is actually necessary.

Proposition 8.20 *Let $\Omega \subset \mathbb{R}^n$ satisfy the multiplier geometrical condition (3.1) and $|\epsilon| > 0$ be small enough. Let $T > 0$ be large enough. Then the adjoint system (8.35) is D-observable if and only if the matrix D satisfies the corresponding Kalman's criterion.*

Proof Since Ω satisfies the multiplier geometrical condition (3.1), following a result in [1], the adjoint system (8.35) is D_0-observable with

$$D_0 = \begin{pmatrix} 1 \\ 0 \end{pmatrix}. \tag{8.83}$$

Noting that

$$D = \begin{pmatrix} a \\ b \end{pmatrix}, \quad A = \begin{pmatrix} 0 & \epsilon \\ \epsilon & 0 \end{pmatrix}, \quad (D, AD) = \begin{pmatrix} a & \epsilon b \\ b & \epsilon a \end{pmatrix}, \tag{8.84}$$

Kalman's criterion (8.28) is satisfied if and only if $a^2 \neq b^2$. On the other hand, the matrix

$$P = \begin{pmatrix} a & b \\ b & a \end{pmatrix} \tag{8.85}$$

is invertible and commutes with A. Thus, by Lemma 8.14, the adjoint problem (8.35) is also PD_0-observable with

$$PD_0 = \begin{pmatrix} a \\ b \end{pmatrix} = D. \tag{8.86}$$

The proof is complete. □

Remark 8.21 Because of the equivalence between the approximate boundary null controllability and the D-observability (cf. Theorem 8.6), in the cases discussed in Theorems 8.16 and 8.18, Kalman's criterion (8.28) is indeed necessary and sufficient to the approximate boundary null controllability for the corresponding system (I). In the one-space-dimensional case, some more general results can be obtained in this direction, and the following two sections are devoted to this end.

8.6 The Unique Continuation for Nonharmonic Series

In this section, we will establish the unique continuation for nonharmonic series, which will be used in the next section to prove the sufficiency of Kalman's criterion (8.28) to the D-observability for $T > 0$ large enough for diagonalizable systems in one-space-dimensional case. The study is based on a generalized Ingham's inequality (cf. [28]).

Let $\mathbb{Z}^*$ denote the set of all nonzero integers and $\{\beta_n^{(l)}\}_{1 \leqslant l \leqslant m, n \in \mathbb{Z}^*}$ be a strictly increasing real sequence:

$$\cdots \beta_{-1}^{(1)} < \cdots < \beta_{-1}^{(m)} < \beta_1^{(1)} < \cdots < \beta_1^{(m)} < \cdots . \tag{8.87}$$

We want to show that the following condition

$$\sum_{n \in \mathbb{Z}^*} \sum_{l=1}^{m} a_n^{(l)} e^{i\beta_n^{(l)} t} = 0 \quad \text{on } [0, T] \tag{8.88}$$

with

$$\sum_{n \in \mathbb{Z}^*} \sum_{l=1}^{m} |a_n^{(l)}|^2 < +\infty \tag{8.89}$$

implies that

$$a_n^{(l)} = 0, \quad n \in \mathbb{Z}^*, \quad 1 \leqslant l \leqslant m, \tag{8.90}$$

provided that $T > 0$ is large enough. If this is true, we say that the sequence $\{e^{i\beta_n^{(l)} t}\}_{1 \leqslant l \leqslant m; n \in \mathbb{Z}^*}$ is ω-**linearly independent** in $L^2(0, T)$.

Theorem 8.22 *Assume that (8.87) holds and that there exist positive constants c, s, and γ such that*

$$\beta_{n+1}^{(l)} - \beta_n^{(l)} \geqslant m\gamma, \tag{8.91}$$

$$\frac{c}{|n|^s} \leqslant \beta_n^{(l+1)} - \beta_n^{(l)} \leqslant \gamma \tag{8.92}$$

for all $1 \leqslant l \leqslant m$ *and all* $n \in \mathbb{Z}^*$ *with* $|n|$ *large enough. Then the sequence* $\{e^{i\beta_n^{(l)}t}\}_{1\leqslant l\leqslant m; n\in\mathbb{Z}^*}$ *is* ω*-linearly independent in* $L^2(0,T)$*, provided that* $T > 2\pi D^+$*, where* D^+ *is the upper density of the sequence* $\{\beta_n^{(l)}\}_{1\leqslant l\leqslant m; n\in\mathbb{Z}^*}$*, defined by*

$$D^+ = \limsup_{R\to+\infty} \frac{N(R)}{2R}, \tag{8.93}$$

where $N(R)$ *denotes the number of* $\{\beta_n^{(l)}\}$ *contained in the interval* $[-R, R]$.

Proof We first define the sequence of difference quotients as follows:

$$e_n^{(l)}(t) = \sum_{p=1}^{l}\Big(\prod_{q=1,q\neq p}^{l}(\beta_n^{(p)}-\beta_n^{(q)})^{-1}\Big)e^{i\beta_n^{(p)}t} \tag{8.94}$$

for $l = 1, \cdots, m$ and $n \in \mathbb{Z}^*$. Since the sequence $\{\beta_n^{(l)}\}_{1\leqslant l\leqslant m, n\in\mathbb{Z}^*}$ can be asymptotically close at the rate $\frac{1}{|n|^s}$, the classical Ingham's theorem does not work. We will use a generalized Ingham-type theorem based on the divided differences, which tolerates asymptotically close frequencies $\{\beta_n^{(l)}\}_{1\leqslant l\leqslant m, n\in\mathbb{Z}^*}$ (cf. Theorem 9.4 in [28]). Namely, under conditions (8.87) and (8.91)–(8.92), the sequence of difference quotients $\{e_n^{(l)}\}_{1\leqslant l\leqslant m, n\in\mathbb{Z}^*}$ is a Riesz sequence in $L^2(0,T)$, provided that $T > 2\pi D^+$.

Let the lower triangular matrix $A_n = (a_n^{(l,p)})$ be defined by

$$a_n^{(1,1)} = 1; \quad a_n^{(l,p)} = \prod_{q=1,q\neq p}^{l}(\beta_n^{(p)}-\beta_n^{(q)})^{-1} \tag{8.95}$$

for $1 < l \leqslant m$ and $1 \leqslant p \leqslant l$. Since the diagonals $a_n^{(l,l)}$ are positive for all $1 \leqslant l \leqslant m$, A_n is invertible. Then, setting $A_n^{-1} = B_n = (b_n^{(l,p)})$, we write (8.94) as

$$e^{i\beta_n^{(l)}t} = \sum_{p=1}^{l} b_n^{(l,p)} e_n^{(p)}(t), \quad l = 1, \cdots, m. \tag{8.96}$$

Inserting (8.96) into (8.88), we get

$$\sum_{n\in\mathbb{Z}^*}\sum_{p=1}^{m}\widetilde{a}_n^{(p)} e_n^{(p)}(t) = 0 \quad \text{on } [0,T], \tag{8.97}$$

where

$$\widetilde{a}_n^{(p)} = \sum_{l=p}^{m} b_n^{(l,p)} a_n^{(l)}. \tag{8.98}$$

Assume for the time being that

$$\sum_{n\in\mathbb{Z}^*}\sum_{p=1}^{m}|\widetilde{a}_n^{(p)}|^2 < +\infty. \tag{8.99}$$

Then by means of the property of Riesz sequence, it follows from (8.97) and (8.99) that

$$\widetilde{a}_n^{(p)} = 0, \quad 1 \leqslant p \leqslant m, \quad n \in \mathbb{Z}^*, \tag{8.100}$$

which implies (8.90).

Now we return to the verification of (8.99). From (8.98), it is sufficient to show that the matrix B_n is uniformly bounded for all n. From (8.92) and (8.95), we have

$$b_n^{(l,l)} = \frac{1}{a_n^{(l,l)}} = \prod_{q=1}^{l-1}(\beta_n^{(l)} - \beta_n^{(q)}) \leqslant c_1\gamma^{l-1}, \quad 1 \leqslant l \leqslant m, \tag{8.101}$$

where c_1 is a positive constant independent of n. Since B_n is also a lower triangular matrix, without loss of generality, we assume that $\gamma < 1$. Then, it follows that the spectral radius $\rho(B_n) \leqslant c_1$. It is well known that for any given $\tilde{\epsilon} > 0$, there exists a vector norm in $\mathbb{R}^m$, such that the subordinate matrix norm satisfies

$$\|B_n\| \leqslant (\rho(B_n) + \tilde{\epsilon}) \leqslant c_1 + 1, \quad \forall n \in \mathbb{Z}^*. \tag{8.102}$$

The proof is then complete. □

Corollary 8.23 *For*

$$\delta_1 < \delta_2 < \cdots < \delta_m, \tag{8.103}$$

we define

$$\begin{cases} \beta_n^{(l)} = \sqrt{n^2 + \epsilon\delta_l}, \; l = 1, 2, \cdots, m, \quad n \geqslant 1, \\ \beta_{-n}^{(l)} = -\beta_n^{(l)}, \qquad l = 1, 2, \cdots, m, \quad n \geqslant 1. \end{cases} \tag{8.104}$$

Then, for $|\epsilon| > 0$ small enough, the sequence $\{e^{i\beta_n^{(l)}t}\}_{1\leqslant l\leqslant m; n\in\mathbb{Z}^*}$ is ω-linearly independent in $L^2(0, T)$, provided that

$$T > 2m\pi. \tag{8.105}$$

Proof First, for $|\epsilon| > 0$ small enough, the sequence $\{\beta_n^{(l)}\}_{1\leqslant l\leqslant m; n\in\mathbb{Z}^*}$ satisfies (8.87). On the other hand, a straightforward computation gives that

$$\beta_{n+1}^{(l)} - \beta_n^{(l)} = O(1) \tag{8.106}$$

and

$$\beta_n^{(l+1)} - \beta_n^{(l)} = \frac{(\delta_{l+1} - \delta_l)\epsilon}{\sqrt{n^2 + \delta_{l+1}\epsilon} + \sqrt{n^2 + \delta_l\epsilon}} = o\left(\left|\frac{\epsilon}{n}\right|\right) \tag{8.107}$$

for $|n|$ large enough. Then the sequence $\{\beta_n^{(l)}\}_{1\leqslant l\leqslant m; n\in\mathbb{Z}^*}$ satisfies all the requirements of Theorem 8.22 with $s = 1$ and $D^+ = m$. Consequently, the sequence $\{e^{i\beta_n^{(l)}t}\}_{1\leqslant l\leqslant m; n\in\mathbb{Z}^*}$ is ω-linearly independent in $L^2(0,T)$, provided that (8.105) holds. □

8.7 Sufficiency of Kalman's Criterion for $T > 0$ Large Enough in the One-Space-Dimensional Case

In this section, under suitable conditions on the coupling matrix A and for $|\epsilon| > 0$ small enough, we will first establish the sufficiency of Kalman's criterion (8.28) for the following one-space-dimensional problem:

$$\begin{cases} \Phi'' - \Phi_{xx} + \epsilon A^T\Phi = 0 & \text{in } (0,+\infty)\times(0,\pi), \\ \Phi(t,0) = \Phi(t,\pi) = 0 & \text{on } (0,+\infty), \\ t = 0: \quad \Phi = \widehat{\Phi}_0, \ \Phi' = \widehat{\Phi}_1 & \text{in } (0,\pi) \end{cases} \tag{8.108}$$

with the observation at the end $x = 0$:

$$D^T\Phi_x(t,0) \equiv 0 \quad \text{on } [0,T]. \tag{8.109}$$

We will next give the optimal time of observation for the coupling matrix A, which has distinct real eigenvalues.

First assume that A^T is diagonalizable with the real eigenvalues:

$$\delta_1 < \delta_2 < \cdots < \delta_m \tag{8.110}$$

and the corresponding eigenvectors $w^{(l,\mu)}$ such that

$$A^T w^{(l,\mu)} = \delta_l w^{(l,\mu)}, \quad 1 \leqslant l \leqslant m, \quad 1 \leqslant \mu \leqslant \mu_l, \tag{8.111}$$

in which

$$\sum_{l=1}^m \mu_l = N. \tag{8.112}$$

Let

$$e_n = \sin nx, \quad n \geqslant 1 \tag{8.113}$$

be the eigenfunctions of $-\Delta$ in $H_0^1(0,\pi)$. Then $e_n w^{(l,\mu)}$ is an eigenvector of $-\Delta + \epsilon A^T$ corresponding to the eigenvalue $n^2 + \epsilon\delta_l$. Furthermore, we define $\{\beta_n^{(l)}\}_{1\leqslant l\leqslant m; n\in\mathbb{Z}^*}$ as in (8.104) and the corresponding eigenvectors of the system given in (8.108) by

$$E_n^{(l,\mu)} = \begin{pmatrix} \dfrac{e_n w^{(l,\mu)}}{i\beta_n^{(l)}} \\ e_n w^{(l,\mu)} \end{pmatrix}, \quad 1 \leqslant l \leqslant m, \quad 1 \leqslant \mu \leqslant \mu_l, \quad n \in \mathbb{Z}^*, \tag{8.114}$$

in which we define $e_{-n} = e_n$ for all $n \geqslant 1$. Since the eigenfunctions $e_n (n \geqslant 1)$ are orthogonal in $L^2(0, \pi)$ as well as in $H_0^1(0, \pi)$, the linear hull $\text{Span}\{E_n^{(l,\mu)}\}_{1\leqslant l\leqslant m, 1\leqslant\mu\leqslant\mu_l}$ of finite dimension N are mutually orthogonal for all $n \geqslant 1$. On the other hand, the system of eigenvectors (8.114) is complete in $(H_0^1(0, \pi))^N \times (L^2(0, \pi))^N$, therefore it forms a Hilbert basis of subspaces, then a Riesz basis in $(H_0^1(0, \pi))^N \times (L^2(0, \pi))^N$ (cf. [15] for the basis of subspaces).

For any given initial data

$$\begin{pmatrix} \widehat{\Phi}_0 \\ \widehat{\Phi}_1 \end{pmatrix} = \sum_{n\in\mathbb{Z}^*} \sum_{l=1}^{m} \sum_{\mu=1}^{\mu_l} \alpha_n^{(l,\mu)} E_n^{(l,\mu)}, \tag{8.115}$$

the corresponding solution of problem (8.108) is given by

$$\begin{pmatrix} \Phi \\ \Phi' \end{pmatrix} = \sum_{n\in\mathbb{Z}^*} \sum_{l=1}^{m} \sum_{\mu=1}^{\mu_l} \alpha_n^{(l,\mu)} e^{i\beta_n^{(l)} t} E_n^{(l,\mu)}. \tag{8.116}$$

In particular, we have

$$\Phi = \sum_{n\in\mathbb{Z}^*} \sum_{l=1}^{m} \sum_{\mu=1}^{\mu_l} \frac{\alpha_n^{(l,\mu)}}{i\beta_n^{(l)}} e^{i\beta_n^{(l)} t} e_n w^{(l,\mu)}, \tag{8.117}$$

and the observation (8.109) becomes

$$\sum_{n\in\mathbb{Z}^*} \sum_{l=1}^{m} D^T \Big(\sum_{\mu=1}^{\mu_l} \frac{n\alpha_n^{(l,\mu)}}{i\beta_n^{(l)}} w^{(l,\mu)} \Big) e^{i\beta_n^{(l)} t} \equiv 0 \quad \text{on } [0, T]. \tag{8.118}$$

Theorem 8.24 *Assume that A and D satisfy Kalman's criterion (8.28). Assume furthermore that A^T is real diagonalizable with (8.110)–(8.111). Then problem (8.108) is D-observable for $|\epsilon| > 0$ small enough, provided that*

$$T > 2m\pi. \tag{8.119}$$

Proof Applying Corollary 8.23 to each line of (8.118), we get

$$D^T \Big(\sum_{\mu=1}^{\mu_l} \frac{n\alpha_n^{(l,\mu)}}{i\beta_n^{(l)}} w^{(l,\mu)} \Big) = 0, \quad 1 \leqslant l \leqslant m, \quad n \in \mathbb{Z}^*. \tag{8.120}$$

By virtue of Proposition 2.12 (ii), because of Kalman's criterion (8.28), $\text{Ker}(D^T)$ does not contain any nontrivial invariant subspace of A^T, then it follows that

$$\sum_{\mu=1}^{\mu_l} \frac{n\alpha_n^{(l,\mu)}}{i\beta_n^{(l)}} w^{(l,\mu)} = 0, \quad 1 \leqslant l \leqslant m, \quad n \in \mathbb{Z}^*. \tag{8.121}$$

Hence

$$\alpha_n^{(l,\mu)} = 0, \quad 1 \leqslant \mu \leqslant \mu_l, \quad 1 \leqslant l \leqslant m, \quad n \in \mathbb{Z}^*. \tag{8.122}$$

The proof is complete. □

We now improve the estimate (8.119) on the observation time in the case that the eigenvalues of A are distinct.

Theorem 8.25 *Under the assumptions of Theorem 8.24, assume furthermore that A^T possesses N distinct real eigenvalues:*

$$\delta_1 < \delta_2 < \cdots < \delta_N. \tag{8.123}$$

Then problem (8.108) is D-observable for $|\epsilon| > 0$ small enough, provided that

$$T > 2\pi(N - rank(D) + 1). \tag{8.124}$$

Proof Let $w^{(1)}, w^{(1)}, \cdots, w^{(N)}$ be the corresponding eigenvectors of A^T. Accordingly, (8.118) becomes

$$\sum_{n\in Z}\sum_{l=1}^{N} D^T \frac{n\alpha_n^{(l)}}{i\beta_n^{(l)}} w^{(l)} e^{i\beta_n^{(l)} t} \equiv 0 \quad \text{on } [0, T]. \tag{8.125}$$

Then, setting $r = \text{rank}(D)$, without loss of generality, we assume that $D^T w^{(1)}, \cdots, D^T w^{(r)}$ are linearly independent. There exists an invertible matrix S of order N, such that

$$SD^T(w^{(1)}, \cdots, w^{(r)}) = (e_1, \cdots e_r), \tag{8.126}$$

where $e_1, \cdots e_r$ are the canonic basis vectors in $\mathbb{R}^N$. Since S is invertible, (8.125) can be equivalently rewritten as

$$\sum_{n\in\mathbb{Z}^*}\Big\{\sum_{l=1}^{r} \frac{n\alpha_n^{(l)}}{i\beta_n^{(l)}} e_l e^{i\beta_n^{(l)} t} + \sum_{l=r+1}^{N} \frac{n\alpha_n^{(l)}}{i\beta_n^{(l)}} SD^T w^{(l)} e^{i\beta_n^{(l)} t}\Big\} \equiv 0 \tag{8.127}$$

for all $0 \leqslant t \leqslant T$. Once again, we apply Corollary 8.23 to each equation of (8.127), but this time, the upper density of the sequence $\{\beta_n^{(1)}, \beta_n^{(l)}\}_{r+1\leqslant l\leqslant N; n\in\mathbb{Z}^*}$ is equal to $(N - r + 1)$. Therefore, we get again (8.122), provided that (8.124) holds. The proof is complete. □

8.8 An Example

Let $|\epsilon| > 0$ be small enough. By Corollary 8.20, we know that the following system

$$\begin{cases} \phi'' - \Delta\phi + \epsilon\psi = 0 \text{ in } (0, +\infty) \times \Omega, \\ \psi'' - \Delta\psi + \epsilon\phi = 0 \text{ in } (0, +\infty) \times \Omega, \\ \phi = \psi = 0 \qquad\qquad \text{on } (0, +\infty) \times \Gamma \end{cases} \tag{8.128}$$

is observable by means of the trace $\partial_\nu\phi|_{\Gamma_1}$ or $\partial_\nu\psi|_{\Gamma_1}$ on all the interval $[0, T]$ for $T > 0$ large enough.

We now consider the following example which is approximately null controllable, but not exactly null controllable:

$$\begin{cases} u'' - \Delta u + \epsilon v = 0 \text{ in } (0, +\infty) \times \Omega, \\ v'' - \Delta v + \epsilon u = 0 \text{ in } (0, +\infty) \times \Omega, \\ u = v = 0 \qquad\qquad \text{on } (0, +\infty) \times \Gamma_0 \\ u = h, \quad v = 0 \qquad \text{on } (0, +\infty) \times \Gamma_1, \end{cases} \tag{8.129}$$

where $N = 2$, $M = 1$, $D = \begin{pmatrix} 1 \\ 0 \end{pmatrix}$ and h is a boundary control.

First, by means of Theorem 3.10, system (8.129) is never exactly null controllable in the space $(\mathcal{H}_0)^2 \times (\mathcal{H}_{-1})^2$ because of the lack of boundary controls.

On the other hand, since its adjoint problem (8.128) is D-observable via the observation of the trace $\partial_\nu\phi|_{\Gamma_1}$, then, applying Theorem 8.6, system (8.129) is approximately null controllable in the space $(L^2(\Omega))^2 \times (H^{-1}(\Omega))^2$ via only one boundary control $h \in L^2(0, T; L^2(\Gamma_1))$, provided that T is large enough.

Chapter 9
Approximate Boundary Synchronization

The approximate boundary synchronization is defined and studied in this chapter for system (I) with Dirichlet boundary controls.

9.1 Definition

We now give the definition on the **approximate boundary synchronization** as follows.

Definition 9.1 System (I) is approximately synchronizable at the time $T > 0$, if for any given initial data $(\widehat{U}_0, \widehat{U}_1) \in (\mathcal{H}_0)^N \times (\mathcal{H}_{-1})^N$, there exists a sequence $\{H_n\}$ of boundary controls in $\mathcal{L}^M$ with compact support in $[0, T]$, such that the corresponding sequence $\{U_n\}$ of solutions to problem (I) and (I0) satisfies

$$u_n^{(k)} - u_n^{(l)} \to 0 \quad \text{as } n \to +\infty \tag{9.1}$$

for all $1 \leqslant k, l \leqslant N$ in the space

$$C^0_{loc}([T, +\infty); \mathcal{H}_0) \cap C^1_{loc}([T, +\infty); \mathcal{H}_{-1}). \tag{9.2}$$

Let C_1 be the synchronization matrix of order $(N-1) \times N$, defined by

$$C_1 = \begin{pmatrix} 1 & -1 & & & \\ & 1 & -1 & & \\ & & \ddots & \ddots & \\ & & & 1 & -1 \end{pmatrix}. \tag{9.3}$$

T. Li and B. Rao, *Boundary Synchronization for Hyperbolic Systems*, Progress in Nonlinear Differential Equations and Their Applications 94,
https://doi.org/10.1007/978-3-030-32849-8_9

Obviously, the approximate boundary synchronization (9.1) can be equivalently rewritten as

$$C_1 U_n \to 0 \quad \text{as } n \to +\infty \tag{9.4}$$

in the space

$$C^0_{loc}([T, +\infty); (\mathcal{H}_0)^{N-1}) \cap C^1_{loc}([T, +\infty); (\mathcal{H}_{-1})^{N-1}). \tag{9.5}$$

9.2 Condition of C_1-Compatibility

Theorem 9.2 *Assume that system (I) is approximately synchronizable, but not approximately null controllable. Then the coupling matrix $A = (a_{ij})$ should satisfy the following condition of compatibility (the* **row-sum condition***):*

$$\sum_{p=1}^{N} a_{kp} := a, \quad k = 1, \cdots, N, \tag{9.6}$$

where a is a constant independent of $k = 1, \cdots, N$.

Proof Let $(\widehat{U}_0, \widehat{U}_1) \in (\mathcal{H}_0)^N \times (\mathcal{H}_{-1})^N$, and let $\{H_n\}$ be a sequence of boundary controls, which realize the approximate boundary synchronization of system (I). Let $\{U_n\}$ be the sequence of the corresponding solutions. We have

$$U_n'' - \Delta U_n + AU_n = 0 \quad \text{in } (T, +\infty) \times \Omega. \tag{9.7}$$

Noting $U_n = (u_n^{(1)}, \cdots, u_n^{(N)})^T$, we have

$$u_n^{(k)''} - \Delta u_n^{(k)} + \sum_{p=1}^{N} a_{kp} u_n^{(p)} = 0 \quad \text{in } (T, +\infty) \times \Omega, \ 1 \leqslant k \leqslant N. \tag{9.8}$$

Let $w_n^{(k)} = u_n^{(k)} - u_n^{(N)}$ for $1 \leqslant k \leqslant N-1$. It follows from (9.8) that

$$w_n^{(k)''} - \Delta w_n^{(k)} + \sum_{p=1}^{N-1} a_{kp} w_n^{(p)} + \Big(\sum_{p=1}^{N} a_{kp} - a\Big) u_n^{(N)} = 0. \tag{9.9}$$

If (9.6) fails, then, noting (9.1), it follows that

$$u_n^{(N)} \to 0 \quad \text{in } \mathcal{D}'\big((T, +\infty) \times \Omega\big) \quad \text{as } n \to +\infty. \tag{9.10}$$

In order to avoid the technical details of the proof, here we claim (*cf.* Corollary 10.15 below) that the convergence (9.10) holds actually in the usual space

$$C^0_{loc}([T, +\infty); \mathcal{H}_0) \cap C^1_{loc}([T, +\infty); \mathcal{H}_{-1}), \tag{9.11}$$

which contradicts the non-approximate boundary null controllability. The proof is complete. □

Remark 9.3 The condition of compatibility (9.6) is just the same as (4.5) for the exact boundary synchronization.

The condition of compatibility (9.6) indicates that $e_1 = (1, \cdots, 1)^T$ is an eigenvector of the coupling matrix A, associated with the eigenvalue a given by (9.6). Moreover, since $\text{Ker}(C_1) = \text{Span}\{e_1\}$, this condition is equivalent to

$$A\text{Ker}(C_1) \subseteq \text{Ker}(C_1), \tag{9.12}$$

namely, $\text{Ker}(C_1)$ is an invariant subspace of A. We call (9.12) the **condition of C_1-compatibility**, which is also equivalent to the existence of a unique matrix $\overline{A}_1$ of order $(N-1)$, such that

$$C_1 A = \overline{A}_1 C_1, \tag{9.13}$$

where the matrix $\overline{A}_1$ is called the **reduced matrix of A by C_1**.

9.3 Fundamental Properties

Under the condition of C_1-compatibility (9.12), let

$$W_1 = (w^{(1)}, \cdots, w^{(N-1)})^T = C_1 U. \tag{9.14}$$

The original problem (I) and (I0) for the variable U can be reduced to the following self-closed problem for the variable W:

$$\begin{cases} W_1'' - \Delta W_1 + \overline{A}_1 W_1 = 0 & \text{in } (0, +\infty) \times \Omega, \\ W_1 = 0 & \text{on } (0, +\infty) \times \Gamma_0, \\ W_1 = C_1 D H & \text{on } (0, +\infty) \times \Gamma_1 \end{cases} \tag{9.15}$$

with the initial data

$$t = 0: \quad W_1 = C_1 \widehat{U}_0, \ W_1' = C_1 \widehat{U}_1 \quad \text{in } \Omega. \tag{9.16}$$

Accordingly, let

$$\Psi_1 = (\psi^{(1)}, \cdots, \psi^{(N-1)})^T. \tag{9.17}$$

Consider the adjoint problem of the **reduced system** (9.15)

$$\begin{cases} \Psi_1'' - \Delta\Psi_1 + \overline{A}_1^T \Psi_1 = 0 & \text{in } (0, +\infty) \times \Omega, \\ \Psi_1 = 0 & \text{on } (0, +\infty) \times \Gamma, \\ t = 0: \quad \Psi_1 = \widehat{\Psi}_0, \ \Psi_1' = \widehat{\Psi}_1 & \text{in } \Omega. \end{cases} \tag{9.18}$$

In what follows, problem (9.18) will be called the **reduced adjoint problem** of system (I).

From Definitions 8.1 and 9.1, we immediately get the following

Lemma 9.4 *Assume that the coupling matrix A satisfies the condition of C_1-compatibility (9.12). Then system (I) is approximately synchronizable at the time $T > 0$ if and only if the reduced system (9.15) is approximately null controllable at the time $T > 0$, or equivalently (cf. Theorem 8.6), the reduced adjoint problem (9.18) is C_1D-observable on the time interval* $[0, T]$.

Moreover, we have

Lemma 9.5 *Under the condition of C_1-compatibility (9.12), assume that system (I) is approximately synchronizable, then we necessarily have*

$$rank(C_1D, C_1AD, \cdots, C_1A^{N-1}D) = N - 1. \tag{9.19}$$

Proof By Lemma 9.4 and Theorem 8.9, we have

$$\text{rank}(C_1D, \overline{A}_1C_1D, \cdots, \overline{A}_1^{N-2}C_1D) = N - 1. \tag{9.20}$$

Then, noting (9.13) and using Proposition 2.16, we get immediately (9.19). □

9.4 Properties Related to the Number of Total Controls

As we have already explained in Sect. 8.3, the rank of the enlarged matrix $(D, AD, \cdots, A^{N-1}D)$ measures the number of total (direct and indirect) controls. So, it is significant to determine the minimal number of total controls, which is necessary to the approximate boundary synchronization of system (I), no matter whether the condition of C_1-compatibility (9.12) is satisfied or not.

Theorem 9.6 *Assume that system (I) is approximately synchronizable under the action of a boundary control matrix D. Then we necessarily have*

$$rank(D, AD, \cdots, A^{N-1}D) \geqslant N - 1. \tag{9.21}$$

In other words, at least $(N - 1)$ total controls are needed in order to realize the approximate boundary synchronization of system (I).

Proof First, assume that A satisfies the condition of C_1-compatibility (9.12). By Lemma 9.5, we have (9.19). Next, assume that A does not satisfy the condition of

C_1-compatibility (9.12). By Proposition 2.15, we have $C_1Ae_1 \neq 0$. So, the matrix $\begin{pmatrix} C_1 \\ C_1A \end{pmatrix}$ is of full column-rank. By (9.4), it is easy to see from problem (I) and (I0) with $U = U_n$ and $H = H_n$ that

$$C_1U_n \to 0 \quad \text{and} \quad C_1AU_n \to 0 \quad \text{in } (\mathcal{D}'((T, +\infty) \times \Omega))^{N-1} \tag{9.22}$$

as $n \to +\infty$. Then we have

$$U_n \to 0 \quad \text{in} \quad (\mathcal{D}'((T, +\infty) \times \Omega))^N \quad \text{as } n \to +\infty. \tag{9.23}$$

We claim that in this case we have

$$\text{rank}(D, AD, \cdots, A^{N-1}D) = N. \tag{9.24}$$

Otherwise, let $d \geqslant 1$ be such that

$$\text{rank}(D, AD, \cdots, A^{N-1}D) = N - d. \tag{9.25}$$

By the assertion (ii) of Proposition 2.12, there exists a nontrivial subspace of A^T, which is contained in $\text{Ker}(D^T)$ and invariant for A^T, therefore, there exists a nontrivial vector E and a number $\lambda \in \mathbb{C}$, such that

$$A^TE_1 = \lambda E_1 \quad \text{and} \quad D^TE_1 = 0. \tag{9.26}$$

Applying E to problem (I) and (I0) with $U = U_n$ and $H = H_n$, and noting $\phi = (E, U_n)$, it follows that

$$\begin{cases} \phi'' - \Delta\phi + \lambda\phi = 0 & \text{in } (0, +\infty) \times \Omega, \\ \phi = 0 & \text{on } (0, +\infty) \times \Gamma, \\ t = 0: \quad \phi = (E_1, \widehat{U}_0), \ \phi' = (E_1, \widehat{U}_1) & \text{in } \Omega. \end{cases} \tag{9.27}$$

Noting that ϕ is obviously independent of n if the initial data is chosen so that $(E_1, \widehat{U}_0) \neq 0$ or $(E_1, \widehat{U}_1) \neq 0$, then $\phi \not\equiv 0$ which contradicts (9.23).

Finally, combining (9.19) and (9.24), we get the rank condition (9.21). The proof is then achieved. □

According to Theorem 9.6, it is natural to consider the approximate boundary synchronization of system (I) with the minimal number $(N - 1)$ of total controls. In this case, the coupling matrix A should possess some fundamental properties related to the synchronization matrix C_1.

Theorem 9.7 *Assume that system (I) is approximately synchronizable under the minimal rank condition*

$$\textit{rank}(D, AD, \cdots, A^{N-1}D) = N - 1. \tag{9.28}$$

Then we have the following assertions:

(i) The coupling matrix A satisfies the condition of C_1-compatibility (9.12).

(ii) There exists a scalar function u as the **approximately synchronizable state,** *such that*

$$u_n^{(k)} \to u \quad as\ n \to +\infty \tag{9.29}$$

for all $1 \leqslant k \leqslant N$ in the space

$$C^0_{loc}([T, +\infty); \mathcal{H}_0) \cap C^1_{loc}([T, +\infty); \mathcal{H}_{-1}). \tag{9.30}$$

Moreover, the approximately synchronizable state u is independent of the sequence $\{H_n\}$ of applied controls.

(iii) The transpose A^T of the coupling matrix A admits an eigenvector E_1 such that $(E_1, e_1) = 1$, where $e_1 = (1, \cdots, 1)^T$ is the eigenvector of A, associated with the eigenvalue a given by (9.6).

Proof (i) If A does not satisfy the condition of C_1-compatibility (9.13), then the minimal rank condition (9.28) makes the rank condition (9.24) impossible.

(ii) By the assertion (ii) of Proposition 2.12 with $d = 1$, the rank condition (9.28) implies the existence of one-dimensional invariant subspace of A^T, contained in $\text{Ker}(D^T)$, therefore, there exist a vector $E_1 \in \text{Ker}(D^T)$ and a number $b \in \mathbb{C}$, such that

$$E_1^T D = 0 \quad \text{and} \quad A^T E_1 = bE_1. \tag{9.31}$$

Applying E_1 to problem (I) and (I0) with $U = U_n$ and $H = H_n$, and setting $\phi = (E_1, U_n)$, it follows that

$$\begin{cases} \phi'' - \Delta\phi + b\phi = 0 & \text{in } (0, +\infty) \times \Omega, \\ \phi = 0 & \text{on } (0, +\infty) \times \Gamma, \\ t = 0: \quad \phi = (E_1, \widehat{U}_0), \ \phi' = (E_1, \widehat{U}_1) & \text{in } \Omega. \end{cases} \tag{9.32}$$

Clearly, ϕ is independent of n and of applied control H_n. Moreover, by (9.4) we have

$$\begin{pmatrix} C_1 \\ E_1^T \end{pmatrix} U_n = \begin{pmatrix} C_1 U_n \\ (E_1, U_n) \end{pmatrix} \to \begin{pmatrix} 0 \\ \phi \end{pmatrix} \quad \text{as } n \to +\infty \tag{9.33}$$

in the space

$$C^0_{loc}([T, +\infty); (\mathcal{H}_0)^N) \cap C^1_{loc}([T, +\infty); (\mathcal{H}_{-1})^N). \tag{9.34}$$

We will show that the matrix $\begin{pmatrix} C_1 \\ E_1^T \end{pmatrix}$ is invertible, then it follows that

$$U_n \to \begin{pmatrix} C_1 \\ E_1^T \end{pmatrix}^{-1} \begin{pmatrix} 0 \\ \phi \end{pmatrix} =: U \quad \text{as } n \to +\infty \tag{9.35}$$

in the space

$$C^0_{loc}([T,+\infty);(\mathcal{H}_0)^N)\cap C^1_{loc}([T,+\infty);(\mathcal{H}_{-1})^N). \tag{9.36}$$

Noting that

$$\mathrm{Ker}(C_1)=\mathrm{Span}(e_1)\ \text{ with }\ e_1=(1,\cdots,1)^T, \tag{9.37}$$

it follows from (9.35) that there exists a scalar function u such that $U=ue_1$, thus, we get (9.29). Since ϕ is independent of applied control H_n, so is u. Moreover, noting (9.35), we have $u\not\equiv 0$ for all initial data $(\widehat{U}_0,\widehat{U}_1)$ such that $(E_1,\widehat{U}_0)\not\equiv 0$ or $(E_1,\widehat{U}_1)\not\equiv 0$.

We now show that the matrix $\begin{pmatrix}C_1\\E_1^T\end{pmatrix}$ is invertible. In fact, assume that there exists a vector $x\in\mathbb{C}^N$, such that

$$x^T\begin{pmatrix}C_1\\E_1^T\end{pmatrix}=0. \tag{9.38}$$

Then, applying x to (9.33), it is easy to see that

$$x^T\begin{pmatrix}0\\\phi\end{pmatrix}=0. \tag{9.39}$$

Clearly, $\phi\not\equiv 0$ at least for an initial data $(\widehat{U}_0,\widehat{U}_1)$. Then the last component of x must be zero. We can thus write

$$x=\begin{pmatrix}\widehat{x}\\0\end{pmatrix}\ \text{ with }\ \widehat{x}\in\mathbb{C}^{N-1}. \tag{9.40}$$

Then, it follows from (9.38) that $\widehat{x}^TC_1=0$. But C_1 is of full row-rank, so $\widehat{x}=0$, therefore $x=0$.

(iii) By (9.33) and noting $U=ue_1$, we get

$$\phi=(E_1,e_1)u. \tag{9.41}$$

Since $\phi\not\equiv 0$ at least for an initial data $(\widehat{U}_0,\widehat{U}_1)$, we get $(E_1,e_1)\neq 0$. Without loss of generality, we may take $(E_1,e_1)=1$. It follows that

$$a(E_1,e_1)=(E_1,Ae_1)=(A^TE_1,e_1)=b(E_1,e_1). \tag{9.42}$$

So $b=a$ is a real number, and E_1 is a real eigenvector of A^T, associated with the eigenvalue a given by (9.6). The proof is then complete. □

Remark 9.8 What is surprising is the existence of the approximately synchronizable state u under the minimal rank condition (9.28). Moreover, by Theorem 8.9, system (I) is not approximately null controllable. Then, at least for an initial data $(\widehat{U}_0,\widehat{U}_1)$, the approximately synchronizable state $u\not\equiv 0$. In this case, system (I) is called to be approximately synchronizable **in the pinning sense**, while that originally given by Definition 9.1 is **in the consensus sense.**

Remark 9.9 The rank condition (9.28) indicates that the number of total controls is equal to $(N-1)$, but the state variable U of system (I) has N components, so, there exists a direction E_1, on which the projection (E_1, U_n) of the solution U_n is independent of $(N-1)$ controls, therefore, (E_1, U_n) converges in the space $C^0_{loc}([0,+\infty);\mathcal{H}_0)\cap C^1_{loc}([0,+\infty);\mathcal{H}_{-1})$ as $n\to+\infty$. Moreover, the necessity of the condition of C_1-compatibility (9.12) becomes also a consequence of the minimal value of the number of applied total controls.

Remark 9.10 By Lemma 9.5, under the condition of C_1-compatibility (9.12), for the approximate boundary synchronization of system (I), we have (9.19), which shows that the reduced system (9.15) is still submitted to $(N-1)$ total controls. In other words, as we transform system (I) into system (9.15), only the number of equations is reduced from N to $(N-1)$, but the number $(N-1)$ of the total controls remains always unchanged, so that we get a reduced system of $(N-1)$ equations still submitted to $(N-1)$ total controls, that is just what we want.

Remark 9.11 By Definition 2.3, the assertion (iii) means that the subspace $\text{Span}\{E_1\}$ is bi-orthonormal to the subspace $\text{Span}\{e_1\}$. Then, by Proposition 2.5 we have

$$\text{Span}\{e_1\}\cap(\text{Span}\{E_1\})^{\perp}=\{0\}. \tag{9.43}$$

Thus, it follows from Proposition 2.2 that $\text{Span}\{E_1\}^{\perp}$ is a supplement of $\text{Span}\{e_1\}$. Moreover, since $\text{Span}\{E_1\}$ is an invariant subspace of A^T, by Proposition 2.7, $\text{Span}\{E_1\}^{\perp}$ is invariant for A. Therefore, A is diagonalizable by blocks according to the decomposition $\text{Span}\{e_1\}\oplus Span\{E_1\}^{\perp}$.

Theorem 9.12 *Let A satisfy the condition of C_1-compatibility (9.12). Assume that A^T admits an eigenvector E_1 such that $(E_1, e_1)=1$ with $e_1=(1,\cdots,1)^T$. Then there exists a boundary control matrix D satisfying the minimal rank condition (9.28), which realizes the approximate boundary synchronization of system (I).*

Proof By Lemma 9.4, under the condition of C_1-compatibility (9.12), the approximate boundary synchronization of system (I) is equivalent to the C_1D-observability of the reduced adjoint problem (9.18).

Let D be the following full column-rank matrix defined by

$$\text{Ker}(D^T)=\text{Span}\{E_1\}. \tag{9.44}$$

Since $\text{Span}\{E_1\}$ is the sole subspace contained in $\text{Ker}(D^T)$ and invariant for A^T, applying the assertion (ii) of Lemma 2.12 with $d=1$, we get the rank condition (9.28). On the other hand, the assumption $(E_1, e_1)=1$ implies that $\text{Ker}(C_1)\cap\text{Im}(D)=\{0\}$, therefore, by Proposition 2.4 we have

$$\text{rank}(C_1D)=\text{rank}(D)=N-1. \tag{9.45}$$

Thus, the $C_1 D$-observation of the reduced adjoint problem (9.18) becomes the complete observation:

$$\partial_\nu \Psi \equiv 0 \quad \text{on } [0, T] \times \Gamma_1, \tag{9.46}$$

which implies well $\Psi \equiv 0$ because of Holmgren's uniqueness Theorem (*cf.* Theorem 8.2 in [62]), provided that $T > 0$ is large enough. The proof is then complete. □

Remark 9.13 Since the matrix D given by (9.44) is of rank $(N-1)$, Theorem 9.12 shows the approximate boundary synchronization of system (I) by means of $(N-1)$ direct boundary controls. However, we are more interested in using fewer direct boundary controls in practice. We will give later in Theorem 10.10 a matrix D with the minimal rank, such that Kalman's criteria (9.28) and (9.19) are simultaneously satisfied. We have pointed out in Chap. 8 that Kalman's criterion (9.19) is indeed sufficient to the approximate boundary null controllability of the reduced system (9.15), then to the approximate boundary synchronization of the original system (I), for some special reduced systems such as the nilpotent system, 2×2 systems with the small spectral gap condition (8.74) and some one-space-dimensional systems, provided that $T > 0$ is large enough.

Remark 9.14 Let $\mathbb{D}_1$ be the set of all boundary control matrices D, which realize the approximate boundary synchronization of system (I). Define the minimal number N_1 of total controls for the approximate boundary synchronization of system (I):

$$N_1 = \inf_{D \in \mathbb{D}_1} \text{rank}(D, AD, \cdots, A^{N-1}D). \tag{9.47}$$

Let V_a denote the subspace of all the eigenvectors of A^T, associated to the eigenvalue a given by (9.6). By the results obtained in Theorems 9.6, 9.7 and 9.12, we have

$$N_1 = \begin{cases} N-1, \text{if } A \text{ is } C_1\text{-compatible and } (E_1, e_1) = 1, \\ N, \text{if } A \text{ is } C_1\text{-compatible but } (E_1, e_1) = 0, \forall E_1 \in V_a. \end{cases} \tag{9.48}$$

Example 9.15 Consider the following system:

$$\begin{cases} u'' - \Delta u + v = 0 & \text{in } (0, +\infty) \times \Omega, \\ v'' - \Delta v - u + 2v = 0 & \text{in } (0, +\infty) \times \Omega, \\ u = v = 0 & \text{on } (0, +\infty) \times \Gamma_0, \\ u = \alpha h, \quad v = \beta h & \text{on } (0, +\infty) \times \Gamma_1. \end{cases} \tag{9.49}$$

Let

$$A = \begin{pmatrix} 0 & 1 \\ -1 & 2 \end{pmatrix}, \quad D = \begin{pmatrix} \alpha \\ \beta \end{pmatrix} \tag{9.50}$$

and

$$C_1 = (1, -1), \quad \text{Ker}(C_1) = \text{Span}\{e_1\} \text{ with } e_1 = (1, 1)^T. \tag{9.51}$$

Clearly, A satisfies the condition of C_1-compatibility with $\overline{A}_1 = 1$.

We point out that the only eigenvector $E_1 = (1, -1)^T$ of A^T satisfies $(E_1, e_1) = 0$. By Theorems 9.6 and 9.7, we have to use two (instead of one!) total controls to realize the approximate boundary synchronization of system (9.49). More precisely, we have

$$(D, AD) = \begin{pmatrix} \alpha & \beta \\ \beta & 2\beta - \alpha \end{pmatrix}, \quad \det(D, AD) = -(\alpha - \beta)^2 \tag{9.52}$$

and

$$(C_1 D, C_1 AD) = (\alpha - \beta, \alpha - \beta). \tag{9.53}$$

It is easy to see that

$$\text{rank}(C_1 D, C_1 AD) = 1 \iff \text{rank}(D, AD) = 2. \tag{9.54}$$

By Corollary 8.7, the corresponding reduced system (9.15) (with $\overline{A}_1 = 1$):

$$\begin{cases} w'' - \Delta w + w = 0 & \text{in } (0, +\infty) \times \Omega, \\ w = 0 & \text{on } (0, +\infty) \times \Gamma_0, \\ w = (\alpha - \beta) h & \text{on } (0, +\infty) \times \Gamma_1 \end{cases} \tag{9.55}$$

is approximately null controllable as $\alpha \neq \beta$, so the original system (9.49) is approximately synchronizable, provided that $\alpha \neq \beta$ and $T > 0$ is large enough.

On the other hand, system (9.49) satisfies well the gap condition (8.74). By Theorem 8.18, Kalman's criterion $\text{rank}(D, CD) = 2$ is sufficient for its approximate boundary null controllability, provided that $T > 0$ is large enough. So, when $\alpha \neq \beta$, we are required to use 2 total controls and then to get a better result that system (9.49) is actually approximately null controllable under the action of the same boundary controls.

Chapter 10
Approximate Boundary Synchronization by p-Groups

The approximate boundary synchronization by p-groups is introduced and studied in this chapter for system (I) with Dirichlet boundary controls.

10.1 Definition

In this chapter, we will consider the **approximate boundary synchronization by p-groups**.

Let $p \geqslant 1$ be an integer and let

$$0 = n_0 < n_1 < n_2 < \cdots < n_p = N \tag{10.1}$$

be integers such that $n_r - n_{r-1} \geqslant 2$ for all $1 \leqslant r \leqslant p$. We divide the components of $U = (u^{(1)}, \cdots, u^{(N)})^T$ into p groups as

$$(u^{(1)}, \cdots, u^{(n_1)}), \ (u^{(n_1+1)}, \cdots, u^{(n_2)}), \cdots, (u^{(n_{p-1}+1)}, \cdots, u^{(n_p)}). \tag{10.2}$$

Definition 10.1 System (I) is approximately synchronizable by p-groups at the time $T > 0$ if for any given initial data $(\widehat{U}_0, \widehat{U}_1) \in (\mathcal{H}_0)^N \times (\mathcal{H}_{-1})^N$, there exists a sequence $\{H_n\}$ of boundary controls in $\mathcal{L}^M$ with compact support in $[0, T]$, such that the corresponding sequence $\{U_n\}$ of solutions to problem (I) and (I0) satisfies the following condition:

$$u_n^{(k)} - u_n^{(l)} \to 0 \quad \text{in } C^0_{loc}([T, +\infty); \mathcal{H}_0) \cap C^1_{loc}([T, +\infty); \mathcal{H}_{-1}) \tag{10.3}$$

as $n \to +\infty$ for all $n_{r-1} + 1 \leqslant k, l \leqslant n_r$ and $1 \leqslant r \leqslant p$.

T. Li and B. Rao, *Boundary Synchronization for Hyperbolic Systems*, Progress in Nonlinear Differential Equations and Their Applications 94,
https://doi.org/10.1007/978-3-030-32849-8_10

Let S_r be the following $(n_r - n_{r-1} - 1) \times (n_r - n_{r-1})$ full row-rank matrix:

$$S_r = \begin{pmatrix} 1 & -1 & & & \\ & 1 & -1 & & \\ & & \ddots & \ddots & \\ & & & 1 & -1 \end{pmatrix}, \quad 1 \leqslant r \leqslant p \tag{10.4}$$

and let C_p be the following $(N - p) \times N$ **matrix of synchronization by p-groups**:

$$C_p = \begin{pmatrix} S_1 & & & \\ & S_2 & & \\ & & \ddots & \\ & & & S_p \end{pmatrix}. \tag{10.5}$$

Clearly, the approximate boundary synchronization by p-groups (10.3) can be equivalently rewritten in the following form:

$$C_p U_n \to 0 \quad \text{as } n \to +\infty \tag{10.6}$$

in the space

$$C^0_{loc}([T, +\infty); (\mathcal{H}_0)^{N-p}) \cap C^1_{loc}([T, +\infty); (\mathcal{H}_{-1})^{N-p}). \tag{10.7}$$

10.2 Fundamental Properties

For $r = 1, \cdots, p$, setting

$$(e_r)_i = \begin{cases} 1, & n_{r-1} + 1 \leqslant i \leqslant n_r, \\ 0, & \text{otherwise,} \end{cases} \tag{10.8}$$

it is clear that

$$\text{Ker}(C_p) = \text{Span}\{e_1, e_2, \cdots, e_p\}. \tag{10.9}$$

We will say that A satisfies the **condition of C_p-compatibility** if $\text{Ker}(C_p)$ is an invariant subspace of A:

$$A\text{Ker}(C_p) \subseteq \text{Ker}(C_p), \tag{10.10}$$

or equivalently, by Proposition 2.15, there exists a matrix $\overline{A}_p$ of order $(N - p)$, such that

$$C_p A = \overline{A}_p C_p. \tag{10.11}$$

$\overline{A}_p$ is called the **reduced matrix of A by C_p**.

Under the condition of C_p-compatibility, setting

$$W_p = C_p U \tag{10.12}$$

and noting (10.11), from problem (I) and (I0) we get the following self-closed **reduced system**:

$$\begin{cases} W_p'' - \Delta W_p + \overline{A}_p W_p = 0 & \text{in } (0, +\infty) \times \Omega, \\ W_p = 0 & \text{on } (0, +\infty) \times \Gamma_0, \\ W_p = C_p D H & \text{on } (0, +\infty) \times \Gamma_1 \end{cases} \tag{10.13}$$

with the initial data:

$$t = 0: \quad W_p = C_p \widehat{U}_0, \ W_p' = C_p \widehat{U}_1 \quad \text{in } \Omega. \tag{10.14}$$

Obviously, we have the following

Lemma 10.2 *Under the condition of C_p-compatibility (10.10), system (I) is approximately synchronizable by p-groups at the time $T > 0$ if and only if the reduced system (10.13) is approximately null controllable at the time $T > 0$, or equivalently (cf. Theorem 8.6) if and only if the adjoint problem of the reduced system (10.13)*

$$\begin{cases} \Psi_p'' - \Delta \Psi_p + \overline{A}_p^T \Psi_p = 0 & \text{in } (0, +\infty) \times \Omega, \\ \Psi_p = 0 & \text{on } (0, +\infty) \times \Gamma, \\ t = 0: \quad \Psi_p = \widehat{\Psi}_{p0}, \ \Psi_p' = \widehat{\Psi}_{p1} & \text{in } \Omega, \end{cases} \tag{10.15}$$

called the **reduced adjoint problem**, *is $C_p D$-observable on the time interval $[0, T]$, namely, the partial observation*

$$(C_p D)^T \partial_\nu \Psi_p \equiv 0 \quad \text{on } [0, T] \times \Gamma_1 \tag{10.16}$$

implies $\widehat{\Psi}_{p0} = \widehat{\Psi}_{p1} \equiv 0$, then $\Psi_p \equiv 0$.

Thus, we have

Lemma 10.3 *Under the condition of C_p-compatibility (10.10), assume that system (I) is approximately synchronizable by p-groups, we necessarily have the following* **Kalman's criterion***:*

$$\mathit{rank}(C_p D, C_p A D, \cdots, C_p A^{N-1} D) = N - p. \tag{10.17}$$

Proof From Lemma 10.2 and Theorem 8.9, we get

$$\text{rank}(C_p D, \overline{A}_p C_p D, \cdots, \overline{A}_p^{N-p-1} C_p A^{N-1} D) = N - p. \tag{10.18}$$

Thus, noting (10.11), (10.17) follows from Proposition 2.16. □

10.3 Properties Related to the Number of Total Controls

The following result indicates the lower bound on the number of total controls necessary to the approximate boundary synchronization by p-groups for system (I), no matter whether the condition of C_p-compatibility (10.10) is satisfied or not.

Theorem 10.4 *Assume that system (I) is approximately synchronizable by p-groups. Then we necessarily have*

$$rank(D, AD, \cdots, A^{N-1}D) \geqslant N - p. \tag{10.19}$$

In other words, at least $(N - p)$ total controls are needed to realize the approximate boundary synchronization by p-groups for system (I).

Proof Keeping in mind that the condition of C_p-compatibility (10.10) is not assumed to be satisfied *a priori*, we define an $(N - \tilde{p}) \times N$ full row-rank matrix $\widetilde{C}_{\tilde{p}}$ $(0 \leqslant \tilde{p} \leqslant p)$ by

$$\mathrm{Im}(\widetilde{C}_{\tilde{p}}^T) = \mathrm{Span}(C_p^T, A^T C_p^T, \cdots, (A^T)^{N-1} C_p^T). \tag{10.20}$$

By Cayley–Hamilton's theorem, we have $A^T \mathrm{Im}(\widetilde{C}_{\tilde{p}}^T) \subseteq \mathrm{Im}(\widetilde{C}_{\tilde{p}}^T)$, or equivalently,

$$A\mathrm{Ker}(\widetilde{C}_{\tilde{p}}) \subseteq \mathrm{Ker}(\widetilde{C}_{\tilde{p}}). \tag{10.21}$$

By Proposition 2.15, there exists a matrix $\widetilde{A}_{\tilde{p}}$ of order $(N - \tilde{p})$, such that

$$\widetilde{C}_{\tilde{p}} A = \widetilde{A}_{\tilde{p}} \widetilde{C}_{\tilde{p}}. \tag{10.22}$$

Then, applying C_p to the equations in system (I) with $U = U_n$ and $H = H_n$, by (10.6) it is easy to see that for any given integer $l \geqslant 0$, we have

$$C_p A^l U_n \to 0 \quad \text{in } (\mathcal{D}'((T, +\infty) \times \Omega))^{N-p} \quad \text{as } n \to +\infty, \tag{10.23}$$

then

$$\widetilde{C}_{\tilde{p}} U_n \to 0 \quad \text{in } (\mathcal{D}'((T, +\infty) \times \Omega))^{N-\tilde{p}} \quad \text{as } n \to +\infty. \tag{10.24}$$

We claim that

$$\mathrm{rank}(\widetilde{C}_{\tilde{p}} D, \widetilde{A}_{\tilde{p}} \widetilde{C}_{\tilde{p}} D, \cdots, \widetilde{A}_{\tilde{p}}^{N-\tilde{p}-1} \widetilde{C}_{\tilde{p}} D) \geqslant N - \tilde{p}. \tag{10.25}$$

Otherwise, noting that $\widetilde{A}_{\tilde{p}}$ is of order $(N - \tilde{p})$, then by the assertion (i) of Proposition 2.12 with $d = 0$, there exists a nontrivial invariant subspace of $\widetilde{A}_{\tilde{p}}^T$, contained in $\mathrm{Ker}(\widetilde{C}_{\tilde{p}} D)^T$. Therefore, there exists a nonzero vector $E \in \mathbb{C}^{N-\tilde{p}}$ and a number $\lambda \in \mathbb{C}$, such that

$$\widetilde{A}_{\tilde{p}}^T E = \lambda E \quad \text{and} \quad (\widetilde{C}_{\tilde{p}} D)^T E = 0. \tag{10.26}$$

Applying $\widetilde{C}_{\tilde{p}}^T E$ to problem (I) and (I0) with $U = U_n$ and $H = H_n$, and noting $\phi = (E, \widetilde{C}_{\tilde{p}} U_n)$, it is easy to see that

$$\begin{cases} \phi'' - \Delta\phi + \lambda\phi = 0 & \text{in } (0, +\infty) \times \Omega, \\ \phi = 0 & \text{on } (0, +\infty) \times \Gamma, \\ t = 0: \ \phi = (E, \widetilde{C}_{\tilde{p}} \widehat{U}_0), \ \phi' = (E, \widetilde{C}_{\tilde{p}} \widehat{U}_1) & \text{in } \Omega. \end{cases} \tag{10.27}$$

If the initial data are chosen such that $(E, \widetilde{C}_{\tilde{p}} \widehat{U}_0) \neq 0$ or $(E, \widetilde{C}_{\tilde{p}} \widehat{U}_1) \neq 0$, then $\phi \not\equiv 0$. Noting that ϕ is independent of n, this contradicts (10.24).

Finally, by Proposition 2.16, under the condition of $\widetilde{C}_{\tilde{p}}$-compatibility (10.22), the rank condition (10.25) yields

$$\begin{aligned} &\text{rank}(D, AD, \cdots, A^{N-1}D) \\ \geqslant &\text{rank}(\widetilde{C}_{\tilde{p}} D, \widetilde{C}_{\tilde{p}} AD, \cdots, \widetilde{C}_{\tilde{p}} A^{N-1} D) \\ = &\text{rank}(\widetilde{C}_{\tilde{p}} D, \widetilde{A}_{\tilde{p}} \widetilde{C}_{\tilde{p}} D, \cdots, \widetilde{A}_{\tilde{p}}^{N-\tilde{p}-1} \widetilde{C}_{\tilde{p}} D) \geqslant N - \tilde{p}, \end{aligned} \tag{10.28}$$

which leads to (10.19) because of $\tilde{p} \leqslant p$. The proof is then complete. □

According to Theorem 10.4, it is natural to consider the approximate boundary synchronization by p-groups for system (I) with the minimal number $(N - p)$ of total controls. In this case, the coupling matrix A should possess some basic properties related to C_p, the matrix of synchronization by p-groups.

Theorem 10.5 *Assume that system (I) is approximately synchronizable by p-groups under the minimal rank condition*

$$\textit{rank}(D, AD, \cdots, A^{N-1}D) = N - p. \tag{10.29}$$

Then we have the following assertions:

(i) There exist some linearly independent scalar functions $u_1, u_2, \cdots, u_p$ *such that*

$$u_n^{(k)} \to u_r \quad \textit{as } n \to +\infty \tag{10.30}$$

for all $n_{r-1} + 1 \leqslant k \leqslant n_r$ *and* $1 \leqslant r \leqslant p$ *in the space*

$$C^0_{loc}([T, +\infty); \mathcal{H}_0) \cap C^1_{loc}([T, +\infty); \mathcal{H}_{-1}). \tag{10.31}$$

(ii) The coupling matrix A satisfies the condition of C_p*-compatibility (10.10).*

(iii) A^T *admits an invariant subspace, which is contained in Ker*(D^T) *and bi-orthonormal to Ker*(C_p)*.*

Proof (i) By the assertion (ii) of Proposition 2.12 with $d = p$, the rank condition (10.29) guarantees the existence of an invariant subspace V of A^T, with dimension p and contained in $\text{Ker}(D^T)$. Let $\{E_1, \cdots, E_p\}$ be a basis of V, such that

$$A^T E_r = \sum_{s=1}^{p} \alpha_{rs} E_s \quad \text{and} \quad D^T E_r = 0, \quad 1 \leqslant r \leqslant p. \tag{10.32}$$

Applying E_r to problem (I) and (I0) with $U = U_n$ and $H = H_n$, and setting $\phi_r = (E_r, U_n)$ for $r = 1, \cdots, p$, we get

$$\begin{cases} \phi_r'' - \Delta\phi_r + \sum_{s=1}^{p} \alpha_{rs}\phi_s = 0 \text{ in } (0, +\infty) \times \Omega, \\ \phi_r = 0 \qquad\qquad\qquad\qquad\quad \text{on } (0, +\infty) \times \Gamma \end{cases} \tag{10.33}$$

with the initial data

$$t = 0: \quad \phi_r = (E_r, \widehat{U}_0), \ \phi_r' = (E_r, \widehat{U}_1) \quad \text{in } \Omega. \tag{10.34}$$

Clearly, $\phi_1, \cdots, \phi_r$ are independent of n and of applied controls H_n. Moreover, by (10.6) we get

$$\begin{pmatrix} C_p \\ E_1^T \\ \vdots \\ E_p^T \end{pmatrix} U_n = \begin{pmatrix} C_p U_n \\ (E_1, U_n) \\ \vdots \\ (E_p, U_n) \end{pmatrix} \to \begin{pmatrix} 0 \\ \phi_1 \\ \vdots \\ \phi_p \end{pmatrix} \quad \text{as } n \to +\infty \tag{10.35}$$

in the space $C^0_{loc}([T, +\infty); (\mathcal{H}_0)^N) \cap C^1_{loc}([T, +\infty); (\mathcal{H}_{-1})^N)$.

We claim that the matrix

$$\begin{pmatrix} C_p \\ E_1^T \\ \vdots \\ E_p^T \end{pmatrix} \tag{10.36}$$

is invertible. Then,

$$U_n \to \begin{pmatrix} C_p \\ E_1^T \\ \vdots \\ E_p^T \end{pmatrix}^{-1} \begin{pmatrix} 0 \\ \phi_1 \\ \vdots \\ \phi_p \end{pmatrix} := U \tag{10.37}$$

in $C^0_{loc}([T, +\infty); (\mathcal{H}_0)^N) \cap C^1_{loc}([T, +\infty); (\mathcal{H}_{-1})^N)$ as $n \to +\infty$. Since $C_p U = 0$ because of (10.35) and (10.37), noting (10.9), there exist $u_r (r = 1, \cdots, p)$ such that

$$U = \sum_{r=1}^{p} u_r e_r. \tag{10.38}$$

Then, by the expression of e_r given in (10.8), the convergence (10.30) follows from (10.37).

Now we go back to show that the matrix (10.36) is indeed invertible. In fact, assume that there exists a vector $x \in \mathbb{R}^N$, such that

$$x^T \begin{pmatrix} C_p \\ E_1^T \\ \vdots \\ E_p^T \end{pmatrix} = 0. \tag{10.39}$$

Let

$$x = \begin{pmatrix} \widehat{x} \\ \widetilde{x} \end{pmatrix} \quad \text{with } \widehat{x} \in \mathbb{C}^{N-p} \text{ and } \widetilde{x} \in \mathbb{C}^p. \tag{10.40}$$

Applying x to (10.35), it is easy to see that

$$t \geqslant T: \quad \widetilde{x}^T \begin{pmatrix} \phi_1 \\ \vdots \\ \phi_p \end{pmatrix} = 0. \tag{10.41}$$

Now for $r = 1, \cdots, p$, let us take the initial data (10.34) as

$$t = 0: \quad \phi_r = \theta_r, \ \phi_r' = 0 \ \text{ in } \Omega. \tag{10.42}$$

Then, it follows from problem (10.33) and (10.34) that

$$(\theta_1, \cdots, \theta_p) \to (\phi_1(T), \cdots, \phi_p(T)) \tag{10.43}$$

defines an isomorphism of $(\mathcal{H}_0)^p$. Thus, as $(\theta_1, \cdots, \theta_p)$ vary in $(\mathcal{H}_0)^p$, the state variable $(\phi_1(T), \cdots, \phi_p(T))$ will run through the space $(\mathcal{H}_0)^p$. Thus, it follows from (10.41) that $\widetilde{x}^T = 0$, then by (10.39) we get $\widehat{x}^T C_p = 0$. But C_p is of full row-rank, we have $\widehat{x} = 0$, then $x = 0$.

(ii) Applying C_p to the equations in system (I) with $U = U_n$ and $H = H_n$, and passing to the limit as $n \to +\infty$, it follows easily from (10.38) that

$$t \geqslant T: \quad \sum_{r=1}^{p} u_r C_p A e_r = 0. \tag{10.44}$$

On the other hand, noting (10.35) and (10.38), we have

$$t \geqslant T: \quad \phi_r = \sum_{s=1}^{p} (E_r, e_s) u_s, \quad r = 1, \cdots, p. \tag{10.45}$$

Since the state variable $(\phi_1(T), \cdots, \phi_p(T))$ of system (10.33) runs through the space $(\mathcal{H}_0)^p$ as $(\theta_1, \cdots, \theta_p)$ given in the initial data (10.42) vary in $(\mathcal{H}_0)^p$, so is the approximately synchronizable state by p-groups $(u_1, \cdots, u_p)^T$. In particular, the functions $u_1, \cdots, u_p$ are linearly independent, then

$$C_p A e_r = 0, \quad 1 \leqslant r \leqslant p, \tag{10.46}$$

namely, $\text{AKer}(C_p) \subseteq \text{AKer}(C_p)$. We get thus (10.10).

(iii) Noting that $\text{Span}\{E_1, \cdots, E_p\}$ and $\text{Ker}(C_p)$ have the same dimension, and that $\{\text{Ker}(C_p)\}^\perp = \text{Im}(C_p^T)$, by Propositions 2.4 and 2.5, in order to show that $\text{Ker}(C_p)$ is bi-orthonormal to $\text{Span}\{E_1, \cdots, E_p\}$, it is sufficient to show that $\text{Span}\{E_1, \cdots, E_p\} \cap \text{Im}(C_p^T) = \{0\}$.

Let $E \in \text{Span}\{E_1, \cdots, E_p\} \cap \text{Im}(C_p^T)$ be a nonzero vector. There exist some coefficients $\alpha_1, \cdots, \alpha_p$ and a nonzero vector $r \in \mathbb{R}^{N-p}$, such that

$$E = \sum_{r=1}^{p} \alpha_r E_r \quad \text{and} \quad E = C_p^T r. \tag{10.47}$$

Let

$$\phi = (E, U_n) = \sum_{r=1}^{p} \alpha_r (E_r, U_n) = \sum_{r=1}^{p} \alpha_r \phi_r, \tag{10.48}$$

where $(\phi_1, \cdots, \phi_p)$ is the solution to problem (10.33) and (10.34) with the homogeneous boundary condition, therefore, independent of n. In particular, for any given nonzero initial data $(\theta_1, \cdots, \theta_p)$, the corresponding solution $\phi \not\equiv 0$. On the other hand, we have

$$\phi = (C_p^T r, U_n) = (r, C_p U_n). \tag{10.49}$$

Then, the convergence (10.6) implies that ϕ goes to zero as $n \to +\infty$ in the space

$$C^0_{loc}([T, +\infty); \mathcal{H}_0) \cap C^1_{loc}([T, +\infty); \mathcal{H}_{-1}). \tag{10.50}$$

We get thus a contradiction, which confirms that $\text{Span}\{E_1, \cdots, E_p\} \cap \text{Im}(C_p^T) = \{0\}$. The proof is then complete. □

Remark 10.6 Under the rank condition (10.29), the requirement (10.6) actually implies the existence of the approximately synchronizable state by p-groups $(u_1, \cdots, u_p)^T$, which are independent of applied boundary controls. In this case, system (I) is approximately synchronizable by p-groups **in the pinning sense**, while that originally given by Definition 10.1 is **in the consensus sense**.

Remark 10.7 The rank condition (10.29) indicates that the number of total controls is equal to $(N - p)$, but the state variable U of system (I) has N independent components, so if system (I) is approximately synchronizable by p-groups, there should exist p directions $E_1, \cdots, E_p$, on which the projections $(E_1, U_n), \cdots, (E_p, U_n)$ of the solution U_n to problem (I) and (I0) are independent of the $(N - p)$ controls (cf.

(10.33)), therefore, converge as $n \to +\infty$. Consequently, we get the necessity of the condition of C_p-compatibility (10.10) in this situation.

Remark 10.8 Noting that the rank condition (10.29) indicates that the reduced system (10.13) of $(N-p)$ equations is still submitted to $(N-p)$ total controls, which is necessary for the corresponding approximate boundary null controllability. In other words, the number of total controls is not reduced. The procedure of reduction from system (I) to the reduced system (10.13) reduces only the number of equations, but not the number of total controls, so that we get a reduced system of $(N-p)$ equations submitted to $(N-p)$ total controls, that is just what we want.

Remark 10.9 Since the invariant subspace $\text{Span}\{E_1, \cdots, E_p\}$ of A^T is bi-orthonormal to the invariant subspace $\text{Span}\{e_1, \cdots, e_p\}$ of A, by Proposition 2.8, the invariant subspace $\text{Span}\{E_1, \cdots, E_p\}^\perp$ of A is a supplement of $\text{Span}\{e_1, \cdots, e_p\}$. Therefore, A is diagonalizable by blocks under the decomposition $\text{Span}\{e_1, \cdots, e_p\} \oplus \text{Span}\{E_1, \cdots, E_p\}^\perp$.

Theorem 10.10 *Assume that the coupling matrix A satisfies the condition of C_p-compatibility (10.10) and that $\text{Ker}(C_p)$ admits a supplement which is also invariant for A. Then there exists a boundary control matrix D which satisfies the minimal rank condition (10.29), and realizes the approximate boundary synchronization by p-groups for system (I) in the pinning sense.*

Proof By Lemma 10.2, under the condition of C_p-compatibility (10.10), the approximate boundary synchronization by p-groups of system (I) is equivalent to the C_pD-observability of the reduced adjoint problem (10.15).

Let $W^\perp$ be a supplement of $\text{Ker}(C_p)$, which is invariant for A. By Proposition 2.8, the subspace W is invariant for A^T and bi-orthonormal to $\text{Ker}(C_p)$.

Define the boundary control matrix D by

$$\text{Ker}(D^T) = W. \tag{10.51}$$

Clearly, W is the only invariant subspace of A^T, with the maximal dimension p and contained in $\text{Ker}(D^T)$. Then, by the assertion (ii) of Proposition 2.12 with $d = p$, the rank condition (10.29) holds for this choice of D. On the other hand, the bi-orthonormality of W with $\text{Ker}(C_p)$ implies that $\text{Ker}(C_p) \cap \text{Im}(D) = \{0\}$. Therefore, by Proposition 2.11 we have

$$\text{rank}(D) = \text{rank}(C_pD) = N - p. \tag{10.52}$$

Then the C_pD-observation (10.16) becomes the complete observation:

$$\partial_\nu \Psi \equiv 0 \quad \text{on } [0, T] \times \Gamma_1, \tag{10.53}$$

which implies well $\Psi \equiv 0$ because of Holmgren's uniqueness Theorem (cf. Theorem 8.2 in [62]). The proof is thus complete. □

Remark 10.11 The matrix D defined by (10.51) is of rank $(N-p)$. So, we realize the approximate boundary synchronization by p-groups for system (I) by means of $(N-p)$ direct boundary controls. But we are more interested in using fewer direct boundary controls to realize the approximate boundary synchronization by p-groups for system (I). We will give later in Proposition 11.14 a matrix D with the minimal rank, such that the rank conditions (10.29) and (10.17) are simultaneously satisfied. We point out that, by the results given in Chap. 8, Kalman's criterion (10.17) is indeed sufficient for the approximate boundary null controllability of the reduced system (10.13), then to the approximate boundary synchronization by p-groups for system (I), for some special reduced systems such as the nilpotent system, 2×2 systems with the gap condition (8.7.4) and some one-space-dimensional systems.

Remark 10.12 Let $\mathbb{D}_p$ be the set of all matrices D, which realize the approximate boundary synchronization by p-groups for system (I). Define the minimal number N_p of total controls for the approximate boundary synchronization by p-groups of system (I) by

$$N_p = \inf_{D\in\mathbb{D}_p} \mathrm{rank}(D, AD, \cdots, A^{N-1}D). \tag{10.54}$$

Then, summarizing the results obtained in Theorems 10.4, 10.5, and 10.10, we have

$$N_p = N - p \tag{10.55}$$

if and only if $\mathrm{Ker}(C_p)$ is an invariant subspace of A, and A^T admits an invariant subspace which is bi-orthonormal to $\mathrm{Ker}(C_p)$, or equivalently, by Proposition 2.8 if and only if $\mathrm{Ker}(C_p)$ admits a supplement V, and both $\mathrm{Ker}(C_p)$ and V are invariant for A.

The following result matches the rank conditions (10.29) and (10.17) with a certain algebraic property of A with respect to the matrix C_p.

Proposition 10.13 *Let C_p be the matrix of synchronization by p-groups, given by* (10.5). *Then the rank conditions (10.29) and (10.17) simultaneously hold for some boundary control matrix D if and only if A^T admits an invariant subspace W which is bi-orthonormal to $\mathrm{Ker}(C_p)$.*

Proof By the assertion (ii) of Proposition 2.12 with $d=p$, the rank condition (10.29) implies the existence of an invariant subspace W of A^T, with the dimension p and contained in $\mathrm{Ker}(D^T)$. It is easy to see that

$$W \subseteq \mathrm{Ker}(D, AD, \cdots, A^{N-1}D)^T \tag{10.56}$$

and

$$\dim(W) = \dim \mathrm{Ker}(D, AD, \cdots, A^{N-1}D)^T = p, \tag{10.57}$$

then we have

$$W = \mathrm{Ker}(D, AD, \cdots, A^{N-1}D)^T. \tag{10.58}$$

By Proposition 2.11, the rank conditions (10.29) and (10.17) imply that

$$\operatorname{Ker}(C_p) \cap W^{\perp} = \operatorname{Ker}(C_p) \cap \operatorname{Im}(D, AD, \cdots, A^{N-1}D) = \{0\}. \tag{10.59}$$

Since $\dim(W) = \dim \operatorname{Ker}(C_p)$, by Propositions 2.4 and 2.5, W is bi-orthonormal to $\operatorname{Ker}(C_p)$.

Conversely, assume that W is an invariant subspace of A^T, and bi-orthonormal to $\operatorname{Ker}(C_p)$, therefore with dimension p. Define a full column-rank matrix D of order $N \times (N - p)$ by

$$\operatorname{Ker}(D^T) = W. \tag{10.60}$$

Clearly, W is an invariant subspace of A^T, with dimension p and contained in $\operatorname{Ker}(D^T)$. Moreover, the dimension of $\operatorname{Ker}(D^T)$ is equal to p. Then by the assertion (ii) of Proposition 2.12 with $d = p$, the rank condition (10.29) holds. Keeping in mind that (10.58) remains true in the present situation, it follows that

$$\operatorname{Ker}(C_p) \cap \operatorname{Im}(D, AD, \cdots, A^{N-1}D) = \operatorname{Ker}(C_p) \cap W^{\perp} = \{0\}, \tag{10.61}$$

which, by Proposition 2.12, implies the rank condition (10.17). The proof is then complete. □

10.4 Necessity of the Condition of C_p-Compatibility

We will continue to discuss the approximate boundary synchronization by p-groups of system (I) and clarify the necessity of the condition of C_p-compatibility.

In Definition 10.1, we should exclude some trivial situations. For this purpose, we assume that each group is not approximately null controllable.

Let us recall that the extension matrix $\widetilde{C}_{\tilde{p}}$ introduced by (10.20) satisfies well the condition of $\widetilde{C}_{\tilde{p}}$-compatibility (10.21). Moreover, the convergence

$$\widetilde{C}_{\tilde{p}} U_n \to 0 \quad \text{in } (\mathcal{D}'((T, +\infty) \times \Omega))^{N-\tilde{p}} \quad \text{as } n \to +\infty \tag{10.62}$$

implies (10.25), then, noting that $\widetilde{C}_{\tilde{p}}$ has only $(N - \tilde{p})$ rows, we have

$$\operatorname{rank}(\widetilde{C}_{\tilde{p}} D, \widetilde{A}_{\tilde{p}} \widetilde{C}_{\tilde{p}} D, \cdots, \widetilde{A}_{\tilde{p}}^{N-\tilde{p}-1} \widetilde{C}_{\tilde{p}} D) = N - \tilde{p}. \tag{10.63}$$

Based on these results, we will introduce a stronger version of the approximate boundary synchronization by p-groups.

Theorem 10.14 *Assume that system (I) is approximately synchronizable by p-groups. Then, for any given initial data $(\widehat{U}_0, \widehat{U}_1)$ in the space $(\mathcal{H}_0)^N \times (\mathcal{H}_{-1})^N$, there exists a sequence of boundary controls $\{H_n\}$ in $L^2(0, +\infty; (L^2(\Gamma_1))^M)$ with compact support in $[0, T]$, such that the corresponding sequence $\{U_n\}$ of solutions* *to problem (I) and (I0) satisfies*

$$\widetilde{C}_{\tilde{p}} U_n \to 0 \quad as\ n \to +\infty \tag{10.64}$$

in the space

$$C^0_{loc}([T, +\infty); (\mathcal{H}_0)^{N-\tilde{p}}) \cap C^1_{loc}([T, +\infty); (\mathcal{H}_{-1})^{N-\tilde{p}}). \tag{10.65}$$

Since the proof of this result is quite long, we will give it at the end of this section.

We first consider the special case $p = 1$.

Corollary 10.15 *In the case $p = 1$, assume that A does not satisfy the condition of C_1-compatibility. If system (I) is approximately synchronizable, then it is approximately null controllable. Consequently if system (I) is approximately synchronizable, but not approximately null controllable, then we necessarily have the condition of C_1-compatibility.*

Proof Since $p = 1$, the extension matrix $\widetilde{C}_{\tilde{p}}$ is of order N and of full row-rank, therefore invertible. Then the convergence (10.64) gives the approximate boundary null controllability. □

Now we consider the general case $p \geqslant 1$.

Theorem 10.16 *Let $\tilde{p}$ be the integer given by* (10.20). *If system (I) is approximately synchronizable by p-groups, then it is approximately synchronizable by $\tilde{p}$-groups under a suitable basis.*

Proof Let

$$0 = \tilde{n}_0 < \tilde{n}_1 < \tilde{n}_2 < \cdots < \tilde{n}_{\tilde{p}} \tag{10.66}$$

be an arbitrarily given partition of the integer set $\{0, 1, \cdots, N\}$ such that $\tilde{n}_r - \tilde{n}_{r-1} \geqslant 2$ for all $1 \leqslant r \leqslant \tilde{p}$.

Denote by $C_{\tilde{p}}$ the corresponding matrix of synchronization by $\tilde{p}$-groups with

$$\mathrm{Ker}(C_{\tilde{p}}) = \{\tilde{e}_1, \cdots, \tilde{e}_{\tilde{p}}\}. \tag{10.67}$$

Since $C_{\tilde{p}}$ and $\widetilde{C}_{\tilde{p}}$ have the same rank, there exists an invertible matrix P of order N, such that

$$\widetilde{C}_{\tilde{p}} = C_{\tilde{p}} P. \tag{10.68}$$

Then, it is easy to see that

$$P\mathrm{Ker}(\widetilde{C}_{\tilde{p}}) = \mathrm{Ker}(C_{\tilde{p}}). \tag{10.69}$$

Now, setting

$$\widetilde{U} = PU \quad \text{and} \quad \widetilde{A} = PAP^{-1}, \tag{10.70}$$

problem (I) and (I0) can be written into the following form:

$$\begin{cases} \widetilde{U}'' - \Delta \widetilde{U} + \widetilde{A}\widetilde{U} = 0 & \text{in } (0, +\infty) \times \Omega, \\ \widetilde{U} = 0 & \text{on } (0, +\infty) \times \Gamma_0, \\ \widetilde{U} = PDH & \text{on } (0, +\infty) \times \Gamma_1 \end{cases} \tag{10.71}$$

with the initial data

$$t = 0: \quad \widetilde{U} = P\widehat{U}_0, \ \widetilde{U}' = P\widehat{U}_1 \quad \text{in } \Omega. \tag{10.72}$$

By Theorem 10.14 and noting (10.68), we have

$$C_{\tilde{p}}\widetilde{U}_n = \widetilde{C}_{\tilde{p}}U_n \to 0 \quad \text{as } n \to +\infty \tag{10.73}$$

in the space

$$C^0_{loc}([T, +\infty); (\mathcal{H}_0)^{N-\tilde{p}}) \cap C^1_{loc}([T, +\infty); (\mathcal{H}_{-1})^{N-\tilde{p}}). \tag{10.74}$$

In other words, under the basis P, system (10.71) is approximately synchronizable by $\tilde{p}$-groups. □

As a direct consequence, we have

Corollary 10.17 *Assume that system (I) is approximately synchronizable by p-groups, but not approximately synchronizable by $\tilde{p}$-groups with $\tilde{p} < p$ under a basis. Then, we necessarily have the condition of C_p-compatibility (10.10).*

Now we return to the proof of Theorem 10.14. We will perceive that the stronger convergence (10.64) follows from the convergence (10.6) and the assumption that A does not satisfy the condition of C_p-compatibility (10.10). We point out that only the rank condition (10.63) can not guarantee this convergence.

Let us first generalize Definition 8.1 to weaker initial data.

Definition 10.18 Let $m \geqslant 0$ be an integer. System (I) is approximately null controllable in the space $(\mathcal{H}_{-2m})^N \times (H_{-(2m+1)})^N$ at the time $T > 0$, if for any given initial data $(\widehat{U}_0, \widehat{U}_1) \in (\mathcal{H}_{-2m})^N \times (H_{-(2m+1)})^N$, there exists a sequence $\{H_n\}$ of boundary controls in $L^2(0, +\infty; (L^2(\Gamma_1)^M)$ with compact support in $[0, T]$, such that the corresponding sequence $\{U_n\}$ of solutions to problem (I) and (I0) satisfies

$$U_n \to 0 \quad \text{as } n \to +\infty \tag{10.75}$$

in the space

$$C^0_{loc}([T, +\infty); (\mathcal{H}_{-2m})^N) \cap C^1_{loc}([T, +\infty); (\mathcal{H}_{-(2m+1)})^N). \tag{10.76}$$

Correspondingly, we give

Definition 10.19 The adjoint problem (8.7) is D-observable in the space $(\mathcal{H}_{2m+1})^N \times (\mathcal{H}_{2m})^N$ on the time interval $[0, T]$ if for $(\widehat{\Phi}_0, \widehat{\Phi}_1) \in (\mathcal{H}_{2m+1})^N \times (\mathcal{H}_{2m})^N$, the observation (8.8) implies $\widehat{\Phi}_0 \equiv \widehat{\Phi}_1 \equiv 0$, then $\Phi \equiv 0$.

Similarly to Theorem 8.6 for the case $m = 0$, we can establish the following

Proposition 10.20 *System (I) is approximately null controllable in the space* $(\mathcal{H}_{-2m})^N \times (H_{-(2m+1)})^N$ *at the time* $T > 0$ *if and only if the adjoint problem (8.7) is D-observable in the space* $(\mathcal{H}_{2m+1})^N \times (\mathcal{H}_{2m})^N$ *on the time interval* $[0, T]$.

Proposition 10.21 *Let* $m \geqslant 0$ *be an integer. System (I) is approximately null controllable in the space* $(\mathcal{H}_{-2m})^N \times (H_{-(2m+1)})^N$ *if and only if it is approximately null controllable in the space* $(\mathcal{H}_0)^N \times (\mathcal{H}_{-1})^N$.

Proof By Proposition 10.20, it is sufficient to show that adjoint problem (8.7) is D-observable in the space $(\mathcal{H}_{2m+1})^N \times (\mathcal{H}_{2m})^N$ if and only if it is D-observable in the space $(\mathcal{H}_1)^N \times (\mathcal{H}_0)^N$.

Assume that adjoint problem (8.7) is D-observable in the space $(\mathcal{H}_{2m+1})^N \times (\mathcal{H}_{2m})^N$, then the following expression

$$\|(\widehat{\Phi}_0, \widehat{\Phi}_1)\|_{\mathcal{F}}^2 = \int_0^T \int_{\Gamma_1} |D^T \partial_\nu \Phi|^2 d\Gamma dt \tag{10.77}$$

defines a Hilbert norm in the space $(\mathcal{H}_{2m+1})^N \times (\mathcal{H}_{2m})^N$. Let $\mathcal{F}$ be the closure of $(\mathcal{H}_{2m+1})^N \times (\mathcal{H}_{2m})^N$ with respect to the F-norm.

By the hidden regularity obtained for problem (8.7) (cf. [62]), we have

$$\int_0^T \int_{\Gamma_1} |D^T \partial_\nu \Phi|^2 d\Gamma dt \leqslant c\|(\widehat{\Phi}_0, \widehat{\Phi}_1)\|^2_{(\mathcal{H}_1)^N \times (\mathcal{H}_0)^N}, \tag{10.78}$$

then

$$(\mathcal{H}_1)^N \times (\mathcal{H}_0)^N \subseteq \mathcal{F}. \tag{10.79}$$

Since problem (8.7) is D-observable in $\mathcal{F}$, therefore, it is still D-observable in its subspace $(\mathcal{H}_1)^N \times (\mathcal{H}_0)^N$. The converse is trivial. The proof is complete. □

Remark 10.22 Similar results on the exact boundary controllability can be found in [12].

Proposition 10.23 *Assume that system (I) is approximately synchronizable by p-groups. Then, for any given integer* $l \geqslant 0$ *and any given initial data* $(\widehat{U}_0, \widehat{U}_1) \in (\mathcal{H}_0)^N \times (\mathcal{H}_{-1})^N$, *we have*

$$C_p A^l U_n \to 0 \text{ in } C^0_{loc}([T, +\infty); (\mathcal{H}_{-2l})^{N-p}) \quad \text{as } n \to +\infty, \tag{10.80}$$

respectively,

$$C_p A^l U_n' \to 0 \text{ in } C^0_{loc}([T, +\infty); (\mathcal{H}_{-(2l+1)})^{N-p}) \quad \text{as } n \to +\infty. \tag{10.81}$$

Proof Note that U_n satisfies the homogeneous system

$$\begin{cases} U_n'' - \Delta U_n + AU_n = 0 \text{ in } (T, +\infty) \times \Omega, \\ U_n = 0 \quad\quad\quad\quad\quad\quad \text{on } (T, +\infty) \times \Gamma. \end{cases} \tag{10.82}$$

Applying $C_p A^{l-1}$ to system (10.82), it follows that

$$\begin{aligned} &\|C_p A^l U_n\|_{C^0_{loc}([T,+\infty);(\mathcal{H}_{-2l})^{N-p})} \\ \leqslant &\|C_p A^{l-1} U_n''\|_{C^0_{loc}([T,+\infty);(\mathcal{H}_{-2l})^{N-p})} \\ &+\|\Delta C_p A^{l-1} U_n\|_{C^0_{loc}([T,+\infty);(\mathcal{H}_{-2l})^{N-p})} \\ \leqslant &c\|C_p A^{l-1} U_n\|_{C^0_{loc}([T,+\infty);(\mathcal{H}_{-2(l-1)})^{N-p})} \\ \leqslant &c^l\|C_p U_n\|_{C^0_{loc}([T,+\infty);(\mathcal{H}_0)^{N-p})}, \end{aligned} \tag{10.83}$$

where $c > 0$ is a positive constant. Similar result can be shown for $C_p A^l U_n'$. The proof is complete. □

We now give the proof of Theorem 10.14.

(i) Let $(\widehat{U}_0, \widehat{U}_1) \in (\mathcal{H}_0)^N \times (\mathcal{H}_{-1})^N$. Then, by (10.20) and noting (10.80)–(10.81) with $0 \leqslant l \leqslant N-1$, we get

$$\widetilde{C}_{\tilde{p}} U_n \to 0 \text{ in } C^0_{loc}([T, +\infty); (\mathcal{H}_{-2(N-1)})^{N-\tilde{p}}) \quad \text{as } n \to +\infty, \tag{10.84}$$

respectively,

$$\widetilde{C}_{\tilde{p}} U_n' \to 0 \text{ in } C^0_{loc}([T, +\infty); (\mathcal{H}_{-(2N-1)})^{N-\tilde{p}}) \quad \text{as } n \to +\infty. \tag{10.85}$$

(ii) Let $(\widehat{U}_0, \widehat{U}_1) \in (\mathcal{H}_{-2(N-1)})^N \times (\mathcal{H}_{-(2N-1)})^N$. By density, there exists a sequence $\{(\widehat{U}_0^m, \widehat{U}_1^m)\}_{m\geqslant 0}$ in $(\mathcal{H}_0)^N \times (\mathcal{H}_{-1})^N$, such that

$$(\widehat{U}_0^m, \widehat{U}_1^m) \to (\widehat{U}_0, \widehat{U}_1) \quad \text{in } (\mathcal{H}_{-2(N-1)})^N \times (\mathcal{H}_{-(2N-1)})^N \tag{10.86}$$

as $m \to +\infty$. For each fixed m, there exists a sequence of boundary controls $\{H_n^m\}_{n\geqslant 0}$, such that the corresponding sequence $\{U_n^m\}_{n\geqslant 0}$ of solutions to problem (I) and (I0) with the initial data $(\widehat{U}_0^m, \widehat{U}_1^m)$ satisfies

$$\widetilde{C}_{\tilde{p}} U_n^m \to 0 \text{ in } C^0_{loc}([T, +\infty); (\mathcal{H}_{-2(N-1)})^{N-\tilde{p}}) \quad \text{as } n \to +\infty. \tag{10.87}$$

(iii) Let $\mathcal{R}$ denote the resolution of problem (I) and (I0) :

$$\mathcal{R}: \quad (\widehat{U}_0, \widehat{U}_1; H_n) \to (U_n, U_n'), \tag{10.88}$$

which is continuous from

$$(\mathcal{H}_{-2(N-1)})^N \times (H_{-(2N-1)})^N \times \mathcal{L}^M \tag{10.89}$$

into

$$C^0_{loc}([T, +\infty); (\mathcal{H}_{-2(N-1)})^N) \cap C^1_{loc}([T, +\infty); (\mathcal{H}_{-(2N-1)})^N). \tag{10.90}$$

Now, for any given $(\widehat{U}_0, \widehat{U}_1) \in (\mathcal{H}_{-2(N-1)})^N \times (\mathcal{H}_{-(2N-1)})^N$, we write

$$\mathcal{R}(\widehat{U}_0, \widehat{U}_1; H^m_n) = \mathcal{R}(\widehat{U}^m_0, \widehat{U}^m_1; H^m_n) + \mathcal{R}(\widehat{U}_0 - \widehat{U}^m_0, \widehat{U}_1 - \widehat{U}^m_1; 0). \tag{10.91}$$

By well-posedness, for all $0 \leqslant t \leqslant T \leqslant S$ we have

$$\|\mathcal{R}(\widehat{U}_0 - \widehat{U}^m_0, \widehat{U}_1 - \widehat{U}^m_1; 0)(t)\| \leqslant c_S \|(\widehat{U}_0 - \widehat{U}^m_0, U_1 - U^m_1)\| \tag{10.92}$$

with respect to the norm of $(\mathcal{H}_{-2(N-1)})^N \times (\mathcal{H}_{-(2N-1)})^N$, where c_S is a positive constant depending only on S. Then, noting (10.86) and (10.87), we can chose a diagonal subsequence $\{H^{m_k}_{n_k}\}_{k \geqslant 0}$ such that

$$\widetilde{C}_{\tilde{p}} \mathcal{R}(\widehat{U}_0, \widehat{U}_1; H^{m_k}_{n_k}) \to 0 \text{ in } C^0_{loc}([T, +\infty); (\mathcal{H}_{-2(N-1)})^{N-\tilde{p}}) \tag{10.93}$$

as $k \to +\infty$.

Hence, the reduced system (10.71) is approximately null controllable in the space $(\mathcal{H}_{-2(N-1)})^{N-\tilde{p}} \times (\mathcal{H}_{-(2N-1)})^{N-\tilde{p}}$, therefore, by Proposition 10.21, it is also approximately null controllable in the space $(\mathcal{H}_0)^{N-\tilde{p}} \times (\mathcal{H}_{-1})^{N-\tilde{p}}$. The proof is complete. □

10.5 Approximate Boundary Null Controllability

Let d be a column vector in D, or more generally, a linear combination of the column vectors of D, namely, $d \in \text{Im}(D)$. If $d \in \text{Ker}(C_p)$, then it will be canceled in the product matrix $C_p D$, therefore it cannot give any effect to the reduced system (10.13). However, the vectors in $\text{Ker}(C_p)$ may play an essential role in the approximate boundary null controllability. More precisely, we have the following

Theorem 10.24 *Let the coupling matrix A satisfy the condition of C_p-compatibility (10.10). Assume that*

$$e_1, \cdots, e_p \in Im(D), \tag{10.94}$$

where $Ker(C_p) = Span\{e_1, \cdots, e_p\}$. If system (I) is approximately synchronizable by p-groups, then it is in fact approximately null controllable.

Proof To prove this theorem, by Theorem 8.6, it is sufficient to show that the adjoint problem (8.7) is D-observable.

For $1 \leqslant r \leqslant p$, applying e_r to the adjoint problem (8.7) and noting $\phi_r = (e_r, \Phi)$, it follows from the condition of C_p-compatibility (10.10) that

$$\begin{cases} \phi_r'' - \Delta\phi_r + \sum_{s=1}^{p} \beta_{rs}\phi_s = 0 & \text{in } (0, T) \times \Omega, \\ \phi_r = 0 & \text{on } (0, T) \times \Gamma, \end{cases} \tag{10.95}$$

where β_{sr} are given by

$$Ae_r = \sum_{s=1}^{p} \beta_{rs} e_s, \quad 1 \leqslant r \leqslant p. \tag{10.96}$$

Noting $e_r \in \text{Im}(D)$, there exists $x_r \in \mathbb{R}^M$, such that $e_r = Dx_r$. Then, the D-observation (8.8) gives

$$\partial_\nu \phi_r = (e_r, \partial_\nu \Phi) = (x_r, D^T \partial_\nu \Phi) \equiv 0 \quad \text{on } (0, T) \times \Gamma_1. \tag{10.97}$$

Hence, by Holmgren's uniqueness theorem, $\phi_r \equiv 0$ for all $1 \leqslant r \leqslant p$. Consequently, we have

$$\Phi \in \{\text{Ker}(C_p)\}^\perp = \text{Im}(C_p^T). \tag{10.98}$$

Hence, the solution of the adjoint problem (8.7) can be written as

$$\Phi = C_p^T \Psi, \tag{10.99}$$

and it follows from adjoint problem (8.7) that

$$\begin{cases} C_p^T \Psi'' - C_p^T \Delta\Psi + A^T C_p^T \Psi = 0 & \text{in } (0, T) \times \Omega, \\ C_p^T \Psi = 0 & \text{on } (0, T) \times \Gamma. \end{cases} \tag{10.100}$$

Noting the condition of C_p-compatibility (10.11), it follows that

$$\begin{cases} C_p^T \Psi'' - C_p^T \Delta\Psi + C_p^T \overline{A}_p^T \Psi = 0 & \text{in } (0, T) \times \Omega, \\ C_p^T \Psi = 0 & \text{on } (0, T) \times \Gamma. \end{cases} \tag{10.101}$$

Since the matrix C_p^T is of full column-rank, we get the reduced adjoint problem (10.15). Accordingly, the D-observation gives

$$D^T \partial_\nu \Phi = D^T C_p^T \partial_\nu \Psi \equiv 0 \quad \text{on } (0, T) \times \Gamma_1. \tag{10.102}$$

By Lemma 10.2, the reduced adjoint problem (10.15) is $C_p D$-observable, it follows that $\Psi \equiv 0$. Then, from (10.99), we have $\Phi \equiv 0$. So, the adjoint problem (8.7) is D-observable. The proof is thus complete. □

Remark 10.25 The extension matrix $\widetilde{C}_{\tilde{p}}$ defined by (10.20) satisfies the condition of $\widetilde{C}_{\tilde{p}}$-compatibility (10.21), and, by Theorem 10.14, the corresponding reduced system (similar to (10.13)) is approximately null controllable. Moreover, we have $Ker(\widetilde{C}_{\tilde{p}}) \subseteq Ker(C_p)$. So, if A does not satisfy the condition of C_p-compatibility (10.10), we can replace C_p by $\widetilde{C}_{\tilde{p}}$, and Theorem 10.24 remains valid in this case.

Chapter 11
Induced Approximate Boundary Synchronization

We introduce the induced approximate boundary synchronization for system (I) with Dirichlet boundary controls and give some examples in this chapter.

11.1 Definition

We have precisely studied the approximate boundary synchronization by p-groups for system (I) under the minimal rank condition (10.29). In this situation, the convergence of the sequence $\{U_n\}$ of solutions to problem (I) and (I0) and the necessity of the condition of C_p-compatibility (10.10) are essentially the consequence of the minimal number of applied total controls.

The objective of this chapter is to investigate the approximate boundary synchronization by p-groups in the case that

$$N_p > N - p, \tag{11.1}$$

where N_p is defined by (10.54).

By the condition of C_p-compatibility (10.10), $\text{Ker}(C_p)$ is invariant for A. However, by Remark 10.12, the situation (11.1) occurs if and only if $\text{Ker}(C_p)$ does not admit any supplement which is invariant for A. So, in order to determine the minimal number N_p of total controls, which are necessary for the approximate boundary synchronization by p-groups of system (I), a natural thought is to extend the matrix C_p of synchronization by p-groups to the **induced extension matrix** C_q^* given in Definition 2.18. Thus, we can bring this situation to the framework of Chap. 10.

Accordingly, we introduce the following

Definition 11.1 System (I) is **induced approximately synchronizable** by C_q^* at the time $T > 0$, if for any given initial data $(\widehat{U}_0, \widehat{U}_1) \in (\mathcal{H}_0)^N \times (\mathcal{H}_{-1})^N$, there exists a

T. Li and B. Rao, *Boundary Synchronization for Hyperbolic Systems*,
Progress in Nonlinear Differential Equations and Their Applications 94,
https://doi.org/10.1007/978-3-030-32849-8_11

sequence $\{H_n\}$ of boundary controls in $\mathcal{L}^M$ with compact support in $[0, T]$, such that the corresponding sequence $\{U_n\}$ of solutions to problem (I) and (I0) satisfies

$$C_q^* U_n \to 0 \quad \text{as } n \to +\infty \tag{11.2}$$

in the space $C^0_{loc}([T, +\infty); (\mathcal{H}_0)^{N-q}) \cap C^1_{loc}([T, +\infty); (\mathcal{H}_{-1})^{N-q})$.

In general, the induced approximate boundary synchronization does not occur. To see this aspect, let $\mathcal{E}_1, \cdots, \mathcal{E}_{m^*}$ be a Jordan chain of A^T associated with an eigenvalue λ:

$$\mathcal{E}_0 = 0, \quad A^T \mathcal{E}_i = \lambda \mathcal{E}_i + \mathcal{E}_{i-1}, \quad i = 1, \cdots, m. \tag{11.3}$$

Let m with $1 \leqslant m < m^*$ be an integer such that $\mathcal{E}_1, \cdots \mathcal{E}_m \in \text{Im}(C_p^T)$. Then, applying $\mathcal{E}_1, \cdots, \mathcal{E}_m$ to system (I) with $U = U_n$ and $H = H_n$, and setting $\phi_i = (\mathcal{E}_i, U_n)(i = 1, \cdots, m)$, we get

$$\begin{cases} \phi''_{1,n} - \Delta\phi_{1,n} + \lambda\phi_{1,n} = 0 & \text{in } (0, +\infty) \times \Omega, \\ \cdots\cdots \\ \phi''_{m,n} - \Delta\phi_{m,n} + \lambda\phi_{m,n} + \phi_{m-1,n} = 0 \text{ in } (0, +\infty) \times \Omega, \\ \cdots\cdots \\ \phi''_{m,n} - \Delta\phi_{m,n} + \lambda\phi_{m,n} + \phi_{m-1,n} = 0 \text{ in } (0, +\infty) \times \Omega. \end{cases} \tag{11.4}$$

Since $\mathcal{E}_1, \cdots, \mathcal{E}_m \in \text{Im}(C_p^T)$, there exist some vectors $r_i \in \mathbb{R}^{N-p}$ for $i = 1, \cdots, m$, such that

$$\mathcal{E}_i = C_p^T r_i, \quad i = 1, \cdots, m. \tag{11.5}$$

By (10.6), for all $i = 1, \cdots, m$ we have

$$\phi_{i,n} = (r_i, C_p U_n) \to 0 \quad \text{as } n \to +\infty \tag{11.6}$$

in the space

$$C^0_{loc}([T, +\infty); \mathcal{H}_0) \cap C^1_{loc}([T, +\infty); \mathcal{H}_{-1}). \tag{11.7}$$

Thus, the first m components $\phi_{1,n}, \cdots, \phi_{m,n}$ converge to zero as $n \to +\infty$. However, except in some specific situations (cf. Theorems 11.11 and 11.12 below), we don't know if the other components $\phi_{m+1,n}, \cdots, \phi_{m,n}$ converge also to zero or not.

However, when system (I) is induced approximately synchronizable, it is not only approximately synchronizable by p-groups, but also possesses some additional properties, which are hidden in the extension matrix C_q^*. That is to say, in this case, when we realize the approximate boundary synchronization by p-groups for system (I), we could additionally get some unexpected results, which would improve the desired approximate boundary synchronization by p-groups!

Since $\mathrm{Im}(C_q^*)$ admits a supplement which is invariant for A^T if system (I) is induced approximately synchronizable, by Remark 10.12, we immediately perceive that

$$N_p = N - q, \tag{11.8}$$

namely, $(N - q)$ total controls are necessary to realize the approximate boundary synchronization by p-groups for system (I). We will show the minimal number (11.8) in the general case. This result improves the estimate (10.19) and deeply reveals that the minimal number of total controls necessary to the approximate boundary synchronization by p-groups depends not only on the number p of groups, but also on the algebraic structure of the coupling matrix A with respect to the matrix C_p of synchronization by p-groups.

11.2 Preliminaries

In what follows, for better understanding the driving idea, *we always assume that* $Im(C_p^T)$ *is* A^T*-marked.* In this framework, the procedure for obtaining the matrix C_q^* is given in Chap. 2. In Remark 11.3 below, we will explain that the non-marked case is senseless for the induced approximate synchronization.

Proposition 11.2 *Assume that the coupling matrix A satisfies the condition of* C_p*-compatibility (10.10). Let D satisfy*

$$rank(C_pD, C_pAD, \cdots, C_pA^{N-1}D) = N - p. \tag{11.9}$$

Then, we necessarily have

$$rank(D, AD, \cdots, A^{N-1}D) \geqslant N - q, \tag{11.10}$$

where q is given by (2.59).

Proof By the assertion (i) of Proposition 2.12, it is sufficient to show that the dimension of any invariant subspace W of A^T, contained in $\mathrm{Ker}(D^T)$, does not exceed q.

By the condition of C_p-compatibility (10.10), there exists a matrix $\overline{A}_p$ of order $(N - p)$, such that $C_pA = \overline{A}_pC_p$. By Proposition 2.16, the rank condition (11.9) is equivalent to

$$\mathrm{rank}(C_pD, \overline{A}_pC_pD, \cdots, \overline{A}_p^{N-p-1}C_pD) = N - p, \tag{11.11}$$

which, by the assertion (ii) of Proposition 2.12, implies that $\mathrm{Ker}(C_pD)^T$ does not contain any nontrivial invariant space of $\overline{A}_p^T$.

Now let W be any given invariant subspace of A^T contained in $\text{Im}(C_p^T)$. The projected subspace

$$\overline{W} = (C_p C_p^T)^{-1} C_p W = \{\overline{x} : \quad C_p^T \overline{x} = x, \quad \forall x \in W\} \tag{11.12}$$

is an invariant subspace of $\overline{A}_p^T$. In particular, we have

$$\overline{W} \cap \text{Ker}(C_p D)^T = \{0\}. \tag{11.13}$$

For any given $x \in W$, there exists $\overline{x} \in \overline{W}$, such that $x = C_p^T \overline{x}$, then we have

$$D^T x = D^T C_p^T \overline{x} = (C_p D)^T \overline{x}. \tag{11.14}$$

Thus we get

$$W \cap \text{Ker}(D^T) = \{0\} \tag{11.15}$$

for any given invariant subspace W of A^T, contained in $\text{Im}(C_p^T)$.

Now let W^* be an invariant subspace of A^T contained in $\text{Im}(C_q^{*T}) \cap \text{Ker}(D^T)$. Since $W^* \cap \text{Im}(C_p^T)$ is also an invariant subspace of A^T, contained in $\text{Im}(C_p^T) \cap \text{Ker}(D^T)$, we have $W^* \cap \text{Im}(C_p^T) = \{0\}$ because of (11.15). Then, it follows that

$$W^* \subseteq \text{Im}(C_q^{*T}) \setminus \text{Im}(C_p^T). \tag{11.16}$$

Since $\text{Im}(C_p^T)$ is A^T-marked, by the construction, $\text{Im}(C_q^{*T}) \setminus \text{Im}(C_p^T)$ does not contain any eigenvector of A^T, then it follows that

$$W^* = \{0\}. \tag{11.17}$$

Finally, let W be an invariant subspace of A^T contained in $\text{Ker}(D^T)$. Since $W \cap \text{Im}(C_q^{*T})$ is an invariant subspace of A^T, contained in $\text{Im}(C_q^{*T}) \cap \text{Ker}(D^T)$, it follows from (11.17) that

$$W \cap \text{Im}(C_q^{*T}) = \{0\}. \tag{11.18}$$

Then, we get

$$\dim \text{Im}(C_q^{*T}) + \dim(W) = N - q + \dim(W) \leqslant N, \tag{11.19}$$

which implies that $\dim(W) \leqslant q$. This achieves the proof. □

Remark 11.3 In the case that $\text{Im}(C_p^T)$ is not A^T-marked, similarly as in §7.3, we can also construct the extension matrix C_q^*. But the set $\text{Im}(C_q^{*T}) \setminus \text{Im}(C_p^T)$ possibly contains eigenvectors of A^T, then (11.17) will not be satisfied. So, instead of (11.10), we can only get a weaker estimation such as

$$\text{rank}(D, AD, \cdots, A^{N-1}D) \geqslant N - q' \tag{11.20}$$

with $q' > q$. In the case of equality, we have

$$\text{rank}(D, AD, \cdots A^{N-1}D) = N - q' < N - q. \tag{11.21}$$

So, system (I) is never induced approximately synchronizable. This explains why we only study the A^T-marked case at the beginning of the section.

Proposition 11.4 *Assume that the coupling matrix A satisfies the condition of C_p-compatibility (10.10). If*

$$rank(D, AD, \cdots, A^{N-1}D) = N - q, \tag{11.22}$$

then we necessarily have the rank condition

$$rank(C_q^* D, C_q^* AD, \cdots C_q^* A^{N-1}D) = N - q. \tag{11.23}$$

Proof First, by (ii) of Proposition 2.12, the rank condition (11.22) implies the existence of an invariant subspace W of A^T contained in $\text{Ker}(D^T)$ with dimension q. On the other hand, noting (11.18), we have

$$\dim \text{Im}(C_q^{*T}) + \dim(W) = (N - q) + q = N, \tag{11.24}$$

then $\text{Im}(C_q^{*T})$ is a supplement of W, which is also invariant for A^T.

Furthermore, by the definition of W, it is easy to see that

$$W \subseteq \text{Ker}(D, AD, \cdots, A^{N-1}D)^T, \tag{11.25}$$

or equivalently,

$$\text{Im}(D, AD, \cdots, A^{N-1}D) \subseteq W^{\perp}, \tag{11.26}$$

which together with the rank condition (11.22) imply

$$\text{Im}(D, AD, \cdots, A^{N-1}D) = W^{\perp}. \tag{11.27}$$

But $W^{\perp}$ is a supplement of $\text{Ker}(C_q^*)$, which is invariant for A, then

$$\text{Im}(D, AD, \cdots, A^{N-1}D) \cap \text{Ker}(C_q^*) = \{0\}. \tag{11.28}$$

By Proposition 2.11, we get

$$\begin{aligned}&\text{rank}(C_q^*D, C_q^*AD, \cdots, C_q^*A^{N-1}D)\\=&\text{rank}(D, AD, \cdots, A^{N-1}D).\end{aligned} \tag{11.29}$$

This achieves the proof. □

11.3 Induced Approximate Boundary Synchronization

We first give a lower bound estimate on the number of total controls which are necessary for the approximate boundary synchronization by p-groups of system (I).

Proposition 11.5 *Assume that the coupling matrix A satisfies the condition of C_p-compatibility (10.10). If system (I) is approximately synchronizable by p-groups under the action of the boundary control matrix D, then we necessarily have*

$$rank(D, AD, \cdots, A^{N-1}D) \geqslant N - q, \tag{11.30}$$

where q is given by (2.59).

Proof By Lemma 10.3, we have the following Kalman's criterion

$$\text{rank}(C_pD, C_pAD, \cdots, C_pA^{N-1}D) = N - p. \tag{11.31}$$

Then, the rank condition (11.30) follows from Proposition 11.2. □

Now we go back to the **induced approximate boundary synchronization**.

Theorem 11.6 *Assume that the coupling matrix A satisfies the condition of C_p-compatibility (10.10). If system (I) is induced approximately synchronizable with the minimal number of total controls (11.22), then, there exist scalar functions $u_1^*, \cdots, u_q^*$ such that*

$$U_n \to \sum_{s=1}^{q} u_s^* e_s^* \quad as\ n \to +\infty \tag{11.32}$$

in the space

$$C^0_{loc}([T, +\infty); (\mathcal{H}_0)^N) \cap (C^1_{loc}([T, +\infty); (\mathcal{H}_{-1})^N), \tag{11.33}$$

where $Span\{e_1^*, \cdots e_q^*\} = Ker(C_q^*)$.

In particular, system (I) is approximately synchronizable by p-groups in the pinning sense.

Proof The proof is similar to that of Theorem 10.5. Here, we only give its sketch as follows.

By Proposition 2.12, the rank condition (11.22) guarantees the existence of a subspace $Span\{E_1, \cdots, E_q\}$, which is invariant for A^T and contained in Ker(D^T). Thus, it is easy to see that the projection of U_n on $Span\{E_1, \cdots, E_q\}$ is independent of boundary controls H_n. Similarly as in Theorem 10.5, we can show that $Span\{E_1, \cdots, E_q\}$ is a supplement of Im(C_q^{*T}). We have

$$U_n = \sum_{r=1}^{q} u_r^* e_r^* + C_q^{*T} R_n, \tag{11.34}$$

where $R_n \in \mathbb{R}^{N-q}$. Since Ker$(C_q^*) = $ Span$\{e_1^*, \cdots e_q^*\}$, the induced approximate boundary synchronization (11.2) implies that $C_q^* C_q^{*T} R_n \to 0$. Noting that $C_q^* C_q^{*T}$ is invertible, we get $R_n \to 0$, which yields (11.32).

Since Ker$(C_q^*) \subseteq$ Ker(C_p), there exist some coefficients α_{sr} such that

$$e_s^* = \sum_{r=1}^{p} \alpha_{sr} e_r, \quad s = 1, \cdots, q. \tag{11.35}$$

Then, setting

$$u_r = \sum_{s=1}^{q} \alpha_{sr} u_s^*, \quad r = 1, \cdots, p \tag{11.36}$$

in (11.32), we get

$$U_n \to \sum_{r=1}^{p} u_r e_r \quad \text{as } n \to +\infty \tag{11.37}$$

in the space

$$C_{loc}^0([T, +\infty); (\mathcal{H}_0)^N) \cap C_{loc}^0([T, +\infty); (\mathcal{H}_{-1})^N). \tag{11.38}$$

Thus, system (I) is approximately synchronizable by p-groups in the pinning sense. □

Remark 11.7 Because of the minimal rank condition (11.22), the functions $u_1^*, \cdots, u_q^*$ in (11.32) are linearly independent, while, since $p > q$, the functions $u_1, \cdots, u_p$ given in (11.36) are linearly dependent. The relationship among $u_1, \cdots, u_p$, which depends on the structure of Ker(C_q^*), will be illustrated in Examples 11.16 and 11.17 below.

Theorem 11.8 *Assume that the coupling matrix A satisfies the condition of C_p-compatibility (10.10). There exists a boundary control matrix D such that the corresponding system (I) is induced approximately synchronizable with the minimal number of total controls (11.22).*

Proof Since Ker(C_q^*) is an invariant subspace of A, by Proposition 2.15, there exists a matrix A_q^* of order $(N-q)$, such that $C_q^* A = A_q^* C_q^*$. Setting $W = C_q^* U$ in problem

(I) and (I0), we get the following reduced system:

$$\begin{cases} W'' - \Delta W + A_q^* W = 0 & \text{in } (0, +\infty) \times \Omega, \\ W = 0 & \text{on } (0, +\infty) \times \Gamma_0, \\ W = C_q^* D H & \text{on } (0, +\infty) \times \Gamma_1 \end{cases} \tag{11.39}$$

with the initial data:

$$t = 0: \quad W = C_q^* \widehat{U}_0, \ W' = C_q^* \widehat{U}_1 \quad \text{in } \Omega. \tag{11.40}$$

Clearly, the induced approximate boundary synchronization by C_q^* for system (I) is equivalent to the approximate boundary null controllability of the reduced system (11.39), then equivalent to the $C_q^* D$-observability of the corresponding reduced adjoint problem

$$\begin{cases} \Psi'' - \Delta \Psi + A_q^{*T} \Psi = 0 & \text{in } (0, +\infty) \times \Omega, \\ \Psi = 0 & \text{on } (0, +\infty) \times \Gamma, \\ t = 0: \quad \Psi = \widehat{\Psi}_0, \ \Psi' = \widehat{\Psi}_1 & \text{in } \Omega. \end{cases} \tag{11.41}$$

Let W be a subspace which is invariant for A^T and bi-orthonormal to $\text{Ker}(C_q^*)$. Clearly, W and $\text{Ker}(C_q^*)$ have the same dimension q. Setting

$$\text{Ker}(D^T) = W, \tag{11.42}$$

W is the largest invariant subspace of A^T contained in $\text{Ker}(D^T)$. By the assertion (ii) of Proposition 2.12 with $d = q$, we have the rank condition (11.22).

On the other hand, since W is bi-orthonormal to $\text{Ker}(C_q^*)$, by Proposition 2.5, we have

$$\text{Im}(D) \cap \text{Ker}(C_q^*) = W^\perp \cap \text{Ker}(C_q^*) = \{0\}. \tag{11.43}$$

Then by Proposition 2.11, we have

$$\text{rank}(C_q^* D) = \text{rank}(D) = N - q. \tag{11.44}$$

Therefore, the $C_q^* D$-observation becomes the complete observation

$$\partial_\nu \Psi \equiv 0 \quad \text{on } [0, T] \times \Gamma_1, \tag{11.45}$$

which, because of Holmgren's uniqueness theorem (cf. Theorem 8.2 in [62]), implies the $C_q^* D$-observability of the reduced adjoint problem (11.41). We get thus the convergence (11.2). The proof is complete. □

As a direct consequence of Theorem 11.8, we have

Corollary 11.9 *Let q with $0 \leqslant q < p$ be given by (2.59). Then we have*

$$N_p = N - q. \tag{11.46}$$

Remark 11.10 Noting that $\text{rank}(D) = N - q$ for the boundary control matrix D given by (11.42), Theorem 11.8 means that the approximate boundary synchronization by p-groups can be realized by means of the minimal number $(N - q)$ of direct controls. However, in practice, we prefer to use a boundary control matrix D which provides $(N - q)$ total controls, and has the minimal $\text{rank}(D)$. To this end, we will replace the rank condition (11.44) by a weaker rank condition (11.23), which is only necessary in general for the induced approximate boundary synchronization. The following results give some confirmative answers under suitable additional conditions.

Theorem 11.11 *Let $\Omega \subset \mathbb{R}^n$ satisfy the multiplier geometrical condition (3.1). Assume that the coupling matrix A satisfies the condition of C_p-compatibility (10.10) and is similar to a nilpotent matrix. If system (I) is approximately synchronizable by p-groups under the rank condition (11.22), then it is actually induced approximately synchronizable.*

Proof As explained at the beginning of the proof of Theorem 11.8, the induced approximate boundary synchronization of system (I) is equivalent to the C_q^*D-observability of the reduced adjoint problem (11.41).

First, by Proposition 11.4, we have the rank condition (11.23). So, by Proposition 2.16, we have

$$\text{rank}(C_q^*D, A_q^*C_q^*D, \cdots, A_q^{*N-q-1}C_q^*D) = N - q. \tag{11.47}$$

On the other hand, the matrix A is similar to a nilpotent matrix. By Proposition 2.21, the projection of the system of root vectors of A provides the system of root vectors of the reduced matrix A_q^*. Thus, it is easy to see that A_q^* is also similar to a nilpotent matrix. Then by Theorem 8.16, the reduced adjoint problem (11.41) is C_q^*D-observable. The proof is thus complete. □

Theorem 11.12 *Let the coupling matrix A satisfy the condition of C_p-compatibility (10.10). Assume that system (I) is approximately synchronizable by p-groups by means of a boundary control matrix D. Let*

$$\widetilde{D} = (\tilde{e}_1, \cdots, \tilde{e}_{p-q}, D), \tag{11.48}$$

where $\tilde{e}_1, \cdots, \tilde{e}_{p-q} \in Ker(C_p) \cap Im(C_q^{*T})$. Then system (I) is induced approximately synchronizable under the action of the new boundary control matrix $\widetilde{D}$.

Proof It is sufficient to show the $C_q^* \widetilde{D}$-observability of the reduced adjoint problem (11.41). Since C_q^{*T} is a bijection from $\mathbb{R}^{N-q}$ onto $\text{Im}(C_q^{*T})$, we can set $\Phi = C_q^{*T} \Psi$. Then, noting $C_q^* A = A_q^* C_q^*$, it follows from system (11.41) that

$$\begin{cases} \Phi'' - \Delta \Phi + A^T \Phi = 0 & \text{in } (0, +\infty) \times \Omega, \\ \Phi = 0 & \text{on } (0, +\infty) \times \Gamma \end{cases} \tag{11.49}$$

with the initial data $(\widehat{\Phi}_0, \widehat{\Phi}_1) \in \text{Im}(C_q^{*T}) \times \text{Im}(C_q^{*T})$. Accordingly, the $C_q^* \widetilde{D}$-observation

$$\widetilde{D}^T C_q^{*T} \Psi \equiv 0 \quad \text{on } (0, T) \times \Gamma \tag{11.50}$$

becomes the $\widetilde{D}$-observation

$$\widetilde{D}^T \Phi \equiv 0 \quad \text{on } (0, T) \times \Gamma. \tag{11.51}$$

Then, the $C_q^* \widetilde{D}$-observability of (11.41) becomes the $\widetilde{D}$-observability of system (11.49) with the initial data $(\widehat{\Phi}_0, \widehat{\Phi}_1) \in \text{Im}(C_q^{*T}) \times \text{Im}(C_q^{*T})$.

Now let $\overline{C}_p$ denote the restriction of C_p to the subspace $\text{Im}(C_q^{*T})$. We have

$$\text{Im}(C_q^{*T}) = \text{Im}(C_p^T) \oplus \text{Ker}(\overline{C}_p). \tag{11.52}$$

Since the induced approximate boundary synchronization can be considered as the approximate boundary null controllability in the subspace $\text{Im}(C_q^{*T}) \times \text{Im}(C_q^{*T})$, so, applying Theorem 10.24, the proof is achieved. □

11.4 Minimal Number of Direct Controls

In this section, we first clarify the relation between the number of total controls and that of direct controls. Then, we give an explicit construction for the boundary control matrix D which has the minimal rank.

Proposition 11.13 *Let A be a matrix of order N and D be an $N \times M$ matrix such that*

$$rank(D, AD, \cdots, A^{N-1}D) = N. \tag{11.53}$$

Then we have the following sharp lower bound estimate

$$rank(D) \geqslant \mu, \tag{11.54}$$

where

$$\mu = \max_{\lambda \in Sp(A^T)} dim\ Ker(A^T - \lambda I) \tag{11.55}$$

is the largest geometrical multiplicity of the eigenvalues of A^T.

Proof Let λ be an eigenvalue of A^T, the geometrical multiplicity of which is equal to μ, and let V be the subspace composed of all the eigenvectors associated with the eigenvalue λ. By the assertion (ii) of Proposition 2.12, the rank condition (11.53) implies that there does not exist any nontrivial invariant subspace of A^T, contained in $\text{Ker}(D^T)$. Therefore, we have

$$\dim \text{Ker}(D^T) + \dim (V) \leqslant N, \tag{11.56}$$

namely,

$$\dim (V) \leqslant N - \dim \text{Ker}(D^T) = \text{rank}(D), \tag{11.57}$$

which yields the lower bound estimate (11.54).

In order to show the sharpness of (11.54), we will construct a matrix D_0 of rank μ, which does not contain any nontrivial invariant subspace of A^T, therefore, satisfies the rank condition (11.53) by Proposition 2.12.

Let $\lambda_1, \cdots, \lambda_d$ be the distinct eigenvalues of A, with geometrical multiplicity $\mu_1, \cdots, \mu_d$, respectively. For any given $r = 1, \cdots, d$ and $k = 1, \cdots, \mu_r$, we denote by $(x_{r,p}^{(k)})$ the following Jordan chain of length $d_r^{(k)}$:

$$x_{r,d_r^{(k)}+1}^{(k)} = 0, \quad Ax_{r,p}^{(k)} = \lambda_r x_{r,p}^{(k)} + x_{r,p+1}^{(k)} \tag{11.58}$$

for $p = 1, \cdots, d_r^{(k)}$.

Accordingly, for any given $s = 1, \cdots, d$ and $l = 1, \cdots, \mu_s$, let $(y_{s,q}^{(l)})$ be the corresponding Jordan chain of length $d_s^{(l)}$:

$$y_{s,0}^{(l)} = 0, \quad A^T y_{s,q}^{(l)} = \lambda_s y_{s,q}^{(l)} + y_{s,q-1}^{(l)}, \quad q = 1, \cdots, d_s^{(l)}. \tag{11.59}$$

We can chose

$$(x_{r,p}^{(k)}, y_{s,q}^{(l)}) = \delta_{kl}\delta_{rs}\delta_{pq}. \tag{11.60}$$

Then we define the matrix D_0 by

$$D_0 = (x_{1,1}^{(1)}, \cdots, x_{1,1}^{(\mu_1)}, \cdots, x_{d,1}^{(1)}, \cdots, x_{d,1}^{(\mu_d)})\overline{D}, \tag{11.61}$$

where $\overline{D}$ is a $(\mu_1 + \cdots + \mu_d) \times \mu$ matrix defined by

$$\overline{D} = \begin{pmatrix} I_{\mu_1} & 0 \\ I_{\mu_2} & 0 \\ \vdots & \vdots \\ I_{\mu_d} & 0 \end{pmatrix}, \tag{11.62}$$

in which, if $\mu_r = \mu$, then the rth zero sub-matrix will disappear. Clearly, $\text{rank}(D_0) = \mu$.

Now for any given $s = 1, \cdots, d$, let

$$y_s = \sum_{l=1}^{\mu_s} \alpha_l y_{s,1}^{(l)} \quad \text{with} \quad (\alpha_1, \cdots, \alpha_{\mu_s}) \neq 0 \tag{11.63}$$

be an eigenvector of A^T, corresponding to the eigenvalue λ_s. Noting (11.60), it is easy to see that

$$y_s^T D_0 = (0, \cdots, 0, \alpha_1, \alpha_2, \cdots, \alpha_{\mu_s}, 0, \cdots, 0) \neq 0. \tag{11.64}$$

Thus, $\text{Ker}(D_0^T)$ does not contain any eigenvector of A^T, then, any nontrivial invariant subspace of A^T either. By Proposition 2.12, the matrix D_0 satisfies well the rank condition (11.53). The proof is complete. □

Proposition 11.14 *There exists a boundary control matrix D with the minimal rank μ, such that the rank conditions (11.22) and (11.23) simultaneously hold.*

Proof Since the coupling matrix A always satisfies the condition of C_q^*-compatibility, there exists a matrix A_q^* of order $(N-q)$, such that $C_q^* A = A_q^* C_q^*$. Then, by Proposition 2.16, the rank condition (11.23) is equivalent to

$$\text{rank}(C_q^* D, A_q^* C_q^* D, \cdots, A_q^{*N-q-1} C_q^* D) = N - q. \tag{11.65}$$

Noting that A_q^* is of order $(N-q)$, as in the proof of Proposition 11.13, there exists a matrix D^* of order $(N-q) \times \mu^*$ with the minimal rank μ^*, such that

$$\text{rank}(D^*, A_q^* D^*, \cdots, A_q^{*N-q-1} D^*) = N - q, \tag{11.66}$$

where μ^* is the largest geometrical multiplicity of the eigenvalues of the reduced matrix A_q^{*T}.

On the other hand, A^T admits an invariant subspace W which is bi-orthonormal to $\text{Ker}(C_q^*)$. Then $\{\text{Ker}(C_q^*)\}^\perp = \text{Im}(C_q^{*T})$ is an invariant subspace of A^T, and $W^\perp = \text{Span}\{e_{q+1}, \cdots, e_N\}$ is an invariant subspace of A and bi-orthonormal to $\text{Im}(C_q^{*T})$, namely, we have

$$C_q^*(e_{q+1}, \cdots, e_N) = I_{N-q}. \tag{11.67}$$

Then, from (11.66) we get that

$$D = (e_{q+1}, \cdots, e_N)D^* \tag{11.68}$$

satisfies (11.65), therefore, (11.23) holds.

Finally, since $\text{Im}(D) \subseteq \text{Span}\{e_{q+1}, \cdots, e_N\}$ which is invariant for A, so

$$\text{Im}(A^k D) \subseteq \text{Span}\{e_{q+1}, \cdots, e_N\}, \quad \forall k \geqslant 0. \tag{11.69}$$

Noting that $\text{Span}\{e_{q+1}, \cdots, e_N\}$ is a supplement of $\text{Ker}(C_q^*)$, it follows that

$$\text{Ker}(C_q^*) \cap \text{Im}(D, AD, \cdots, A^{N-1}D) = \{0\}, \tag{11.70}$$

which, thanks to Proposition 2.11 and noting (11.23), implies the equality (11.22) and the sharpness of the rank of D:

$$\text{rank}(C_q^* D) = \text{rank}(D) = \mu^*. \tag{11.71}$$

The proof is thus complete. □

11.5 Examples

Example 11.15 As shown in Proposition 11.13, a "good" coupling matrix A should have distinct eigenvalues, or the geometrical multiplicity of its eigenvalues should be as small as possible!

In particular, if all the eigenvalues $\lambda_i (i = 1, \cdots, N)$ of A are simple, we can take a boundary control matrix D of rank one:

$$D = x, \tag{11.72}$$

where

$$x = \sum_{i=1}^{N} x_i \text{ with } Ax_i = \lambda_i x_i \ (i = 1, \cdots, N). \tag{11.73}$$

Let $y_j (j = 1, \cdots, N)$ be the eigenvectors of A^T, such that $(x_i, y_j) = \delta_{ij}$. Then for all $j = 1, \cdots N$, we have

$$D^T y_j = x^T y_j = \sum_{i=1}^{N} (x_i, y_j) = 1. \tag{11.74}$$

So, $\text{Ker}(D^T)$ does not contain any eigenvector of A^T, by (ii) of Proposition 2.12 (with $d = 0$), D satisfies Kalman's criterion (11.53).

If A has one double eigenvalue:

$$\lambda_1 = \lambda_2 < \lambda_3 \cdots < \lambda_N \text{ with } Ax_i = \lambda_i x_i \ (i = 1, \cdots, N), \tag{11.75}$$

we can take a boundary control matrix D of rank two:

$$D = (x_1, x), \tag{11.76}$$

where

$$x = \sum_{i=2}^{N} x_i. \tag{11.77}$$

Let $y_j (j = 1, \cdots N)$ be the eigenvectors of A^T, such that $(x_i, y_j) = \delta_{ij}$. Then we have

$$D^T y_1 = \begin{pmatrix} (x_1, y_1) \\ (x, y_1) \end{pmatrix} = \begin{pmatrix} 1 \\ 0 \end{pmatrix}, \tag{11.78}$$

and for $j = 2, \cdots, N$, we have

$$D^T y_j = \begin{pmatrix} (x_1, y_j) \\ (x, y_j) \end{pmatrix} = \begin{pmatrix} 0 \\ 1 \end{pmatrix}. \tag{11.79}$$

So, $\text{Ker}(D^T)$ does not contain any eigenvector of A^T. Once again, by (ii) of Proposition 2.12 (with $d = 0$), D satisfies Kalman's criterion (11.53).

Example 11.16 Let $N = 4$, $M = 1$, $p = 2$,

$$A = \begin{pmatrix} 2 & -2 & -1 & 1 \\ 1 & -1 & 0 & 0 \\ 1 & -1 & -1 & 1 \\ 0 & 0 & 0 & 0 \end{pmatrix} \tag{11.80}$$

and

$$C_2 = \begin{pmatrix} 1 & -1 & 0 & 0 \\ 0 & 0 & 1 & -1 \end{pmatrix} \quad \text{with} \quad e_1 = \begin{pmatrix} 1 \\ 1 \\ 0 \\ 0 \end{pmatrix}, \quad e_2 = \begin{pmatrix} 0 \\ 0 \\ 1 \\ 1 \end{pmatrix}. \tag{11.81}$$

First, noting that $Ae_1 = Ae_2 = 0$, we have the condition of C_2-compatibility: $A\text{Ker}(C_2) \subseteq \text{Ker}(C_2)$, and the corresponding reduced matrix

$$\overline{A}_2 = \begin{pmatrix} 1 & -1 \\ 1 & -1 \end{pmatrix} \tag{11.82}$$

is similar to the cascade matrix $\begin{pmatrix} 0 & 1 \\ 0 & 0 \end{pmatrix}$ of order 2.

In order to determine the minimal number of total controls necessary to the approximate boundary synchronization by 2-groups, we exhibit the system of root vectors of the matrix A^T:

$$\mathcal{E}_1^{(1)} = \begin{pmatrix} 0 \\ 0 \\ 0 \\ 1 \end{pmatrix}, \quad \mathcal{E}_1^{(2)} = \begin{pmatrix} 1 \\ -1 \\ -1 \\ 1 \end{pmatrix}, \quad \mathcal{E}_2^{(2)} = \begin{pmatrix} 0 \\ 0 \\ 1 \\ -1 \end{pmatrix}, \quad \mathcal{E}_3^{(2)} = \begin{pmatrix} 0 \\ 1 \\ -1 \\ 0 \end{pmatrix}. \tag{11.83}$$

Since $\mathcal{E}_1^{(2)}, \mathcal{E}_2^{(2)} \in \mathrm{Im}(C_2^T)$, $\mathrm{Im}(C_2^T)$ is A^T-marked. Then, the induced matrix C_1^* can be chosen as

$$C_1^* = C_1 = \begin{pmatrix} 1 & -1 & 0 & 0 \\ 0 & 1 & -1 & 0 \\ 0 & 0 & 1 & -1 \end{pmatrix}. \tag{11.84}$$

This justifies well that we have to use 3 (instead of 2!) total controls to realize the approximate boundary synchronization by 2-groups.

On the other hand, all the one-column matrices D satisfying the following rank conditions:

$$\mathrm{rank}(C_2 D, \overline{A}_2 C_2 D) = 2 \quad \text{and} \quad \mathrm{rank}(D, AD, A^2 D, A^3 D) = 3 \tag{11.85}$$

are given either by

$$D = \begin{pmatrix} \alpha + \beta \\ \alpha \\ \beta \\ 1 \end{pmatrix}, \quad \forall \alpha, \beta \in \mathbb{R} \tag{11.86}$$

or by

$$D = \begin{pmatrix} \gamma \\ \alpha \\ \beta \\ 0 \end{pmatrix}, \quad \forall \alpha, \beta, \gamma \in \mathbb{R} \text{ such that } \gamma \neq \alpha + \beta. \tag{11.87}$$

Since the above one-column matrix D provides only 3 total controls, by Theorems 8.6 and 8.9, system (I) is not approximately boundary null controllable. However, the corresponding Kalman's criterion $\mathrm{rank}(C_2 D, \overline{A}_2 C_2 D) = 2$ is sufficient for the approximate boundary null controllability of the corresponding reduced system (10.13) of cascade type, then sufficient for the approximate boundary synchronization by 2-groups of the original system (I) in the consensus sense:

$$u_n^{(1)} - u_n^{(2)} \to 0 \quad \text{and} \quad u_n^{(3)} - u_n^{(4)} \to 0 \quad \text{as } n \to +\infty \tag{11.88}$$

in the space

$$C^0_{loc}([T,+\infty);\mathcal{H}_0)\cap C^1_{loc}([T,+\infty);\mathcal{H}_{-1}). \tag{11.89}$$

Furthermore, the reduced matrix A_1^* given by

$$A_1^* = \begin{pmatrix} 1 & 0 & -1 \\ 0 & 0 & 1 \\ 1 & 0 & -1 \end{pmatrix} \tag{11.90}$$

is similar to a cascade matrix. Moreover, noting that $q=1$ and $C_1^*A = A_1^*C_1^*$, by Proposition 11.4, we necessarily have the Kalman's criterion

$$\text{rank}(C_1^*D, A_1^*C_1^*D, A_1^{*2}C_1^*D, A_1^{*3}C_1^*D) = 3 \tag{11.91}$$

for the boundary control matrix D in the both cases (11.86) and (11.87). Since the rank condition (11.91) is sufficient for the approximate boundary null controllability of the corresponding reduced system (11.39), system (I) is induced approximately synchronizable. By Theorem 11.6, there exists a nontrivial scalar function u^* such that

$$U_n \to u^*e_1^* \quad \text{as } n\to+\infty, \tag{11.92}$$

where $\text{Ker}(C_1^*) = \{e_1^*\}$ with $e_1^* = (1,1,1,1)^T$, or equivalently,

$$u_n^{(1)} \to u,\ u_n^{(2)} \to u,\ u_n^{(3)} \to u,\ u_n^{(4)} \to u \quad \text{as } n\to+\infty \tag{11.93}$$

in the space

$$C^0_{loc}([T,+\infty);\mathcal{H}_0)\cap C^1_{loc}([T,+\infty);\mathcal{H}_{-1}). \tag{11.94}$$

The induced approximate boundary synchronization (11.93) reveals the hidden information and makes clear the situation of the approximate boundary synchronization by 2-groups (11.88) in the consensus sense.

Example 11.17 Let $N=4$, $M=1$, $p=2$,

$$A = \begin{pmatrix} 0 & 0 & 1 & -1 \\ 0 & 0 & -1 & 1 \\ 1 & -1 & 0 & 0 \\ 1 & -1 & 0 & 0 \end{pmatrix}, \quad D = \begin{pmatrix} 0 \\ 0 \\ 1 \\ -1 \end{pmatrix} \tag{11.95}$$

and

$$C_2 = \begin{pmatrix} 1 & -1 & 0 & 0 \\ 0 & 0 & 1 & -1 \end{pmatrix} \text{ with } e_1 = \begin{pmatrix} 1 \\ 1 \\ 0 \\ 0 \end{pmatrix}, \quad e_2 = \begin{pmatrix} 0 \\ 0 \\ 1 \\ 1 \end{pmatrix}. \tag{11.96}$$

First, the rank of Kalman's matrix

$$(D, AD, A^2D, A^3D) = \begin{pmatrix} 0 & 2 & 0 & 0 \\ 0 & -2 & 0 & 0 \\ 1 & 0 & 4 & 0 \\ -1 & 0 & 4 & 0 \end{pmatrix} \tag{11.97}$$

is equal to 3. So, by Theorems 8.6 and 8.9, system (I) is not approximately null controllable under the action of the boundary control matrix D.

Next, noting $Ae_1 = Ae_2 = 0$, we have the condition of C_2-compatibility (10.10): $A\mathrm{Ker}(C_2) \subseteq \mathrm{Ker}(C_2)$. The reduced matrix is

$$\overline{A}_2 = C_2AC_2^T(C_2C_2^T)^{-1} = \begin{pmatrix} 0 & 2 \\ 0 & 0 \end{pmatrix}. \tag{11.98}$$

Moreover,

$$(C_2D, \overline{A}_2C_2D) = \begin{pmatrix} 0 & 4 \\ 2 & 0 \end{pmatrix}. \tag{11.99}$$

The reduced matrix $\overline{A}_2$ is of cascade type and then Kalman's criterion: $\mathrm{rank}(C_2D, \overline{A}_2C_2D) = 2$ is sufficient for the approximate boundary null controllability of the corresponding reduced system (10.13), hence, the original system (I) is approximately synchronizable by 2-groups in the consensus sense:

$$u_n^{(1)} - u_n^{(2)} \to 0 \quad \text{and} \quad u_n^{(3)} - u_n^{(4)} \to 0 \tag{11.100}$$

as $n \to +\infty$ in the space

$$C^0_{loc}([T, +\infty); \mathcal{H}_0) \cap C^1_{loc}([T, +\infty); \mathcal{H}_{-1}). \tag{11.101}$$

Now, we exhibit the system of root vectors of the matrix A^T:

$$\mathcal{E}_1^{(1)} = \begin{pmatrix} 1 \\ 1 \\ 0 \\ 0 \end{pmatrix}, \ \mathcal{E}_1^{(2)} = \begin{pmatrix} 0 \\ 0 \\ 2 \\ -2 \end{pmatrix}, \ \mathcal{E}_2^{(2)} = \begin{pmatrix} 1 \\ -1 \\ 0 \\ 0 \end{pmatrix}, \ \mathcal{E}_3^{(2)} = \begin{pmatrix} 0 \\ 0 \\ 0 \\ 1 \end{pmatrix}. \tag{11.102}$$

Since $\mathcal{E}_1^{(2)}, \mathcal{E}_2^{(2)} \in \mathrm{Im}(C_2^T)$, $\mathrm{Im}(C_2^T)$ is A^T-marked. Then, the induced matrix C_1^* can be chosen as

$$C_1^* = \begin{pmatrix} 0 & 0 & 1 & -1 \\ 1 & -1 & 0 & 0 \\ 0 & 0 & 0 & 1 \end{pmatrix}. \tag{11.103}$$

Furthermore, the reduced matrix A_1^* given by

$$A_1^* = \begin{pmatrix} 0 & 0 & 0 \\ 1 & 0 & 0 \\ 0 & 1 & 0 \end{pmatrix} \tag{11.104}$$

is a cascade matrix. Moreover, by Proposition 11.4, we necessarily have the following Kalman's criterion

$$\text{rank}(C_1^* D, A_1^* C_1^* D, A_1^{*2} C_1^* D, A_1^{*3} C_1^* D) = 3,$$

which is sufficient for the approximate boundary null controllability of the corresponding reduced system (11.39). Then system (I) is induced approximately synchronizable. By Theorem 11.6, there exists a nontrivial scalar function u^*, such that

$$U_n \to u^* e_1^* \quad \text{as } n \to +\infty, \tag{11.105}$$

where $\text{Ker}(C_1^*) = \{e_1^*\}$ with $e_1^* = (1, 1, 0, 0)^T$, or equivalently,

$$u_n^{(1)} \to u^*, \; u_n^{(2)} \to u^* \quad \text{and} \quad u_n^{(3)} \to 0, \; u_n^{(4)} \to 0 \tag{11.106}$$

as $n \to +\infty$ in the space

$$C_{loc}^0([T, +\infty); \mathcal{H}_0) \cap C_{loc}^1([T, +\infty); \mathcal{H}_{-1}). \tag{11.107}$$

Once again, the induced approximate boundary synchronization (11.106) clarifies the approximate boundary synchronization by 2-groups (11.100) in the consensus sense.

Remark 11.18 The nature of the induced approximate boundary synchronization is determined by the structure of $\text{Ker}(C_q^*)$. In Example 11.16, since $C_1^* = C_1$, the induced approximate boundary synchronization becomes the approximate boundary synchronization. This is purely a matter of chance. In fact, in Example 11.17, since $C_1^* \neq C_1$, the induced approximate boundary synchronization implies the approximate boundary synchronization for the first group and the approximate boundary null controllability for the second one, namely, the approximate boundary null controllability and synchronization by 2-groups, which can be treated in a way similar to the approximate boundary synchronization by groups (cf. Chap. 10).

Other examples can be constructed to illustrate more complicated situations. However, the next example shows that the induced approximate boundary synchronization can fail in the general case.

Example 11.19 Let $N = 2$, $M = 1$, $p = 1$, and

$$A = \begin{pmatrix} 0 & \epsilon \\ \epsilon & 0 \end{pmatrix}, \quad D = \begin{pmatrix} 1 \\ 0 \end{pmatrix}, \tag{11.108}$$

where $\epsilon \neq 0$ is a real number. We consider the following system:

$$\begin{cases} u'' - \Delta u + \epsilon v = 0 \text{ in } (0, +\infty) \times \Omega, \\ v'' - \Delta v + \epsilon u = 0 \text{ in } (0, +\infty) \times \Omega, \\ u = v = 0 \qquad\qquad \text{on } (0, +\infty) \times \Gamma_0, \\ u = h, \quad v = 0 \qquad \text{on } (0, +\infty) \times \Gamma_1. \end{cases} \tag{11.109}$$

Setting $w = u - v$, we get the reduced system

$$\begin{cases} w'' - \Delta w - \epsilon w = 0 \text{ in } (0, +\infty) \times \Omega, \\ w = 0 \qquad\qquad\qquad \text{on } (0, +\infty) \times \Gamma_0, \\ w = h \qquad\qquad\qquad \text{on } (0, +\infty) \times \Gamma_1, \end{cases} \tag{11.110}$$

which is approximately null controllable, therefore system (11.109) is approximately synchronizable. Moreover, it is easy to see that the Kalman's criterion $\text{rank}(D, AD) = 2$ holds for all $\epsilon \neq 0$, namely, the number of total controls is equal to $2 > N - p = 1$. However, as shown in Theorem 8.11, the adjoint system (8.35) is not D-observable for some values of ϵ. So, system (11.109) is not approximately null controllable in general.

This example shows that for the approximate boundary synchronization by p-groups, even $\text{rank}(D, AD, \cdots, A^{N-1}D)$, the number of total controls, is bigger than $N - p$, we cannot always get more information from the point of view of the induced approximate boundary synchronization.

Part III
Synchronization for a Coupled System of Wave Equations with Neumann Boundary Controls: Exact Boundary Synchronization

We consider the following coupled system of wave equations with Neumann boundary controls:

$$\begin{cases} U'' - \Delta U + AU = 0 & in \quad (0, +\infty) \times \Omega, \\ U = 0 & on \quad (0, +\infty) \times \Gamma_0, \\ \partial_\nu U = DH & on \quad (0, +\infty) \times \Gamma_1 \end{cases} \tag{II}$$

with the initial condition

$$t = 0: \quad U = \widehat{U}_0, \ U' = \widehat{U}_1 \quad \text{in } \Omega, \tag{II0}$$

where $\Omega \subset \mathbb{R}^n$ is a bounded domain with smooth boundary $\Gamma = \Gamma_1 \cup \Gamma_0$ such that $\overline{\Gamma}_1 \cap \overline{\Gamma}_0 = \emptyset$ and $\text{mes}(\Gamma_1) > 0$; "$'$" stands for the time derivative; $\Delta = \sum_{k=1}^n \frac{\partial^2}{\partial x_k^2}$ is the Laplacian operator; ∂_ν denotes the outward normal derivative on the boundary; $U = \left(u^{(1)}, \cdots, u^{(N)}\right)^T$ and $H = \left(h^{(1)}, \cdots, h^{(M)}\right)^T (M \leqslant N)$ stand for the state variables and the boundary controls, respectively; the coupling matrix $A = (a_{ij})$ is of order N, and D as the boundary control matrix is a full column-rank matrix of order $N \times M$, both with constant elements.

The exact boundary synchronization and the exact boundary synchronization by groups for system (II) will be presented and discussed in this part, while, correspondingly, the approximate boundary synchronization and the approximate boundary synchronization by groups for system (II) will be introduced and considered in the next part (Part IV).

Chapter 12
Exact Boundary Controllability and Non-exact Boundary Controllability

In this chapter, we will consider the exact boundary controllability and the non-exact boundary controllability for system (II) of wave equations with Neumann boundary controls.

12.1 Introduction

In this part, we will consider system (II) of wave equations with Neumann boundary controls, together with the initial data (II0).

Let us denote

$$\mathcal{H}_0 = L^2(\Omega) \quad \text{and} \quad \mathcal{H}_1 = H^1_{\Gamma_0}(\Omega), \tag{12.1}$$

where $H^1_{\Gamma_0}(\Omega)$ is the subspace of $H^1(\Omega)$, composed of functions with the null trace on Γ_0. We denote by $\mathcal{H}_{-1}$ the dual space of $\mathcal{H}_1$. When $\text{mes}(\Gamma_0) = 0$, namely, $\Gamma_1 = \Gamma$, instead of (12.1) we denote

$$\begin{cases} \mathcal{H}_0 = \left\{\phi \in L^2(\Omega), \int_\Omega \phi dx = 0\right\}, \\ \mathcal{H}_1 = \left\{\phi \in H^1(\Omega), \int_\Omega \phi dx = 0\right\}. \end{cases} \tag{12.2}$$

We will show the exact boundary controllability of system (II) for any given initial data $(\widehat{U}_0, \widehat{U}_1) \in (\mathcal{H}_1)^N \times (\mathcal{H}_0)^N$ via the HUM approach.

To this end, let

$$\Phi = (\phi^{(1)}, \cdots, \phi^{(N)})^T \tag{12.3}$$

T. Li and B. Rao, *Boundary Synchronization for Hyperbolic Systems*, Progress in Nonlinear Differential Equations and Their Applications 94,
https://doi.org/10.1007/978-3-030-32849-8_12

denote the adjoint variables. We consider the following **adjoint problem**:

$$\begin{cases} \Phi'' - \Delta\Phi + A^T\Phi = 0 & \text{in } (0, +\infty) \times \Omega, \\ \Phi = 0 & \text{on } (0, +\infty) \times \Gamma_0, \\ \partial_\nu \Phi = 0 & \text{on } (0, +\infty) \times \Gamma_1, \\ t = 0: \quad \Phi = \widehat{\Phi}_0, \ \Phi' = \widehat{\Phi}_1 & \text{in } \Omega. \end{cases} \tag{12.4}$$

We will establish the following

Theorem 12.1 *There exist positive constants $T > 0$ and $C > 0$, independent of initial data, such that the following observability inequality*

$$\|(\widehat{\Phi}_0, \widehat{\Phi}_1)\|^2_{(\mathcal{H}_0)^N \times (\mathcal{H}_{-1})^N} \leqslant C \int_0^T \int_{\Gamma_1} |\Phi|^2 d\Gamma dt \tag{12.5}$$

holds for the solution Φ to the adjoint problem (12.4) with any given $(\widehat{\Phi}_0, \widehat{\Phi}_1)$ belonging to a subspace $\mathcal{F} \subset (\mathcal{H}_0)^N \times (\mathcal{H}_{-1})^N$ (cf. (12.64) below).

Recall that without any assumption on the coupling matrix A, the usual multiplier method cannot be applied directly. The absorption of coupling lower terms is a delicate issue even for a single wave equation (cf. [62] and [26]). In order to deal with the lower order terms, we propose a method based on the compactness-uniqueness argument that we formulate in the following

Lemma 12.2 *Let $\mathcal{F}$ be a Hilbert space endowed with the p-norm. Assume that*

$$\mathcal{F} = \mathcal{N} \bigoplus \mathcal{L}, \tag{12.6}$$

where $\oplus$ denotes the direct sum and $\mathcal{L}$ is a finite co-dimensional closed subspace of $\mathcal{F}$. Assume that there is another norm—the q-norm in $\mathcal{F}$, such that the projection from $\mathcal{F}$ into $\mathcal{N}$ is continuous with respect to the q-norm. Assume furthermore that

$$q(y) \leqslant p(y), \quad \forall y \in \mathcal{L}. \tag{12.7}$$

Then there exists a positive constant $C > 0$, such that

$$q(z) \leqslant Cp(z), \quad \forall z \in \mathcal{F}. \tag{12.8}$$

Following the above Lemma, we have to first show the observability inequality for the initial data with higher frequencies in $\mathcal{L}$. In order to extend this inequality to the whole space $\mathcal{F}$, it is sufficient to verify the continuity of the projection from $\mathcal{F}$ into $\mathcal{N}$ for the q-norm. In many situations, it often occurs that the subspaces $\mathcal{N}$ and $\mathcal{L}$ are mutually orthogonal with respect to the q-inner product, and this is true in the present case. This new approach turns out to be particularly simple and efficient for getting the observability of some distributed systems with lower order terms.

As for the problem with Dirichlet boundary controls in Chap. 3, we show the exact boundary controllability and the non-exact boundary controllability for system (II) with Neumann boundary controls in the case $M = N$ or with fewer boundary controls ($M < N$) (cf. Theorems 12.14 and 12.15), respectively. Roughly speaking, in the framework that all the components of initial data are in the same energy space, a coupled system of wave equations with Dirichlet or Neumann boundary controls is exactly controllable if and only if one applies the same number of boundary controls as the number of state variables of wave equations.

12.2 Proof of Lemma 12.2

Assume that (12.8) fails, then there exists a sequence $z_n \in \mathcal{F}$, such that

$$q(z_n) = 1 \quad \text{and} \quad p(z_n) \to 0 \quad \text{as } n \to +\infty. \tag{12.9}$$

Using (12.6), we write $z_n = x_n + y_n$ with $x_n \in \mathcal{N}$ and $y_n \in \mathcal{L}$. Since the projection from $\mathcal{F}$ into $\mathcal{N}$ is continuous with respect to the q-norm, there exists a positive constant $c > 0$, such that

$$q(x_n) \leqslant cq(z_n) = c, \quad \forall n \geq 1. \tag{12.10}$$

Noting that $\mathcal{N}$ is of finite dimension, we may assume that there exists $x \in \mathcal{N}$, such that $x_n \to x$ in $\mathcal{N}$. Then, since the second relation of (12.9) means that $z_n \to 0$ in $\mathcal{F}$, we deduce that $y_n \to -x$ in $\mathcal{L}$ for the p-norm. Therefore, we get $x \in \mathcal{L} \cap \mathcal{N}$, which leads to $x = 0$. Then, we have

$$q(x_n) \to 0 \quad \text{and} \quad p(y_n) \to 0 \tag{12.11}$$

as $n \to +\infty$. Then using (12.7), we get a contradiction:

$$1 = q(z_n) \leqslant q(x_n) + q(y_n) \leqslant q(x_n) + p(y_n) \to 0 \tag{12.12}$$

as $n \to +\infty$. The proof is then complete.

Remark 12.3 Noting that $\mathcal{L}$ is not necessarily closed with respect to the weaker q-norm, so, a priori, the projection $z \to x$ is not continuous with respect to the q-norm (cf. [8]).

12.3 Observability Inequality

In order to give the proof of Theorem 12.1, we first start with some useful preliminary results. Assume that $\Omega \subset \mathbb{R}^n$ is a bounded domain with smooth boundary Γ. Let $\Gamma = \Gamma_1 \cup \Gamma_0$ be a partition of Γ such that $\overline{\Gamma}_1 \cap \overline{\Gamma}_0 = \emptyset$. Throughout this chapter, we assume that Ω satisfies the usual multiplier geometrical condition (cf. [7, 26, 62]). More precisely, assume that there exists $x_0 \in \mathbb{R}^n$, such that setting $m = x - x_0$, we have

$$(m, \nu) \leqslant 0, \quad \forall x \in \Gamma_0; \quad (m, \nu) > 0, \quad \forall x \in \Gamma_1, \tag{12.13}$$

where $(\cdot, \cdot)$ denotes the inner product in $\mathbb{R}^n$.

We define the linear unbounded operator $-\Delta$ in $\mathcal{H}_0$ by

$$D(-\Delta) = \{\phi \in H^2(\Omega) : \quad \phi|_{\Gamma_0} = 0, \quad \partial_\nu \phi|_{\Gamma_1} = 0\}. \tag{12.14}$$

Clearly, $-\Delta$ is a densely defined self-adjoint and coercive operator with a compact resolvent in $\mathcal{H}_0$. Then we can define the power operator $(-\Delta)^{s/2}$ for any given $s \in \mathbb{R}$ (cf. [64]). Moreover, the domain $\mathcal{H}_s = D((-\Delta)^{s/2})$ endowed with the norm $\|\phi\|_s = \|(-\Delta)^{s/2}\phi\|_{\mathcal{H}_0}$ is a Hilbert space, and its dual space with respect to the pivot space $\mathcal{H}_0$ is $\mathcal{H}_s' = \mathcal{H}_{-s}$. In particular, we have

$$\mathcal{H}_1 = D(\sqrt{-\Delta}) = \{\phi \in H^1(\Omega) : \quad \phi = 0 \quad \text{on } \Gamma_0\}. \tag{12.15}$$

Then we formulate the adjoint problem (12.4) into an abstract evolution problem in the space $(\mathcal{H}_s)^N \times (\mathcal{H}_{s-1})^N$ for any given $s \in \mathbb{R}$:

$$\begin{cases} \Phi'' - \Delta\Phi + A^T\Phi = 0, \\ t = 0 : \quad \Phi = \widehat{\Phi}_0, \ \ \Psi' = \widehat{\Phi}_1. \end{cases} \tag{12.16}$$

Moreover, we have the following result (cf. [60, 64, 70]).

Proposition 12.4 *For any given initial data* $(\widehat{\Phi}_0, \widehat{\Phi}_1) \in (\mathcal{H}_s)^N \times (\mathcal{H}_{s-1})^N$ *with* $s \in \mathbb{R}$*, the adjoint problem (12.16) admits a unique weak solution* Φ *in the sense of* C^0*-semigroups, such that*

$$\Phi \in C^0([0, +\infty); (\mathcal{H}_s)^N) \cap C^1([0, +\infty); (\mathcal{H}_{s-1})^N). \tag{12.17}$$

Now let e_m be the normalized eigenfunction defined by

$$\begin{cases} -\Delta e_m = \mu_m^2 e_m & \text{in } \Omega, \\ e_m = 0 & \text{on } \Gamma_0, \\ \partial_\nu e_m = 0 & \text{on } \Gamma_1, \end{cases} \tag{12.18}$$

where the positive sequence $\{\mu_m\}_{m\geqslant 1}$ is increasing such that $\mu_m \to +\infty$ as $m \to +\infty$.

For each $m \geqslant 1$, we define the subspace Z_m by

$$Z_m = \{\alpha e_m : \quad \alpha \in \mathbb{R}^N\}. \tag{12.19}$$

Since A is a matrix with constant coefficients, for any given $m \geqslant 1$ the subspace Z_m is invariant for A^T. Moreover, for any given integers m, n with $m \neq n$ and any given vectors $\alpha, \beta \in \mathbb{R}^N$, we have

$$\begin{aligned}
&(\alpha e_m, \beta e_n)_{(\mathcal{H}_s)^N} \\
=&(\alpha, \beta)_{\mathbb{R}^N}\big((-\Delta)^{s/2} e_m, (-\Delta)^{s/2} e_n\big)_{\mathcal{H}_0} \\
=&(\alpha, \beta)_{\mathbb{R}^N}\mu_m^s \mu_n^s \big(e_m, e_n\big)_{\mathcal{H}_0} \\
=&(\alpha, \beta)_{\mathbb{R}^N}\mu_m^s \mu_n^s \delta_{mn}.
\end{aligned} \tag{12.20}$$

Then, the subspaces $Z_m (m \geqslant 1)$ are mutually orthogonal in the Hilbert space $(\mathcal{H}_s)^N$ for any given $s \in \mathbb{R}$, and in particular, we have

$$\|\Phi\|_{(\mathcal{H}_s)^N} = \frac{1}{\mu_m}\|\Phi\|_{(\mathcal{H}_{s+1})^N}, \quad \forall \Phi \in Z_m. \tag{12.21}$$

Let $m_0 \geqslant 1$ be an integer. We denote by $\bigoplus_{m \geqslant m_0}(Z_m \times Z_m)$ the linear hull of the subspaces $Z_m \times Z_m$ for $m \geqslant m_0$. In other words, $\bigoplus_{m \geqslant m_0}(Z_m \times Z_m)$ is composed of all finite linear combinations of elements of $Z_m \times Z_m$ for $m \geqslant m_0$.

Proposition 12.5 *Let Φ be the solution to the adjoint problem (12.16) with the initial data $(\widehat{\Phi}_0, \widehat{\Phi}_1) \in \bigoplus_{m \geqslant 1}(Z_m \times Z_m)$, and satisfy an additional condition*

$$\Phi \equiv 0 \quad \text{on } [0, T] \times \Gamma_1 \tag{12.22}$$

for $T > 0$ large enough. Then, we have $\widehat{\Phi}_0 \equiv \widehat{\Phi}_1 \equiv 0$.

Proof By Schur's Theorem, we may assume that $A = (a_{ij})$ is an upper triangular matrix so that problem (12.16) with the additional condition (12.22) can be rewritten as for $k = 1, \cdots, N$,

$$\begin{cases}
(\phi^{(k)})'' - \Delta\phi^{(k)} + \sum_{p=1}^{k} a_{pk}\phi^{(p)} = 0 & \text{in } (0, +\infty) \times \Omega, \\
\phi^{(k)} = 0 & \text{on } (0, +\infty) \times \Gamma, \\
\partial_\nu \phi^{(k)} = 0 & \text{on } (0, +\infty) \times \Gamma_1, \\
t = 0: \quad \phi^{(k)} = \widehat{\phi}_0^{(k)}, \ (\phi^{(k)})' = \widehat{\phi}_1^{(k)} & \text{in } \Omega,
\end{cases} \tag{12.23}$$

where corresponding to (12.3) we have

$$\widehat{\Phi}_0 = (\widehat{\phi}_0^{(1)}, \cdots, \widehat{\phi}_0^{(N)})^T \quad \text{and} \quad \widehat{\Phi}_1 = (\widehat{\phi}_1^{(1)}, \cdots, \widehat{\phi}_1^{(N)})^T. \tag{12.24}$$

Then, using Holmgren's uniqueness theorem (cf. Theorem 8.2 in [62]), there exists a positive constant $T > 0$ large enough and independent of the initial data $(\widehat{\phi}_0^{(1)}, \widehat{\phi}_1^{(1)})$, such that $\phi^{(1)} \equiv 0$. Then, we get successively $\phi^{(k)} \equiv 0$ for $k = 1, \cdots, N$. The proof is then complete. □

Proposition 12.6 *Let $m_0 \geqslant 1$ be an integer and Φ be the solution to the adjoint problem (12.16) with the initial data $(\widehat{\Phi}_0, \widehat{\Phi}_1) \in \bigoplus_{m \geqslant m_0}(Z_m \times Z_m)$. Define the energy by*

$$E(t) = \frac{1}{2}\int_{\Omega}(|\Phi'|^2 + |\nabla\Phi|^2)dx. \tag{12.25}$$

Let σ denote the Euclidian norm of the matrix A. Then we have the following energy estimates:

$$e^{\frac{-\sigma t}{\mu_{m_0}}}E(0) \leqslant E(t) \leqslant e^{\frac{\sigma t}{\mu_{m_0}}}E(0), \quad t \geqslant 0 \tag{12.26}$$

for all $(\widehat{\Phi}_0, \widehat{\Phi}_1) \in \bigoplus_{m \geqslant m_0}(Z_m \times Z_m)$, where the sequence $(\mu_m)_{m \geqslant 1}$ is defined by problem (12.18).

Proof First, a straightforward computation yields

$$E'(t) = -\int_{\Omega}(A\Phi', \Phi)dx. \tag{12.27}$$

Then, using (12.21) we get

$$\begin{aligned} &\left|\int_{\Omega}(A\Phi', \Phi)dx\right| \\ \leqslant &\sigma\|\Phi'\|_{\mathcal{H}_0}\|\Phi\|_{\mathcal{H}_0} \leqslant \frac{\sigma}{\mu_{m_0}}\|\Phi'\|_{\mathcal{H}_0}\|\Phi\|_{\mathcal{H}_1} \leqslant \frac{\sigma}{\mu_{m_0}}E(t). \end{aligned} \tag{12.28}$$

It then follows that

$$-\frac{\sigma}{\mu_{m_0}}E(t) \leqslant E'(t) \leqslant \frac{\sigma}{\mu_{m_0}}E(t). \tag{12.29}$$

Therefore, the function $E(t)e^{\frac{\sigma t}{\mu_{m_0}}}$ is increasing with respect to the variable t; while, the function $E(t)e^{\frac{-\sigma t}{\mu_{m_0}}}$ is decreasing with respect to the variable t. Thus we get (12.26) and then the proof is complete. □

Proposition 12.7 *There exist an integer $m_0 \geqslant 1$ and positive constants $T > 0$ and $C > 0$ independent of initial data, such that the following observability inequality*

$$\|(\widehat{\Phi}_0, \widehat{\Phi}_1)\|^2_{(\mathcal{H}_1)^N \times (\mathcal{H}_0)^N} \leqslant C\int_0^T\int_{\Gamma_1}|\Phi'|^2 d\Gamma dt \tag{12.30}$$

holds for all solutions Φ to adjoint problem (12.16) with the initial data $(\widehat{\Phi}_0, \widehat{\Phi}_1) \in \bigoplus_{m \geqslant m_0}(Z_m \times Z_m)$.

Proof First we write the adjoint problem (12.16) as

$$\begin{cases} (\phi^{(k)})'' - \Delta\phi^{(k)} + \sum_{p=1}^{N} a_{pk}\phi^{(p)} = 0 \text{ in } (0, +\infty) \times \Omega, \\ \phi^{(k)} = 0 \qquad \text{on } (0, +\infty) \times \Gamma_0, \\ \partial_\nu \phi^{(k)} = 0 \qquad \text{on } (0, +\infty) \times \Gamma_1, \\ t = 0: \quad \phi^{(k)} = \widehat{\phi}_0^{(k)}, \ (\phi^{(k)})' = \widehat{\phi}_1^{(k)} \quad \text{in } \Omega \end{cases} \tag{12.31}$$

for $k = 1, 2, \cdots, N$. Then, multiplying the k-th equation of (12.31) by

$$M^{(k)} := 2m \cdot \nabla\phi^{(k)} + (N-1)\phi^{(k)} \quad \text{with} \quad m = x - x_0 \tag{12.32}$$

and integrating by parts, we get easily the following identities (cf. [26, 62]):

$$\begin{aligned} &\int_0^T \int_\Gamma \big(\partial_\nu \phi^{(k)} M^{(k)} + (m, \nu)(|(\phi^{(k)})'|^2 - |\nabla\phi^{(k)}|^2\big) d\Gamma dt \\ =&\Big[\int_\Omega (\phi^{(k)})' M^{(k)} dx\Big]_0^T + \int_0^T \int_\Omega \big(|(\phi^{(k)})'|^2 + |\nabla\phi^{(k)}|^2\big) dx dt \\ &+ \sum_{p=1}^N \int_0^T \int_\Omega a_{kp}\phi^{(p)} M^{(k)} dx dt, \quad k = 1, \cdots, N. \end{aligned} \tag{12.33}$$

Noting the multiplier geometrical condition (12.13), we have

$$\begin{aligned} &\partial_\nu \phi^{(k)} M^{(k)} + (m, \nu)(|(\phi^{(k)})'|^2 - |\nabla\phi^{(k)}|^2) \\ =&(m, \nu)|\partial_\nu \phi^{(k)}|^2 \leqslant 0 \qquad \text{on } (0, T) \times \Gamma_0 \end{aligned} \tag{12.34}$$

and

$$\begin{aligned} &\partial_\nu \phi^{(k)} M^{(k)} + (m, \nu)(|(\phi^{(k)})'|^2 - |\nabla\phi^{(k)}|^2) \\ =&(m, \nu)(|(\phi^{(k)})'|^2 - |\nabla\phi^{(k)}|^2) \\ \leqslant&(m, \nu)|\phi'^{(k)}|^2 \qquad \text{on } (0, T) \times \Gamma_1. \end{aligned} \tag{12.35}$$

Then, it follows from (12.33) that

$$\begin{aligned} &\int_0^T \int_\Omega \big(|(\phi^{(k)})'|^2 + |\nabla\phi^{(k)}|^2\big) dx dt \\ \leqslant&\int_0^T \int_{\Gamma_1} (m, \nu)|(\phi^{(k)})'|^2 d\Gamma dt - \Big[\int_\Omega (\phi^{(k)})' M^{(k)} dx\Big]_0^T \\ &- \sum_{p=1}^N \int_0^T \int_\Omega a_{kp}\phi^{(p)} M^{(k)} dx dt, \quad k = 1, \cdots, N. \end{aligned} \tag{12.36}$$

Taking the summation of (12.36) with respect to $k = 1, \cdots, N$, we get

$$2\int_0^T E(t)dt \leqslant \int_0^T \int_{\Gamma_1} (m, \nu)|\Phi'|^2 d\Gamma dt \tag{12.37}$$
$$-\Big[\int_\Omega (\Phi', M)dx\Big]_0^T - \int_0^T \int_\Omega (\Phi, AM)dxdt,$$

where M is the vector composed of $M^{(k)} (k = 1, \cdots, N)$ given by (12.32).

Next, we estimate the last two terms on the right-hand side of (12.37).

First, it follows from (12.32) that

$$\|M\|_{(\mathcal{H}_0)^N} \leqslant 2R \sum_{k=1}^N \|\nabla\phi^{(k)}\|_{(\mathcal{H}_0)^n} \tag{12.38}$$
$$+ (N-1)\|\Phi\|_{(\mathcal{H}_0)^N} \leqslant \gamma\|\Phi\|_{(\mathcal{H}_1)^N},$$

where $R = \|m\|_\infty$ is the diameter of Ω and

$$\gamma = \sqrt{4R^2 + (N-1)^2}. \tag{12.39}$$

On the other hand, since Z_m is invariant for A^T, for any given $(\widehat{\Phi}_0, \widehat{\Phi}_1) \in \bigoplus_{m \geqslant m_0}(Z_m \times Z_m)$, the corresponding solution Φ to the adjoint problem (12.16) belongs to $\bigoplus_{m \geqslant m_0} Z_m$ for any given $t \geqslant 0$. Thus, using (12.21) and (12.38), we have

$$\Big|\int_\Omega (\Phi, AM)dx\Big| \leqslant \sigma\|\Phi\|_{(\mathcal{H}_0)^N}\|M\|_{(\mathcal{H}_0)^N} \tag{12.40}$$
$$\leqslant \gamma\sigma\|\Phi\|_{(\mathcal{H}_0)^N}\|\Phi\|_{(\mathcal{H}_1)^N} \leqslant \frac{2\gamma\sigma}{\mu_{m_0}} E(t).$$

Similarly, we have

$$\Big|\int_\Omega (\Phi', M)dx\Big| \leqslant \|\Phi'\|_{(\mathcal{H}_0)^N}\|M\|_{(\mathcal{H}_0)^N} \tag{12.41}$$
$$\leqslant \gamma\|\Phi'\|_{(\mathcal{H}_0)^N}\|\Phi\|_{(\mathcal{H}_1)^N} \leqslant \gamma E(t). \tag{12.42}$$

Thus, setting

$$T = \frac{\mu_{m_0}}{\sigma} \tag{12.43}$$

and noting (12.26), we get

$$\Big|\Big[\int_\Omega \Phi' M dx\Big]_0^T\Big| \leqslant \gamma(E(T) + E(0)) \leqslant \gamma(1+e)E(0). \tag{12.44}$$

Inserting (12.40) and (12.44) into (12.37) gives

$$
\begin{aligned}
2\int_0^T E(t)dt &\leqslant \int_0^T \int_{\Gamma_1} (m,\nu)|\Phi'|^2 d\Gamma dt \\
&+\gamma(1+e)E(0) + \frac{2\sigma\gamma}{\mu_{m_0}} \int_0^T E(t)dt.
\end{aligned}
\tag{12.45}
$$

Thus, we have

$$
\int_0^T E(t)dt \leqslant R\int_0^T \int_{\Gamma_1} |\Phi'|^2 d\Gamma dt + \gamma(1+e)E(0), \tag{12.46}
$$

provided that m_0 is so large that

$$
\mu_{m_0} \geqslant 2\sigma\gamma. \tag{12.47}
$$

Now, integrating the inequality on the left-hand side of (12.26) over $[0, T]$, we get

$$
\frac{\mu_{m_0}}{\sigma}(1 - e^{\frac{-\sigma T}{\mu_{m_0}}})E(0) \leqslant \int_0^T E(t)dt, \tag{12.48}
$$

then, noting (12.26), we get

$$
T(1-e^{-1})E(0) \leqslant \int_0^T E(t)dt. \tag{12.49}
$$

Thus, it follows from (12.46) and (12.49) that

$$
E(0) \leqslant \frac{R}{T(1-e^{-1}) - \gamma(1+e)} \int_0^T \int_{\Gamma_1} |\Phi'|^2 d\Gamma dt \tag{12.50}
$$

holds for any given $(\widehat{\Phi}_0, \widehat{\Phi}_1) \in \bigoplus_{m \geqslant m_0} (Z_m \times Z_m)$, provided that

$$
T > \frac{\gamma e(1+e)}{e-1}, \tag{12.51}
$$

which is guaranteed by the following choice:

$$
\mu_{m_0} > \frac{2\sigma\gamma e(1+e)}{e-1} \tag{12.52}
$$

(cf. (12.43), (12.47), and (12.51)). The proof is then complete. □

Proposition 12.8 *There exist an integer $m_0 \geqslant 1$ and positive constants $T > 0$ and $C > 0$ independent of initial data, such that the following observability inequality*

$$\|(\widehat{\Phi}_0, \widehat{\Phi}_1)\|^2_{(\mathcal{H}_0)^N \times (\mathcal{H}_{-1})^N} \leqslant C \int_0^T \int_{\Gamma_1} |\Phi|^2 d\Gamma dt \tag{12.53}$$

holds for all solutions Φ to the adjoint problem (12.16) with the initial data $(\widehat{\Phi}_0, \widehat{\Phi}_1) \in \bigoplus_{m \geqslant m_0}(Z_m \times Z_m)$.

Proof Noting that $\mathrm{Ker}(-\Delta + A^T)$ is of finite dimension, there exists an integer $m_0 \geqslant 1$ so large that

$$\mathrm{Ker}(-\Delta + A^T) \bigcap \bigoplus_{m \geqslant m_0} Z_m = \{0\}. \tag{12.54}$$

Let

$$\mathcal{W} = \overline{\{\bigoplus_{m \geqslant m_0} Z_m\}}^{(\mathcal{H}_0)^N} \subseteq (\mathcal{H}_0)^N. \tag{12.55}$$

Since $\bigoplus_{m \geqslant m_0} Z_m$ is an invariant subspace of $(-\Delta + A^T)$, by Fredholm's alternative, $(-\Delta + A^T)^{-1}$ is an isomorphism from $\mathcal{W}$ onto its dual space $\mathcal{W}'$. Moreover, we have

$$\|(-\Delta + A^T)^{-1}\Psi\|^2_{(\mathcal{H}_0)^N} \sim \|\Psi\|^2_{(\mathcal{H}_{-1})^N}, \quad \forall \Psi \in \mathcal{W}. \tag{12.56}$$

For any given $(\widehat{\Phi}_0, \widehat{\Phi}_1) \in \bigoplus_{m \geqslant m_0}(Z_m \times Z_m)$, let

$$\Psi_0 = (\Delta - A^T)^{-1}\widehat{\Phi}_1, \quad \Psi_1 = \widehat{\Phi}_0. \tag{12.57}$$

We have

$$\|\Psi_0\|^2_{(\mathcal{H}_1)^N} + \|\Psi_1\|^2_{(\mathcal{H}_0)^N} \sim \|\widehat{\Phi}_1\|^2_{(\mathcal{H}_{-1})^N} + \|\widehat{\Phi}_0\|^2_{(\mathcal{H}_0)^N}. \tag{12.58}$$

Next let Ψ be the solution to the adjoint problem (12.16) with the initial data (Ψ_0, Ψ_1) given by (12.57). We have

$$t = 0: \quad \Psi' = \widehat{\Phi}_0, \ \Psi'' = \widehat{\Phi}_1. \tag{12.59}$$

By well-posedness, we get

$$\Psi' = \Phi. \tag{12.60}$$

On the other hand, since the subspace $\bigoplus_{m \geqslant m_0} Z_m$ is invariant for $(-\Delta + A^T)$, we have

$$(\Psi_0, \Psi_1) \in \bigoplus_{m \geqslant m_0}(Z_m \times Z_m). \tag{12.61}$$

Then, applying (12.30) to Ψ, we get

$$\|(\Psi_0, \Psi_1)\|^2_{(\mathcal{H}_1)^N \times (\mathcal{H}_0)^N} \leqslant C \int_0^T \int_{\Gamma_1} |\Psi'|^2 d\Gamma dt. \tag{12.62}$$

Thus, using (12.58) and (12.60), we get immediately (12.53). The proof is finished. □

We are now ready to give the proof of Theorem 12.1.

For any given $(\widehat{\Phi}_0, \widehat{\Phi}_1) \in \bigoplus_{m \geqslant 1}(Z_m \times Z_m)$, define

$$p(\widehat{\Phi}_0, \widehat{\Phi}_1) = \sqrt{\int_0^T \int_{\Gamma_1} |\Phi|^2 d\Gamma dt}, \tag{12.63}$$

where Φ is the solution to the adjoint problem (12.16). By Proposition 12.5, for $T > 0$ large enough, $p(\cdot, \cdot)$ defines well a norm in $\bigoplus_{m \geqslant 1}(Z_m \times Z_m)$. Then, we denote by $\mathcal{F}$ the completion of $\bigoplus_{m \geqslant 1}(Z_m \times Z_m)$ with respect to the p-norm. Clearly, $\mathcal{F}$ is a Hilbert space.

We next write

$$\mathcal{F} = \mathcal{N} \bigoplus \mathcal{L} \tag{12.64}$$

with

$$\mathcal{N} = \bigoplus_{1 \leqslant m < m_0} (Z_m \times Z_m), \quad \mathcal{L} = \overline{\{ \bigoplus_{m \geqslant m_0} (Z_m \times Z_m)\}}^p. \tag{12.65}$$

Clearly, $\mathcal{N}$ is a finite-dimensional subspace and $\mathcal{L}$ is a closed subspace in $\mathcal{F}$. In particular, the observability inequality (12.53) can be extended to all the initial data in the whole subspace $\mathcal{L}$.

Now we introduce the q-norm by

$$q(\widehat{\Phi}_0, \widehat{\Phi}_1) = \|(\widehat{\Phi}_0, \widehat{\Phi}_1)\|_{(\mathcal{H}_0)^N \times (\mathcal{H}_{-1})^N}, \quad \forall (\widehat{\Phi}_0, \widehat{\Phi}_1) \in \mathcal{F}. \tag{12.66}$$

By (12.20), the subspaces $(Z_m \times Z_m)$ for all $m \geqslant 1$ are mutually orthogonal in $(\mathcal{H}_0)^N \times (\mathcal{H}_{-1})^N$, then the subspace $\mathcal{N}$ is an orthogonal complement of $\mathcal{L}$ in $(\mathcal{H}_0)^N \times (\mathcal{H}_{-1})^N$. In particular, the projection from $\mathcal{F}$ into $\mathcal{N}$ is continuous with respect to the q-norm.

On the other hand, (12.53) means that (12.7) holds. Then, applying Lemma 12.2, we get the inequality (12.8), namely, (12.5) in the present situation. In particular, we have

$$\mathcal{F} \subset (\mathcal{H}_0)^N \times (\mathcal{H}_{-1})^N. \tag{12.67}$$

The proof of Theorem 12.1 is now completed. □

Remark 12.9 Without any additional assumptions on the coupling matrix A, the adjoint problem (12.16) is not conservative and the usual multiplier method cannot be applied directly. However, since each subspace Z_m is invariant for the matrix A^T, for any given initial data $(\widehat{\Phi}_0, \widehat{\Phi}_1) \in Z_m \times Z_m$, the corresponding solution Φ to the adjoint problem (12.16) lies still in the subspace Z_m. Then, because of the identity (12.21), the coupling term $\|A^T \Phi\|_{(\mathcal{H}_0)^N}$ is negligible comparing with $\frac{1}{\mu_m}\|\Phi\|_{(\mathcal{H}_1)^N}$. Therefore, we first expect the observability inequality (12.30) only for the initial data $(\widehat{\Phi}_0, \widehat{\Phi}_1)$ with higher frequencies lying in the sublinear hull $\bigoplus_{m \geqslant m_0}(Z_m \times Z_m)$ with an integer $m_0 \geqslant 1$ large enough. We next extend it to the whole linear hull $\bigoplus_{m \geqslant 1}(Z_m \times Z_m)$ by an argument of compact perturbation as shown in Lemma 12.2.

Remark 12.10 The compactness-uniqueness arguments are frequently used in the study of the observability of distributed parameter systems. It turns out that this method is particularly simple and efficient for dealing with some systems with lower order terms. A natural formulation is to consider the problem as a compact perturbation of a skew-adjoint operator (cf. [27, 69]). This approach requests that the eigensystem of the underlying system forms a Riesz basis in the energy space. However, since the Riesz basis is not stable even for the compact perturbation, this may cause serious problems. By contrast, the method proposed here does not require any spectral condition on the underlying system. In particular, instead of Riesz basis property, we assume only that the projection from $\mathcal{F}$ into $\mathcal{N}$ is continuous with respect to the q-norm. Moreover, it often occurs that the subspaces $\mathcal{N}$ and $\mathcal{L}$ are mutually orthogonal with respect to the q-inner product, hence, the continuity of the projection from $\mathcal{F}$ into $\mathcal{N}$ is much easier to be checked than the Riesz basis property.

Other approaches by energy method for systems with variable coefficients can be found in [11, 82, 84] and the references therein.

12.4 Exact Boundary Controllability

We will first show the **exact boundary null controllability** of system (II) by a standard application of the HUM method proposed by Lions in [61, 62].

Obviously, we have

$$\mathcal{H}_s \subset H^s(\Omega), \quad s \geqslant 0. \tag{12.68}$$

On the other hand, by (12.67) and the trace embedding $H^s(\Omega) \to L^2(\Gamma_1)$ for all $s > 1/2$, noting (12.63), we get the following continuous embeddings:

$$(\mathcal{H}_s)^N \times (\mathcal{H}_{s-1})^N \subset \mathcal{F} \subset (\mathcal{H}_0)^N \times (\mathcal{H}_{-1})^N, \quad s > 1/2. \tag{12.69}$$

Multiplying the equations in system (II) by a solution Φ to the adjoint problem (12.4) and integrating by parts, we get

$$\begin{aligned} &(U'(t), \Phi(t))_{(\mathcal{H}_0)^N} - (U(t), \Phi'(t))_{(\mathcal{H}_0)^N} \\ &= (\widehat{U}_1, \widehat{\Phi}_0)_{(\mathcal{H}_0)^N} - (\widehat{U}_0, \widehat{\Phi}_1)_{(\mathcal{H}_0)^N} + \int_0^t \int_{\Gamma_1} (DH(\tau), \Phi(\tau)) d\Gamma d\tau. \end{aligned} \tag{12.70}$$

Taking $(\mathcal{H}_0)^N$ as the pivot space and noting (12.69), (12.70) can be written as

$$\begin{aligned} &\langle (U'(t), -U(t)), (\Phi(t), \Phi'(t))\rangle \\ = &\langle (\widehat{U}_1, -\widehat{U}_0), (\widehat{\Phi}_0, \widehat{\Phi}_1)\rangle + \int_0^t \int_{\Gamma_1} (DH(\tau), \Phi(\tau)) d\Gamma d\tau, \end{aligned} \tag{12.71}$$

where $\langle \cdot, \cdot \rangle$ denotes the duality between the spaces $(\mathcal{H}_{-s})^N \times (\mathcal{H}_{1-s})^N$ and $(\mathcal{H}_s)^N \times (\mathcal{H}_{s-1})^N$.

Definition 12.11 U is a weak solution to problem (II) and (II0), if

$$(U, U') \in C^0([0, T]; (\mathcal{H}_{1-s})^N \times (\mathcal{H}_{-s})^N) \tag{12.72}$$

such that the variational Eq. (12.71) holds for any given $(\widehat{\Phi}_0, \widehat{\Phi}_1) \in (\mathcal{H}_s)^N \times (\mathcal{H}_{s-1})^N$ with $s > 1/2$.

Proposition 12.12 *For any given $H \in L^2(0, T; (L^2(\Gamma_1))^M)$ and any given $(\widehat{U}_0, \widehat{U}_1) \in (\mathcal{H}_{1-s})^N \times (\mathcal{H}_{-s})^N$ with $s > 1/2$, problem (II) and (II0) admits a unique weak solution U. Moreover, the linear map*

$$\mathcal{R}: \quad (\widehat{U}_0, \widehat{U}_1, H) \to (U, U') \tag{12.73}$$

is continuous with respect to the corresponding topologies.

Proof Define the linear form

$$\begin{aligned} &L_t(\widehat{\Phi}_0, \widehat{\Phi}_1) \\ = &\langle (\widehat{U}_1, -\widehat{U}_0), (\widehat{\Phi}_0, \widehat{\Phi}_1)\rangle + \int_0^t \int_{\Gamma_1} (DH(\tau), \Phi(\tau)) d\Gamma d\tau. \end{aligned} \tag{12.74}$$

By the definition (12.63) of the p-norm and the continuous embedding (12.69), the linear form L_t is bounded in $(\mathcal{H}_s)^N \times (\mathcal{H}_{s-1})^N$ for any given $t \geqslant 0$. Let S_t be the semigroup associated to the adjoint problem (12.16) on the Hilbert space $(\mathcal{H}_s)^N \times (\mathcal{H}_{s-1})^N$, which is an isomorphism on $(\mathcal{H}_s)^N \times (\mathcal{H}_{s-1})^N$. The composed linear form $L_t \circ S_t^{-1}$ is bounded in $(\mathcal{H}_s)^N \times (\mathcal{H}_{s-1})^N$. Then, by Riesz–Fréchet's representation theorem, there exists a unique element $(U'(t), -U(t)) \in (\mathcal{H}_{-s})^N \times (\mathcal{H}_{1-s})^N$, such that

$$L_t \circ S_t^{-1}(\Phi(t), \Phi'(t)) = \langle (U'(t), -U(t)), (\Phi(t), \Phi'(t))\rangle \tag{12.75}$$

for any given $(\widehat{\Phi}_0, \widehat{\Phi}_1) \in (\mathcal{H}_s)^N \times (\mathcal{H}_{s-1})^N$. Since

$$L_t \circ S_t^{-1}(\Phi(t), \Phi'(t)) = L_t(\widehat{\Phi}_0, \widehat{\Phi}_1), \tag{12.76}$$

we get (12.71) for any given $(\widehat{\Phi}_0, \widehat{\Phi}_1) \in (\mathcal{H}_s)^N \times (\mathcal{H}_{s-1})^N$. Moreover, we have

$$\begin{aligned}&\|(U'(t), -U(t))\|_{(\mathcal{H}_{-s})^N \times (\mathcal{H}_{1-s})^N} \\ \leqslant\ & C_T\big(\|(\widehat{U}_1, \widehat{U}_0)\|_{(\mathcal{H}_{-s})^N \times (\mathcal{H}_{1-s})^N} + \|H\|_{L^2(0,T;(L^2(\Gamma_1))^M)}\big).\end{aligned} \tag{12.77}$$

Then, by a classic argument of density, we get the regularity (12.72). The proof is thus complete. □

Definition 12.13 System (II) is **exactly null controllable** at the time T in the space $(\mathcal{H}_1)^N \times (\mathcal{H}_0)^N$, if there exists a positive constant $T > 0$, such that for any given $(\widehat{U}_1, \widehat{U}_0) \in (\mathcal{H}_1)^N \times (\mathcal{H}_0)^N$, there exists a boundary control $H \in L^2(0, T; (L^2(\Gamma_1))^M)$, such that problem (II) and (II0) admits a unique weak solution U satisfying the final condition

$$t = T : \quad U = U' = 0. \tag{12.78}$$

Theorem 12.14 *Assume that $M = N$. Then there exists a positive constant $T > 0$, such that system (II) is exactly null controllable at the time T for any given initial data $(\widehat{U}_0, \widehat{U}_1) \in (\mathcal{H}_1)^N \times (\mathcal{H}_0)^N$. Moreover, we have the continuous dependence*

$$\|H\|_{L^2(0,T;(L^2(\Gamma_1))^N)} \leqslant c\|(\widehat{U}_0, \widehat{U}_1)\|_{(\mathcal{H}_1)^N \times (\mathcal{H}_0)^N}. \tag{12.79}$$

Proof Let Φ be the solution to the adjoint problem (12.4) in $(\mathcal{H}_s)^N \times (\mathcal{H}_{s-1})^N$ with $s > 1/2$. Let

$$H = D^{-1}\Phi|_{\Gamma_1}. \tag{12.80}$$

Because of the first inclusion in (12.69), we have $H \in L^2(0, T; (L^2(\Gamma_1))^N)$. Then, by Proposition 12.12, the corresponding backward problem

$$\begin{cases} U'' - \Delta U + AU = 0 & \text{in } (0, T) \times \Omega, \\ U = 0 & \text{on } (0, T) \times \Gamma_0, \\ \partial_\nu U = \Phi & \text{on } (0, T) \times \Gamma_1, \\ t = T : U = U' = 0 & \text{in } \Omega \end{cases} \tag{12.81}$$

admits a unique weak solution U with (12.72). Accordingly, we define the linear map

$$\Lambda(\widehat{\Phi}_0, \widehat{\Phi}_1) = (-U'(0), U(0)). \tag{12.82}$$

Clearly, Λ is a continuous map from $(\mathcal{H}_s)^N \times (\mathcal{H}_{s-1})^N$ into $(\mathcal{H}_{-s})^N \times (H_{1-s})^N$.

Next, noting (12.78), it follows from (12.71) that

$$\langle \Lambda(\widehat{\Phi}_0, \widehat{\Phi}_1), (\widehat{\Psi}_0, \widehat{\Psi}_1)\rangle = \int_0^T \int_{\Gamma_1} \Phi(\tau)\Psi(\tau)d\Gamma d\tau, \tag{12.83}$$

where Ψ is the solution to the adjoint problem (12.16) with the initial data $(\widehat{\Psi}_0, \widehat{\Psi}_1)$. It then follows that

$$\langle \Lambda(\widehat{\Phi}_0, \widehat{\Phi}_1), (\widehat{\Psi}_0, \widehat{\Psi}_1)\rangle \leqslant \|(\widehat{\Phi}_0, \widehat{\Phi}_1)\|_{\mathcal{F}} \|(\widehat{\Psi}_0, \widehat{\Psi}_1)\|_{\mathcal{F}}. \tag{12.84}$$

By definition, $(\mathcal{H}_s)^N \times (\mathcal{H}_{s-1})^N$ is dense in $\mathcal{F}$, then the linear form

$$(\widehat{\Psi}_0, \widehat{\Psi}_1) \to \langle \Lambda(\widehat{\Phi}_0, \widehat{\Phi}_1), (\widehat{\Psi}_0, \widehat{\Psi}_1)\rangle \tag{12.85}$$

can be continuously extended to $\mathcal{F}$, so that $\Lambda(\Phi_0, \Phi_1) \in \mathcal{F}'$. Moreover, we have

$$\|\Lambda(\widehat{\Phi}_0, \widehat{\Phi}_1)\|_{\mathcal{F}'} \leqslant \|(\widehat{\Phi}_0, \widehat{\Phi}_1)\|_{\mathcal{F}}. \tag{12.86}$$

Then, Λ is a continuous linear map from $\mathcal{F}$ to $\mathcal{F}'$. Therefore, the symmetric bilinear form

$$\langle \Lambda(\widehat{\Phi}_0, \widehat{\Phi}_1), (\widehat{\Psi}_0, \widehat{\Psi}_1)\rangle, \tag{12.87}$$

where $\langle \cdot, \cdot\rangle$ denotes the duality between the spaces $(\mathcal{H}_{-s})^N \times (\mathcal{H}_{1-s})^N$ and $(\mathcal{H}_s)^N \times (\mathcal{H}_{s-1})^N$, is continuous and coercive in the product space $\mathcal{F} \times \mathcal{F}$. By Lax–Milgram's Lemma, Λ is an isomorphism from $\mathcal{F}$ onto $\mathcal{F}'$. Then for any given $(-\widehat{U}_1, \widehat{U}_0) \in \mathcal{F}'$, there exists an element $(\widehat{\Phi}_0, \widehat{\Phi}_1) \in \mathcal{F}$, such that

$$\Lambda(\widehat{\Phi}_0, \widehat{\Phi}_1) = (-\widehat{U}_1, \widehat{U}_0). \tag{12.88}$$

This is precisely the exact boundary controllability of system (II) for any given initial data $(\widehat{U}_1, -\widehat{U}_0) \in \mathcal{F}'$, in particular, for any given initial data $(\widehat{U}_1, -\widehat{U}_0) \in (\mathcal{H}_0)^N \times (\mathcal{H}_1)^N \subset \mathcal{F}'$, because of the second inclusion in (12.69).

Finally, from the definitions (12.80) and (12.63), we have

$$\|H\|_{L^2(0,T;(L^2(\Gamma_1))^N)} \leqslant C\|\Phi\|_{L^2(0,T;(L^2(\Gamma_1))^N)} = C\|(\widehat{\Phi}_0, \widehat{\Phi}_1)\|_{\mathcal{F}} \tag{12.89}$$

which together with (12.88) implies the continuous dependence:

$$\|H\|_{L^2(0,T;(L^2(\Gamma_1))^N)} \leqslant C\|\Lambda^{-1}\|_{\mathcal{L}(\mathcal{F}',\mathcal{F})}\|(\widehat{U}_0, \widehat{U}_1)\|_{(\mathcal{H}_1)^N \times (\mathcal{H}_0)^N}. \tag{12.90}$$

The proof is thus complete. □

12.5 Non-exact Boundary Controllability

In the case of fewer boundary controls, we have the following negative result.

Theorem 12.15 *Assume that $M < N$. Then system (II) is not exactly boundary controllable for all initial data $(\widehat{U}_0, \widehat{U}_1) \in (\mathcal{H}_1)^N \times (\mathcal{H}_0)^N$.*

Proof Since $M < N$, there exists a non-null vector $e \in \mathbb{R}^N$, such that $D^T e = 0$. We choose a special initial data as

$$\widehat{U}_0 = \theta e, \quad \widehat{U}_1 = 0, \tag{12.91}$$

where $\theta \in \mathcal{D}(\Omega)$ is arbitrarily given. If system (II) is exactly boundary controllable, noting that Theorem 3.9 is still valid here, there exists a boundary control H such that

$$\|H\|_{L^2(0,T;(L^2(\Gamma_1))^M)} \leqslant C\|\theta\|_{H^1(\Omega)}. \tag{12.92}$$

Then, by Proposition 12.12 we have

$$\|U\|_{L^2(0,T;(\mathcal{H}_{1-s}(\Omega))^N)} \leqslant C\|\theta\|_{H^1(\Omega)}, \quad \forall s > 1/2. \tag{12.93}$$

Now, taking the inner product of e with problem (II) and (II0) and noting $\phi = (e, U)$, we get

$$\begin{cases} \phi'' - \Delta\phi = -(e, AU) & \text{in } (0,T)\times\Omega, \\ \phi = 0 & \text{on } (0,T)\times\Gamma_0, \\ \partial_\nu\phi = 0 & \text{on } (0,T)\times\Gamma_1, \\ t = 0: \quad \phi = \theta, \ \phi' = 0 \text{ in } \Omega, \\ t = T: \quad \phi = 0, \ \phi' = 0 \text{ in } \Omega. \end{cases} \tag{12.94}$$

Noting (12.68), by well-posedness and noting (12.93), we get

$$\|\theta\|_{H^{2-s}(\Omega)} \leqslant C\|U\|_{L^2(0,T;(H_{1-s}(\Omega))^N)} \leqslant C'\|\theta\|_{H^1(\Omega)} \tag{12.95}$$

Choosing s such that $1 > s > 1/2$, then $2 - s > 1$, which gives a contradiction. The proof is then complete. □

Remark 12.16 As shown in the proof of Theorem 12.14, a weaker regularity such as $(U, U') \in C^0([0,T]; (\mathcal{H}_0)^N \times (\mathcal{H}_{-1})^N)$ is sufficient to make sense to the value $(U(0), U'(0))$, therefore, sufficient for proving the exact boundary controllability. At this stage, it is not necessary to pay much attention to the regularity of the weak solution with respect to the space variable. However, in order to establish the non-exact boundary controllability in Theorem 12.15, this regularity becomes indispensable for the argument of compact perturbation. In the case with Dirichlet boundary controls, the weak solution has the same regularity as the controllable initial data. This regularity yields the non-exact boundary controllability in the case of fewer boundary controls (cf. Chap. 3). However, for Neumann boundary controls, the direct inequality is much weaker than the inverse inequality. For example, in Proposition 12.12, we can get only $(U, U') \in C^0([0,T]; (\mathcal{H}_{1-s})^N \times (H_{-s})^N)$ for any given $s > 1/2$, while, the controllable initial data $(\widehat{U}_0, \widehat{U}_1)$ lies in the space $(\mathcal{H}_1)^N \times (\mathcal{H}_0)^N$. Even though this regularity is not sharp in general, it is already sufficient for the proof of the non-exact boundary controllability of system (II). Nevertheless, the gap of regularity between the solution and its initial data could cause some problems for the non-exact boundary controllability with coupled Robin boundary controls (cf. Part V and Part VI below).

Remark 12.17 As for the system with Dirichlet boundary controls discussed in Part 1, we have shown in Theorems 12.14 and 12.15 that system (II) with Neumann boundary controls is exactly boundary controllable if and only if the boundary controls have the same number as the state variables or the wave equations. Of course, the non-exact boundary controllability is valid only in the framework that all the components of the initial data are in the same energy space. For example, the authors of [65] considered the exact boundary controllability by means of only one boundary control for a system of two wave equations with initial data of different levels of finite energy. More specifically, the exact boundary controllability by means of only one boundary control for the a cascade system of N wave equations was established in [1, 2]. On the other hand, in contrast with the exact boundary controllability, the approximate boundary controllability is more flexible with respect to the number of boundary controls, and is closely related to the so-called Kalman's criterion on the rank of an enlarged matrix composed of the coupling matrix A and the boundary control matrix D (cf. Part IV below).

Chapter 13
Exact Boundary Synchronization and Non-exact Boundary Synchronization

In the case of partial lack of boundary controls, we consider the exact boundary synchronization and the non-exact boundary synchronization in this chapter for system (II) with Neumann boundary controls.

13.1 Definition

Let

$$U = \big(u^{(1)}, \cdots, u^{(N)}\big)^T \text{and} \big(h^{(1)}, \cdots, h^{(M)}\big)^T \tag{13.1}$$

with $M \leqslant N$. Consider the coupled system (II) with the initial condition (II0).

For simplifying the statement and without loss of generality, in what follows we always suppose that $\text{mes}(\Gamma_0) \neq 0$ and denote

$$\mathcal{H}_0 = L^2(\Omega), \quad \mathcal{H}_1 = H^1_{\Gamma_0}(\Omega), \quad \mathcal{L} = L^2_{loc}(0, +\infty; L^2(\Gamma_1)), \tag{13.2}$$

where $H^1_{\Gamma_0}(\Omega)$ is the subspace of $H^1(\Omega)$ composed of all the functions with the null trace on Γ_0.

According to Theorem 12.14, when $M = N$, there exists a constant $T > 0$, such that system (II) is exactly null controllable at the time T for any given initial data $(\widehat{U}_0, \widehat{U}_1) \in (\mathcal{H}_1)^N \times (\mathcal{H}_0)^N$.

On the other hand, according to Theorem 12.15 if there is a lack of boundary controls, namely, when $M < N$, no matter how large $T > 0$ is, system (II) is not exactly null controllable at the time T for any given initial data $(\widehat{U}_0, \widehat{U}_1) \in (\mathcal{H}_1)^N \times (\mathcal{H}_0)^N$.

Thus, it is necessary to discuss whether system (II) is controllable in some weaker senses when there is a lack of boundary controls, namely, when $M < N$. Although the results are similar to those for the coupled system of wave equations with Dirichlet

T. Li and B. Rao, *Boundary Synchronization for Hyperbolic Systems*,
Progress in Nonlinear Differential Equations and Their Applications 94,
https://doi.org/10.1007/978-3-030-32849-8_13

boundary controls, discussed in Part I, since the solution to a coupled system of wave equations with Neumann boundary conditions has a relatively weaker regularity, in order to realize the desired result, we need stronger function spaces, and the corresponding adjoint problem is also different.

First, we give the following

Definition 13.1 System (II) is **exactly synchronizable** at the time $T > 0$ in the space $(\mathcal{H}_1)^N \times (\mathcal{H}_0)^N$ if for any given initial data $(\widehat{U}_0, \widehat{U}_1) \in (\mathcal{H}_1)^N \times (\mathcal{H}_0)^N$, there exists a boundary control $H \in \mathcal{L}^M$ with compact support in $[0, T]$, such that the weak solution $U = U(t, x)$ to the mixed initial-boundary value problem (II) and (II0) satisfies

$$t \geqslant T: \quad u^{(1)} \equiv \cdots \equiv u^{(N)} := u, \tag{13.3}$$

where $u = u(t, x)$, being unknown a priori, is called the corresponding **exactly synchronizable state.**

The above definition requires that system (II) maintains the exactly synchronizable state even though the boundary control is canceled after the time T.

Let

$$C_1 = \begin{pmatrix} 1 & -1 & & & \\ & 1 & -1 & & \\ & & \ddots & \ddots & \\ & & & 1 & -1 \end{pmatrix}_{(N-1)\times N} \tag{13.4}$$

be the corresponding **matrix of synchronization.** C_1 is a full row-rank matrix, and $\text{Ker}(C_1) = \text{Span}\{e_1\}$, where $e_1 = (1, \cdots, 1)^T$. Clearly, the **exact boundary synchronization** (13.3) can be equivalently written as

$$t \geqslant T: \quad C_1 U(t, x) \equiv 0 \quad \text{in } \Omega. \tag{13.5}$$

13.2 Condition of C_1-Compatibility

We have

Theorem 13.2 *Assume that $M < N$. If system (II) is exactly synchronizable in the space $(\mathcal{H}_1)^N \times (\mathcal{H}_0)^N$, then the coupling matrix $A = (a_{ij})$ should satisfy the following condition of compatibility (**row-sum condition**: the sum of elements in every row is equal to each other):*

$$\sum_{j=1}^{N} a_{ij} := a, \tag{13.6}$$

where a is a constant independent of $i = 1, \cdots, N$.

Proof By Theorem 12.15, since $M < N$, system (II) is not exactly null controllable, then there exists an initial data $(\widehat{U}_0, \widehat{U}_1) \in (\mathcal{H}_1)^N \times (\mathcal{H}_0)^N$, such that for any given boundary control H, the corresponding exactly synchronizable state $u(t,x) \not\equiv 0$. Then, noting (13.3), the solution to problem (II) and (II0) corresponding to this initial data satisfies

$$u'' - \Delta u + \Big(\sum_{j=1}^{N} a_{ij}\Big)u = 0 \quad \text{in } \mathcal{D}'((T, +\infty) \times \Omega) \tag{13.7}$$

for all $i = 1, \cdots, N$. Then, we have

$$\Big(\sum_{j=1}^{N} a_{kj} - \sum_{j=1}^{N} a_{ij}\Big)u = 0 \quad \text{in } \mathcal{D}'((T, +\infty) \times \Omega) \tag{13.8}$$

for $i, k = 1, \cdots, N$. It follows that

$$\sum_{j=1}^{N} a_{kj} = \sum_{j=1}^{N} a_{ij}, \quad i, k = 1, \cdots, N, \tag{13.9}$$

which is just the required condition of compatibility (13.6). □

By Proposition 2.15, it is easy to get

Lemma 13.3 *The following properties are equivalent:*

(i) The condition of compatibility (13.6) *holds;*

(ii) $e = (1, \cdots, 1)^T$ is an eigenvector of A corresponding to the eigenvalue a given by (13.6)*;*

(iii) $Ker(C_1)$ is a one-dimensional invariant subspace of A:

$$AKer(C_1) \subseteq Ker(C_1); \tag{13.10}$$

(iv) There exists a unique matrix $\overline{A}_1$ of order $(N-1)$, such that

$$C_1 A = \overline{A}_1 C_1. \tag{13.11}$$

$\overline{A}_1 = (\overline{a}_{ij})$ is called the **reduced matrix of** *A by C_1, where*

$$\overline{a}_{ij} = \sum_{p=j+1}^{N} (a_{i+1,p} - a_{ip}) = \sum_{p=1}^{j} (a_{ip} - a_{i+1,p}) \tag{13.12}$$

for $i, j = 1, \cdots, N-1$.

In what follows we will call the row-sum condition (13.6) to be the **condition of** C_1**-compatibility**.

13.3 Exact Boundary Synchronization and Non-exact Boundary Synchronization

We have

Theorem 13.4 *Under the condition of* C_1*-compatibility* (13.10)*, system (II) is exactly synchronizable at some time* $T > 0$ *in the space* $(\mathcal{H}_1)^N \times (\mathcal{H}_0)^N$ *if and only if* $\text{rank}(C_1 D) = N - 1$*. Moreover, we have the following continuous dependence:*

$$\|H\|_{L^2(0,T;(L^2(\Gamma_1))^{N-1})} \leqslant c\|C_1(\widehat{U}_0, \widehat{U}_1)\|_{(\mathcal{H}_1)^N \times (\mathcal{H}_0)^{N-1}}, \tag{13.13}$$

where c *is a positive constant.*

Proof Under the condition of C_1-compatibility (13.10), let

$$W = C_1 U, \quad \widehat{W}_0 = C_1 \widehat{U}_0, \quad \widehat{W}_1 = C_1 \widehat{U}_1. \tag{13.14}$$

Noting (13.11), it is easy to see that the original problem (II) and (II0) for U can be reduced to the following self-closed system for $W = (w^{(1)}, \cdots, w^{(N-1)})^T$:

$$\begin{cases} W'' - \Delta W + \overline{A}_1 W = 0 \text{ in } (0, +\infty) \times \Omega, \\ W = 0 \qquad\qquad\qquad \text{on } (0, +\infty) \times \Gamma_0, \\ \partial_\nu W = \overline{D} H \qquad\qquad \text{on } (0, +\infty) \times \Gamma_1 \end{cases} \tag{13.15}$$

with the initial condition

$$t = 0: \quad W = \widehat{W}_0, \ W' = \widehat{W}_1 \quad \text{in } \Omega, \tag{13.16}$$

where $\overline{D} = C_1 D$.

Noting that C_1 is a surjection from $(\mathcal{H}_1)^N \times (\mathcal{H}_0)^N$ onto $(\mathcal{H}_1)^{N-1} \times (\mathcal{H}_0)^{N-1}$, we easily check that the exact boundary synchronization of system (II) for U is equivalent to the exact boundary null controllability of the **reduced system** (13.15) for W. Thus, by means of Theorems 12.14 and 12.15, the exact boundary synchronization of system (II) is equivalent to the rank condition $\text{rank}(\overline{D}) = \text{rank}(C_1 D) = N - 1$. Moreover, the continuous dependence (13.13) comes directly from (12.79). □

13.4 Attainable Set of Exactly Synchronizable States

In the case that system (II) possesses the exact boundary synchronization at the time $T > 0$, under the condition of C_1-compatibility (13.6), it is easy to see that for $t \geqslant T$, the exactly synchronizable state $u = u(t, x)$ defined by (13.3) satisfies the following wave equation with homogenous boundary conditions:

$$\begin{cases} u'' - \Delta u + au = 0 \text{ in } (T, +\infty) \times \Omega, \\ u = 0 \qquad\qquad\quad \text{on } (T, +\infty) \times \Gamma_0, \\ \partial_\nu u = 0 \qquad\qquad \text{on } (T, +\infty) \times \Gamma_1, \end{cases} \tag{13.17}$$

where a is given by (13.3). Hence, the evolution of the exactly synchronizable state $u = u(t, x)$ with respect to t is completely determined by the values of (u, u') at the time $t = T$:

$$t = T: \quad u = \widehat{u}_0, \quad u' = \widehat{u}_1 \quad \text{in } \Omega. \tag{13.18}$$

We have

Theorem 13.5 *Under the condition of* C_1*-compatibility* (13.6), *the attainable set of the values* (u, u') *at the time* $t = T$ *of the exactly synchronizable state* $u = u(t, x)$ *is the whole space* $\mathcal{H}_1 \times \mathcal{H}_0$ *as the initial data* $(\widehat{U}_0, \widehat{U}_1)$ *vary in the space* $(\mathcal{H}_1)^N \times (\mathcal{H}_0)^N$.

Proof For any given $(\widehat{u}_0, \widehat{u}_1) \in \mathcal{H}_1 \times \mathcal{H}_0$, by solving the following backward problem

$$\begin{cases} u'' - \Delta u + au = 0 \text{ in } (0, T) \times \Omega, \\ u = 0 \qquad\qquad\quad \text{on } (0, T) \times \Gamma_0, \\ \partial_\nu u = 0 \qquad\qquad \text{on } (0, T) \times \Gamma_1 \end{cases} \tag{13.19}$$

with the final condition

$$t = T: \quad u = \widehat{u}_0, \quad u' = \widehat{u}_1 \quad \text{in } \Omega, \tag{13.20}$$

we get the corresponding solution $u = u(t, x)$. Then, under the condition of C_1-compatibility (13.6), the function

$$U(t, x) = u(t, x)e_1 \tag{13.21}$$

with $e_1 = (1, \cdots, 1)^T$ is the solution to problem (II) and (II0) with the null boundary control $H \equiv 0$ and the initial condition

$$t = 0: \quad U = u(0, x)e_1, \; U' = u'(0, x)e_1. \tag{13.22}$$

Therefore, by solving problem (II) and (II0) with the null boundary control and the initial condition (13.22), we can reach any given exactly synchronizable state $(\widehat{u}_0, \widehat{u}_1)$

at the time $t = T$. This fact shows that any given state $(\widehat{u}_0, \widehat{u}_1) \in \mathcal{H}_1 \times \mathcal{H}_0$ can be expected to be a exactly synchronizable state. Consequently, the set of the values $(u(T), u'(T))$ of the exactly synchronizable state $u = (t, x)$ at the time T is the whole space $\mathcal{H}_1 \times \mathcal{H}_0$ as the initial data $(\widehat{U}_0, \widehat{U}_1)$ vary in the space $(\mathcal{H}_1)^N \times (\mathcal{H}_0)^N$. The proof is complete. □

The determination of the exactly synchronizable state u for each given initial data $(\widehat{U}_0, \widehat{U}_1)$ will be considered in Chap. 15.

Chapter 14
Exact Boundary Synchronization by p-Groups

The exact boundary synchronization by p-groups will be considered in this chapter for system (II) with further lack of Neumann boundary controls.

14.1 Definition

When there is a further lack of boundary controls, we consider the **exact boundary synchronization by p-groups** with $p \geqslant 1$. This indicates that the components of U are divided into p groups:

$$(u^{(1)}, \cdots, u^{(n_1)}),\ (u^{(n_1+1)}, \cdots, u^{(n_2)}), \cdots, (u^{(n_{p-1}+1)}, \cdots, u^{(n_p)}), \tag{14.1}$$

where $0 = n_0 < n_1 < n_2 < \cdots < n_p = N$ are integers such that $n_r - n_{r-1} \geqslant 2$ for all $1 \leqslant r \leqslant p$.

Definition 14.1 System (II) is exactly synchronizable by p-groups at the time $T > 0$ in the space $(\mathcal{H}_1)^N \times (\mathcal{H}_0)^N$ if for any given initial data $(\widehat{U}_0, \widehat{U}_1) \in (\mathcal{H}_1)^N \times (\mathcal{H}_0)^N$, there exists a boundary control $H \in \mathcal{L}^M$ with compact support in $[0, T]$, such that the weak solution $U = U(t, x)$ to problem (II) and (II0) satisfies

$$t \geqslant T: \quad u^{(i)} = u_r, \quad n_{r-1} + 1 \leqslant i \leqslant n_r, \quad 1 \leqslant r \leqslant p, \tag{14.2}$$

where, $u = (u_1, \cdots, u_p)^T$, being unknown a priori, is called the corresponding **exactly synchronizable state by p-groups**.

Let S_r be a $(n_r - n_{r-1} - 1) \times (n_r - n_{r-1})$ full row-rank matrix:

T. Li and B. Rao, *Boundary Synchronization for Hyperbolic Systems*, Progress in Nonlinear Differential Equations and Their Applications 94,
https://doi.org/10.1007/978-3-030-32849-8_14

$$S_r = \begin{pmatrix} 1 & -1 & & & \\ & 1 & -1 & & \\ & & \ddots & \ddots & \\ & & & 1 & -1 \end{pmatrix}, \quad 1 \leqslant r \leqslant p, \tag{14.3}$$

and let C_p be the following $(N-p) \times N$ **matrix of synchronization by** p**-groups**:

$$C_p = \begin{pmatrix} S_1 & & & \\ & S_2 & & \\ & & \ddots & \\ & & & S_p \end{pmatrix}. \tag{14.4}$$

Obviously, we have

$$\text{Ker}(C_p) = \text{Span}\{e_1, \cdots, e_p\}, \tag{14.5}$$

where for $1 \leqslant r \leqslant p$,

$$(e_r)_i = \begin{cases} 1, & n_{r-1} + 1 \leqslant i \leqslant n_r, \\ 0, & \text{otherwise.} \end{cases} \tag{14.6}$$

Thus, the exact boundary synchronization by p-groups (14.2) can be equivalently written as

$$t \geqslant T: \quad C_p U \equiv 0 \quad \text{in } \Omega, \tag{14.7}$$

or equivalently,

$$t \geqslant T: \quad U = \sum_{r=1}^{p} u_r e_r \quad \text{in } \Omega. \tag{14.8}$$

14.2 Condition of C_p-Compatibility

We have

Theorem 14.2 *Assume that system (II) is exactly synchronizable by p-groups. Then we necessarily have $M \geqslant N - p$. In particular, when $M = N - p$, the coupling matrix $A = (a_{ij})$ should satisfy the following* **condition of C_p-compatibility**:

$$AKer(C_p) \subseteq Ker(C_p). \tag{14.9}$$

Proof Noting that Lemma 6.3 is proved in a way independent of applied boundary conditions, it can be still used in the case with Neumann boundary controls.

Since we have (14.7), if $A\mathrm{Ker}(C_p) \not\subseteq \mathrm{Ker}(C_p)$, by Lemma 6.3, we can construct an enlarged full row-rank $(N-p+1)\times N$ matrix $\widetilde{C}_{p-1}$ such that

$$t \geqslant T: \quad \widetilde{C}_{p-1}U \equiv 0 \quad \text{in } \Omega.$$

If $A\mathrm{Ker}(\widetilde{C}_{p-1}) \not\subseteq \mathrm{Ker}(\widetilde{C}_{p-1})$, still by Lemma 6.3, we can construct another enlarged full row-rank $(N-p+2)\times N$ matrix $\widetilde{C}_{p-2}$ such that

$$t \geqslant T: \quad \widetilde{C}_{p-2}U \equiv 0 \quad \text{in } \Omega,$$

$\cdots\cdots$. This procedure should stop at the rth step with $0 \leqslant r \leqslant p$. Thus, we get an enlarged full row-rank $(N-p+r)\times N$ matrix $\widetilde{C}_{p-r}$ such that

$$t \geqslant T: \quad \widetilde{C}_{p-r}U \equiv 0 \quad \text{in } \Omega \tag{14.10}$$

and

$$A\mathrm{Ker}(\widetilde{C}_{p-r}) \subseteq \mathrm{Ker}(\widetilde{C}_{p-r}). \tag{14.11}$$

Then, by Proposition 2.15, there exists a unique matrix $\tilde{A}$ of order $(N-p+r)$, such that

$$\widetilde{C}_{p-r}A = \tilde{A}\widetilde{C}_{p-r}. \tag{14.12}$$

Setting $W = \widetilde{C}_{p-r}U$ in problem (II) and (II0), we get the following reduced problem for $W = (w^{(1)}, \cdots, w^{(N-p+r)})^T$:

$$\begin{cases} W'' - \Delta W + \tilde{A}W = 0 \text{ in } (0,+\infty)\times\Omega, \\ W = 0 \qquad\qquad\qquad \text{on } (0,+\infty)\times\Gamma_0, \\ \partial_\nu W = \tilde{D}H \qquad\qquad \text{on } (0,+\infty)\times\Gamma_1 \end{cases} \tag{14.13}$$

with the initial condition

$$t = 0: \quad W = \widetilde{C}_{p-r}\widehat{U}_0, \ W' = \widetilde{C}_{p-r}\widehat{U}_1 \quad \text{in } \Omega, \tag{14.14}$$

where $\tilde{D} = \widetilde{C}_{p-r}D$. Moreover, by (14.10) we have

$$t \geqslant T: \quad W \equiv 0. \tag{14.15}$$

Noting that $\widetilde{C}_{p-r}$ is a $(N-p+r)\times N$ full row-rank matrix, the linear mapping

$$(\widehat{U}_0, \widehat{U}_1) \to (\widetilde{C}_{p-r}\widehat{U}_0, \widetilde{C}_{p-r}\widehat{U}_1) \tag{14.16}$$

is a surjection from $(\mathcal{H}_1)^N \times (\mathcal{H}_0)^N$ onto $(\mathcal{H}_1)^{N-p+r} \times (\mathcal{H}_0)^{N-p+r}$, then system (14.13) is exactly null controllable at the time T in the space $(\mathcal{H}_1)^{N-p+r} \times (\mathcal{H}_0)^{N-p+r}$. By Theorems 12.14 and 12.15, we necessarily have

$$\text{rank}(\widetilde{C}_{p-r}D) = N - p + r,$$

then we get

$$M = \text{rank}(D) \geqslant \text{rank}(\widetilde{C}_{p-r}D) = N - p + r \geqslant N - p. \tag{14.17}$$

In particular, when $M = N - p$, we have $r = 0$, namely, the condition of C_p-compatibility (14.9) holds. □

Remark 14.3 The condition of C_p-compatibility (14.9) is equivalent to the fact that there exist some constants α_{rs} $(1 \leqslant r, s \leqslant p)$ such that

$$Ae_r = \sum_{s=1}^{p} \alpha_{sr} e_s, \quad 1 \leqslant r \leqslant p, \tag{14.18}$$

or, noting (14.6), A satisfies the following **row-sum condition by blocks**:

$$\sum_{j=n_{s-1}+1}^{n_s} a_{ij} = \alpha_{rs}, \quad n_{r-1} + 1 \leqslant i \leqslant n_r, \quad 1 \leqslant r, s \leqslant p. \tag{14.19}$$

Especially, this condition of compatibility becomes the row-sum condition 13.6 when $p = 1$.

14.3 Exact Boundary Synchronization by p-Groups and Non-exact Boundary Synchronization by p-Groups

Theorem 14.4 *Assume that the condition of C_p-compatibility* (14.9) *holds. Then system (II) is exactly synchronizable by p-groups if and only if rank(C_pD) $= N - p$. Moreover, we have the continuous dependence:*

$$\|H\|_{L^2(0;T;(L^2(\Gamma_1))^{N-p})} \leqslant c\|C_p(\widehat{U}_0, \widehat{U}_1)\|_{(\mathcal{H}_1)^{N-p}\times(\mathcal{H}_0)^{N-p}}, \tag{14.20}$$

where c is a positive constant.

Proof Assume that the coupling matrix $A = (a_{ij})$ satisfies the condition of C_p-compatibility (14.9). By Proposition 2.15, there exists a unique matrix $\overline{A}_p$ of order $(N - p)$, such that

$$C_p A = \overline{A}_p C_p. \tag{14.21}$$

Setting

$$W = C_p U, \quad \overline{D} = C_p D, \tag{14.22}$$

we get the following reduced system for $W = (w^{(1)}, \cdots, w^{(N-p)})^T$:

$$\begin{cases} W'' - \Delta W + \overline{A}_p W = 0 \text{ in } (0, +\infty) \times \Omega, \\ W = 0 \qquad\qquad\qquad\quad \text{on } (0, +\infty) \times \Gamma_0, \\ \partial_\nu W = \overline{D} H \qquad\qquad\quad \text{on } (0, +\infty) \times \Gamma_1 \end{cases} \tag{14.23}$$

with the initial condition

$$t = 0: \quad W = C_p \widehat{U}_0, \ W' = C_p \widehat{U}_1 \quad \text{in } \Omega. \tag{14.24}$$

Noting that C_p is an $(N - p) \times N$ full row-rank matrix, it is easy to see that system (II) is exactly synchronizable by p-groups if and only if the reduced system (14.23) is exactly null controllable, or equivalently, by Theorem 12.14 and Theorem 12.15 if and only if $\text{rank}(C_p D) = N - p$. Moreover, the continuous dependence (14.20) comes from (12.79). □

14.4 Attainable Set of Exactly Synchronizable States by p-Groups

Under the condition of C_p-compatibility (14.9), if system (II) is exactly synchronizable by p-groups at the time $T > 0$, then it is easy to see that for $t \geqslant T$, the exactly synchronizable state by p-groups $u = (u_1, \cdots u_p)^T$ satisfies the following coupled system of wave equations with homogenous boundary conditions:

$$\begin{cases} u'' - \Delta u + \widetilde{A} u = 0 \text{ in } (T, +\infty) \times \Omega, \\ u = 0 \qquad\qquad\qquad \text{on } (T, +\infty) \times \Gamma_0, \\ \partial_\nu u = 0 \qquad\qquad\quad \text{on } (T, +\infty) \times \Gamma_1, \end{cases} \tag{14.25}$$

where $\widetilde{A} = (\alpha_{rs})$ is given by (14.18). Hence, the evolution of the exactly synchronizable state by p-groups $u = (u_1, \cdots u_p)^T$ with respect to t is completely determined by the values of (u, u') at the time $t = T$:

$$t = T: \quad u = \widehat{u}_0, \quad u' = \widehat{u}_1. \tag{14.26}$$

As in Theorem 13.5 for the case $p = 1$, we can show that the attainable set of all possible values of (u, u') at $t = T$ is the whole space $(\mathcal{H}_1)^p \times (\mathcal{H}_0)^p$, when the initial data $(\widehat{U}_0, \widehat{U}_1)$ vary in the space $(\mathcal{H}_1)^N \times (\mathcal{H}_0)^N$.

The determination of the exactly synchronizable state by p-groups $u = (u_1, \cdots, u_p)^T$ for each given initial data $(\widehat{U}_0, \widehat{U}_1)$ will be considered in Chap. 15.

Chapter 15
Determination of Exactly Synchronizable States by p-Groups

When system (II) possesses the exact boundary synchronization by p-groups, the corresponding exactly synchronizable states by p-groups will be studied in this chapter.

15.1 Introduction

Now, under the condition of C_p-compatibility (14.9), we are going to discuss the determination of exactly synchronizable states by p-groups $u = (u_1, \cdots, u_p)^T$ with $p \geqslant 1$ for system (II). Generally speaking, exactly synchronizable states by p-groups should depend on the initial data $(\widehat{U}_0, \widehat{U}_1)$ and applied boundary controls H. However, when the coupling matrix A possesses some good properties, exactly synchronizable states by p-groups are independent of applied boundary controls, and can be determined entirely by the solution to a system of wave equations with homogeneous boundary conditions for any given initial data $(\widehat{U}_0, \widehat{U}_1)$.

First, as in the case of Dirichlet boundary controls, we have the following

Theorem 15.1 *Let $\mathcal{U}_{ad}$ denote the set of all boundary controls $H = (h^{(1)}, \cdots, h^{(N-p)})^T$ which can realize the exact boundary synchronization by p-groups at the time $T > 0$ for system (II). Assume that the condition of C_p-compatibility (14.9) holds. Then for $\epsilon > 0$ small enough, the value of $H \in \mathcal{U}_{ad}$ on $(0, \epsilon) \times \Gamma_1$ can be arbitrarily chosen.*

Proof Under the condition of C_p-compatibility (14.9), the exact boundary synchronization by p-groups of system (II) is equivalent to the exact boundary null controllability of the reduced system (14.23). Let $T_0 > 0$ be a number independent of the initial data, such that the reduced system (14.23) is exactly null controllable at any given time T with $T > T_0$. Therefore, taking an $\epsilon > 0$ so small that $T - \epsilon > T_0$,

T. Li and B. Rao, *Boundary Synchronization for Hyperbolic Systems*, Progress in Nonlinear Differential Equations and Their Applications 94, https://doi.org/10.1007/978-3-030-32849-8_15

system (II) is still exactly synchronizable by p-groups at the time T for the initial data given at $t = \epsilon$.

For any given initial data $(\widehat{U}_0, \widehat{U}_1) \in (\mathcal{H}_1)^N \times (\mathcal{H}_0)^N$, we arbitrarily give

$$\widehat{H}_\epsilon \in (C_0^\infty([0, \epsilon] \times \Gamma_1))^{N-p} \tag{15.1}$$

and solve problem (II) and (II0) on the time interval $[0, \epsilon]$ with $H = \widehat{H}_\epsilon$. Let $(\widehat{U}_\epsilon, \widehat{U}_\epsilon')$ denote the corresponding solution. We check easily that

$$(\widehat{U}_\epsilon, \widehat{U}_\epsilon') \in C^0([0, \epsilon]; (\mathcal{H}_1)^N \times (\mathcal{H}_0)^N). \tag{15.2}$$

By Theorem 14.4, we can find a boundary control

$$\widetilde{H}_\epsilon \in L^2(\epsilon, T; (L^2(\Gamma_1))^{N-p}) \tag{15.3}$$

which realizes the exact boundary synchronization by p-groups at the time $t = T$ for system (II) with the initial condition

$$t = \epsilon : \quad \widetilde{U}_\epsilon = \widehat{U}_\epsilon, \quad \widetilde{U}_\epsilon' = \widehat{U}_\epsilon'. \tag{15.4}$$

Let

$$H = \begin{cases} \widehat{H}_\epsilon, & t \in (0, \epsilon), \\ \widetilde{H}_\epsilon, & t \in (\epsilon, T), \end{cases} \quad U = \begin{cases} \widehat{U}_\epsilon, & t \in (0, \epsilon), \\ \widetilde{U}_\epsilon, & t \in (\epsilon, T). \end{cases} \tag{15.5}$$

It is easy to check that U is the solution to problem (II) and (II0) with the boundary control H. Then, system (II) is exactly synchronizable by p-groups still at the time T for the initial data $(\widehat{U}_0, \widehat{U}_1)$ given at $t = 0$. □

15.2 Determination and Estimation of Exactly Synchronizable States by p-Groups

Exactly synchronizable states by p-groups are closely related to the properties of the coupling matrix A.

Let

$$\mathcal{D}_{N-p} = \{D \in \mathbb{M}^{N\times(N-p)}(\mathbb{R}) : \text{rank}(D) = \text{rank}(C_p D) = N - p\}. \tag{15.6}$$

By Proposition 6.12, $D \in \mathcal{D}_{N-p}$ if and only if it can be expressed by

$$D = C_p^T D_1 + (e_1, \cdots, e_p) D_0, \tag{15.7}$$

where $e_1, \cdots, e_p$ are given by (14.6), D_1 is an invertible matrix of order $(N-p)$, and D_0 is a $p \times (N-p)$ matrix.

We have

Theorem 15.2 *Under the condition of C_p-compatibility (14.9), assume that A^T possesses an invariant subspace Span$\{E_1, \cdots, E_p\}$ which is bi-orthonormal to Ker$(C_p) = $ Span$\{e_1, \cdots, e_p\}$:*

$$(E_r, e_s) = \delta_{rs}, \quad 1 \leqslant r, s \leqslant p. \tag{15.8}$$

Then there exists a boundary control matrix $D \in \mathcal{D}_{N-p}$, such that each exactly synchronizable state by p-groups $u = (u_1, \cdots, u_p)^T$ is independent of applied boundary controls, and can be determined as follows:

$$t \geqslant T: \quad u = \psi, \tag{15.9}$$

where $\psi = (\psi_1, \cdots, \psi_p)^T$ is the solution to the following problem with homogeneous boundary conditions for $r = 1, \cdots, p$:

$$\begin{cases} \psi_r'' - \Delta\psi_r + \sum_{s=1}^{p} \alpha_{rs}\psi_s = 0 & \text{in } (0, +\infty) \times \Omega, \\ \psi_r = 0 & \text{on } (0, +\infty) \times \Gamma_0, \\ \partial_\nu \psi_r = 0 & \text{on } (0, +\infty) \times \Gamma_1, \\ t = 0: \quad \psi_r = (E_r, \widehat{U}_0), \ \psi_r' = (E_r, \widehat{U}_1) & \text{in } \Omega, \end{cases} \tag{15.10}$$

where α_{rs} $(1 \leqslant r, s \leqslant p)$ are given by (14.19).

Proof Noting that Span$\{E_1, \cdots, E_p\}$ is bi-orthonormal to Ker$(C_p) = $ Span $\{e_1, \cdots, e_p\}$, and taking

$$D_1 = I_{N-p}, \quad D_0 = -E^T C_p^T \quad \text{with} \quad E = (E_1, \cdots, E_p) \tag{15.11}$$

in (15.7), we get a boundary control matrix $D \in \mathcal{D}_{N-p}$ and it is easy to see that

$$E_s \in \text{Ker}(D^T), \quad 1 \leqslant s \leqslant p. \tag{15.12}$$

On the other hand, since Span$\{E_1, \cdots, E_p\}$ is an invariant subspace of A^T, noting (14.18) and the bi-orthonormality (15.8), we get easily that

$$A^T E_s = \sum_{r=1}^{p} \alpha_{sr} E_r, \quad 1 \leqslant s \leqslant p. \tag{15.13}$$

Let $\psi_s = (E_s, U)$ for $s = 1, \cdots, p$. Taking the inner product with E_s on both sides of problem (II) and (II0), we get (15.10). Finally, for the exactly synchronizable state by p-groups $u = (u_1, \cdots, u_p)^T$, by (14.8) we have

$$t \geqslant T: \quad \psi_s = (E_s, U) = \sum_{r=1}^{p} (E_s, e_r) u_r = u_s, \quad 1 \leqslant s \leqslant p. \tag{15.14}$$

The proof is complete. □

When A^T does not possess any invariant subspace Span$\{E_1, \cdots, E_p\}$ which is bi-orthonormal to Ker(C_p) = Span$\{e_1, \cdots, e_p\}$, we can use the solution of problem (15.10) to give an estimate on each exactly synchronizable state by p-groups.

Theorem 15.3 *Under the condition of C_p-compatibility (14.9), assume that there exists a subspace Span$\{E_1, \cdots, E_p\}$ that is bi-orthonormal to the subspace Span $\{e_1, \cdots, e_p\}$. Then there exist a boundary control matrix $D \in \mathcal{D}_{N-p}$, such that each exactly synchronizable state by p-groups $u = (u_1, \cdots, u_p)^T$ satisfies the following estimate:*

$$\begin{aligned} &\|(u, u')(t) - (\psi, \psi')(t)\|_{(\mathcal{H}_{2-s})^p \times (\mathcal{H}_{1-s})^p} \\ \leqslant &c \|C_p(\widehat{U}_0, \widehat{U}_1)\|_{(\mathcal{H}_1)^{N-p} \times (\mathcal{H}_0)^{N-p}}, \end{aligned} \tag{15.15}$$

for all $t \geqslant T$, where $\psi = (\psi_1, \cdots, \psi_p)^T$ is the solution to problem (15.10)*, $s > \frac{1}{2}$, and c is a positive constant independent of the initial data.*

Proof Since Span$\{E_1, \cdots, E_p\}$ is bi-orthonormal to Span$\{e_1, \cdots, e_p\}$, still by (15.11), there exists a boundary control matrix $D \in \mathcal{D}_{N-p}$, such that (15.12) holds. Noting (14.18), it is easy to see that

$$A^T E_s - \sum_{r=1}^{p} \alpha_{sr} E_r \in \{Ker(C_p)\}^{\perp} = \text{Im}(C_p^T),$$

then there exists a vector $R_s \in \mathbb{R}^{N-p}$, such that

$$A^T E_s - \sum_{r=1}^{p} \alpha_{sr} E_r = C_p^T R_s. \tag{15.16}$$

Thus, taking the inner product with E_s on both sides of problem (II) and (II0), and setting $\phi_s = (E_s, U)$ for $s = 1, \cdots, p$, we have

$$\begin{cases} \phi_s'' - \Delta\phi_s + \sum_{r=1}^{p} \alpha_{sr}\phi_r = -(R_s, C_p U) & \text{in } (0, +\infty) \times \Omega, \\ \phi_s = 0 & \text{on } (0, +\infty) \times \Gamma_0, \\ \partial_\nu \phi_s = 0 & \text{on } (0, +\infty) \times \Gamma_1, \\ t = 0: \quad \phi_r = (E_s, \widehat{U}_0), \ \phi_s' = (E_s, \widehat{U}_1) & \text{in } \Omega, \end{cases} \tag{15.17}$$

where α_{sr} $(1 \leqslant r, s \leqslant p)$ are defined by (14.19), and $U = U(t, x) \in C_{loc}(0, +\infty; (\mathcal{H}_{1-s})^N) \cap C^1_{loc}(0, +\infty; (\mathcal{H}_{-s})^N)$ with $s > 1/2$ is the solution to problem (II) and (II0). Moreover, we have

$$t \geqslant T: \quad \phi_s = (E_s, U) = \sum_{r=1}^{p} (E_s, e_r) u_r = u_s \tag{15.18}$$

for $s = 1, \cdots, p$. Noting that (15.10) and (15.17) possess the same initial data and the same boundary conditions, by the well-posedness for a system of wave equations with Neumann boundary condition, we have (cf. Chap. 3 in [70]) that

$$\begin{aligned} &\|(\psi, \psi')(t) - (\phi, \phi')(t)\|^2_{(\mathcal{H}_{2-s})^p \times (\mathcal{H}_{1-s})^p} \\ \leqslant &c \int_0^T \|C_p U\|^2_{(\mathcal{H}_{1-s})^{N-p}} ds, \qquad \forall\, 0 \leqslant t \leqslant T, \end{aligned} \tag{15.19}$$

where c is a positive constant independent of the initial data. Noting that $W = C_p U$, by the well-posedness of the reduced problem (14.23), it is easy to get that

$$\begin{aligned} &\int_0^T \|C_p U\|^2_{(\mathcal{H}_{1-s})^{N-p}} ds \\ \leqslant &\, c(\|C_p(\widehat{U}_0, \widehat{U}_1)\|^2_{(\mathcal{H}_1)^{N-p} \times (\mathcal{H}_0)^{N-p}} + \|\overline{D}H\|^2_{L^2(0,T;(L^2(\Gamma_1))^{N-p})}). \end{aligned} \tag{15.20}$$

Moreover, by (14.20) we have

$$\|\overline{D}H\|_{L^2(0,T;(L^2(\Gamma_1))^{N-p})} \leqslant c\|C_p(\widehat{U}_0, \widehat{U}_1)\|_{(\mathcal{H}_1)^{N-p} \times (\mathcal{H}_0)^{N-p}}. \tag{15.21}$$

Substituting it into (15.20), we have

$$\int_0^T \|C_p U\|^2_{(\mathcal{H}_{1-s})^{N-p}} ds \leqslant c\|C_p(\widehat{U}_0, \widehat{U}_1)\|^2_{(\mathcal{H}_1)^{N-p} \times (\mathcal{H}_0)^{N-p}}, \tag{15.22}$$

then, by (15.19) we get

$$\begin{aligned} &\|(\psi, \psi')(t) - (\phi, \phi')(t)\|^2_{(\mathcal{H}_{2-s})^p \times (\mathcal{H}_{1-s})^p} \\ \leqslant &c\|C_p(\widehat{U}_0, \widehat{U}_1)\|^2_{(\mathcal{H}_1)^{N-p} \times (\mathcal{H}_0)^{N-p}}. \end{aligned} \tag{15.23}$$

Finally, substituting (15.18) into (15.23), we get (15.15). □

Remark 15.4 Differently from the case with Dirichlet boundary controls, although the solution to problem (II) and (II0) with Neumann boundary controls possesses a weaker regularity, the solution to problem (15.10), which is used to estimate the exactly synchronizable state by p-groups, possesses a higher regularity than the original problem (II) and (II0) itself. This improved regularity makes it possible to approach the exactly synchronizable state by p-groups by a solution to a relatively smoother problem.

If there does not exist any subspace $\text{Span}\{E_1, \cdots, E_p\}$, which is invariant for A^T and bi-orthonormal to $\text{Ker}(C_p) = \text{Span}\{e_1, \cdots, e_p\}$, in order to express the exactly synchronizable state by p-groups of system (II), by the same procedure as in Sect. 7.3, we can always extend the subspace $\text{Span}\{e_1, \cdots, e_p\}$ to the minimal invariant subspace $\text{Span}\{e_1, \cdots, e_q\}$ of A with $q \geqslant p$, so that A^T possesses an invariant subspace $\text{Span}\{E_1, \cdots, E_q\}$, which is bi-orthonormal to $\text{Span}\{e_1, \cdots, e_q\}$.

Let P be the projection on the subspace $\text{Span}\{e_1, \cdots, e_q\}$ given by

$$P = \sum_{s=1}^{q} e_s \otimes E_s, \tag{15.24}$$

in which the tensor product $\otimes$ is defined by

$$(e \otimes E)U = (E, U)e = E^T U e, \quad \forall\, U \in \mathbb{R}^N.$$

P can be represented by a matrix of order N, such that

$$\text{Im}(P) = \text{Span}\{e_1, \cdots, e_q\} \tag{15.25}$$

and

$$\text{Ker}(P) = \big(\text{Span}\{E_1, \cdots, E_q\}\big)^{\perp}. \tag{15.26}$$

Moreover, it is easy to check that

$$PA = AP. \tag{15.27}$$

Let $U = U(t, x)$ be the solution to problem (II) and (II0). We define its **synchronizable part** U_s and **controllable part** U_c, respectively, as follows:

$$U_s := PU, \quad U_c := (I - P)U. \tag{15.28}$$

If system (II) is exactly synchronizable by p-groups, then

$$U \in \text{Span}\{e_1, \cdots, e_p\} \subseteq \text{Span}\{e_1, \cdots, e_q\} = \text{Im}(P), \tag{15.29}$$

hence we have

$$t \geqslant T: \quad U_s = PU = U, \quad U_c = (I - P)U = 0. \tag{15.30}$$

Noting (15.27), multiplying P and $(I - P)$ from the left on both sides of problem (II) and (II0), respectively, we see that the synchronizable part U_s of U satisfies the following problem:

$$\begin{cases} U_s'' - \Delta U_s + AU_s = 0 & \text{in } (0, +\infty) \times \Omega, \\ U_s = 0 & \text{on } (0, +\infty) \times \Gamma_0, \\ \partial_\nu U_s = PDH & \text{on } (0, +\infty) \times \Gamma_1, \\ t = 0: \quad U_s = P\widehat{U}_0, \ U_s' = P\widehat{U}_1 & \text{in } \Omega, \end{cases} \tag{15.31}$$

while the controllable part U_c of U satisfies the following problem:

$$\begin{cases} U_c'' - \Delta U_c + AU_c = 0 & \text{in } (0, +\infty) \times \Omega, \\ U_c = 0 & \text{on } (0, +\infty) \times \Gamma_0, \\ \partial_\nu U_c = (I - P)DH & \text{on } (0, +\infty) \times \Gamma_1, \\ t = 0: \quad U_c = (I - P)\widehat{U}_0, \ U_c' = (I - P)\widehat{U}_1 & \text{in } \Omega. \end{cases} \tag{15.32}$$

In fact, by (15.30), under the boundary control H, U_c with the initial data $((I - P)\widehat{U}_0, (I - P)\widehat{U}_1) \in \text{Ker}(P) \times \text{Ker}(P)$ is exactly null controllable, while U_s with the initial data $(P\widehat{U}_0, P\widehat{U}_1) \in \text{Im}(P) \times \text{Im}(P)$ is exactly synchronizable.

Theorem 15.5 *Assume that the condition of C_p-compatibility (14.9) holds. If system (II) is exactly synchronizable by p-groups, and the synchronizable part U_s is independent of applied boundary controls H for $t \geqslant 0$, then A^T possesses an invariant subspace which is bi-orthonormal to Ker(C_p).*

Proof It is sufficient to show that $p = q$. Then Span$\{E_1, \cdots, E_p\}$ will be the desired subspace.

Suppose that the synchronizable part U_s is independent of applied boundary controls H for all $t \geqslant 0$. Let H_1 and H_2 be two boundary controls which realize simultaneously the exact boundary synchronization by p-groups for system (II). It follows from (15.31) that

$$PD(H_1 - H_2) = 0 \quad \text{on } (0, \epsilon) \times \Gamma_1.$$

By Theorem 15.1, the value of H_1 on $(0, \epsilon) \times \Gamma_1$ can be arbitrarily taken, then

$$PD = 0,$$

hence

$$\text{Im}(D) \subseteq \text{Ker}(P).$$

Noting (15.25), we have

$$\dim \text{Ker}(P) = N - q \quad \text{and} \quad \dim \text{Im}(D) = N - p,$$

then $p = q$. The proof is complete. □

Remark 15.6 In particular, for the initial data $(\widehat{U}_0, \widehat{U}_1)$ such that $P\widehat{U}_0 = P\widehat{U}_1 = 0$, system (II) is exactly null controllable.

15.3 Determination of Exactly Synchronizable States

In the case of exact boundary synchronization, by the condition of C_1-compatibility (13.10), $e_1 = (1, \cdots, 1)^T$ is an eigenvector of A, corresponding to the eigenvalue a defined by (13.6). Let $\epsilon_1, \cdots, \epsilon_r$ and $E_1, \cdots, E_r$ with $r \geqslant 1$ be the Jordan chains of A and A^T, respectively, corresponding to the eigenvalue a, and Span$\{\epsilon_1, \cdots, \epsilon_r\}$ is bi-orthonormal to Span$\{E_1, \cdots, E_r\}$. Thus, we have

$$\begin{cases} A\epsilon_l = a\epsilon_l + \epsilon_{l+1}, & 1 \leqslant l \leqslant r, \\ A^T E_k = aE_k + E_{k-1}, & 1 \leqslant k \leqslant r, \\ (E_k, \epsilon_l) = \delta_{kl}, & 1 \leqslant k, l \leqslant r \end{cases} \tag{15.33}$$

with

$$\epsilon_r = e_1 = (1, \cdots, 1)^T, \quad \epsilon_{r+1} = 0, \quad E_0 = 0. \tag{15.34}$$

Let $U = U(t, x)$ be the solution to problem (II) and (II0). If system (II) is exactly synchronizable, then

$$t \geqslant T: \quad U = u\epsilon_r, \tag{15.35}$$

where $u = u(t, x)$ is the corresponding exactly synchronizable state. For the synchronizable part U_s and the controllable part U_c, we have

$$t \geqslant T: \quad U_s = u\epsilon_r, \quad U_c = 0. \tag{15.36}$$

Setting

$$\psi_k = (E_k, U), \quad 1 \leqslant k \leqslant r, \tag{15.37}$$

and noting (15.24) and (15.28), similarly we have

$$U_s = \sum_{k=1}^{r} (E_k, U)\epsilon_k = \sum_{k=1}^{r} \psi_k \epsilon_k. \tag{15.38}$$

Thus, $(\psi_1, \cdots, \psi_r)$ can be regarded as the coordinates of U_s under the basis $\{\epsilon_1, \cdots, \epsilon_r\}$.

Taking the inner product with E_k on both sides of (15.31), and noting (15.35), we have

$$t \geqslant T: \quad \psi_k = (E_k, U) = (E_k, \epsilon_r) = \delta_{kr}, \quad 1 \leqslant k \leqslant r, \tag{15.39}$$

then we easily get the following

Theorem 15.7 *Let $\epsilon_1, \cdots, \epsilon_r$ and $E_1, \cdots, E_r$ be the Jordan chains of A and A^T, respectively, corresponding to the eigenvalue a given by (13.6), in which $\epsilon_r = e_1 = (1, \cdots, 1)^T$. Then, the exactly synchronizable state u is determined by*

$$t \geqslant T: \quad u = \psi_r, \tag{15.40}$$

where the synchronizable part $U_s = (\psi_1, \cdots, \psi_r)$ is determined by the following system ($1 \leqslant k \leqslant r$):

$$\begin{cases} \psi_k'' - \Delta\psi_k + a\psi_k + \psi_{k-1} = 0, & \text{in } (0, +\infty) \times \Omega, \\ \psi_k = 0 & \text{on } (0, +\infty) \times \Gamma_0, \\ \partial_\nu \psi_k = h_k & \text{on } (0, +\infty) \times \Gamma_1, \\ t = 0: \quad \psi_k = (E_k, \widehat{U}_0), \ \psi_k' = (E_k, \widehat{U}_1) & \text{in } \Omega, \end{cases} \tag{15.41}$$

where

$$\psi_0 = 0, \quad h_k = E_k^T PDH. \tag{15.42}$$

Remark 15.8 When $r = 1$, A^T possesses an eigenvector E_1 such that

$$(E_1, \epsilon_1) = 1. \tag{15.43}$$

By Theorem 15.2, we can choose a boundary control matrix D such that the exactly synchronizable state is independent of applied boundary controls H.

When $r \geqslant 1$, the exactly synchronizable state depends on applied boundary controls H. Moreover, in order to get the exactly synchronizable state, we must solve the whole coupled problem (15.41)–(15.42).

15.4 Determination of Exactly Synchronizable States by 3-Groups

In this section, for an example, we will give the details on the determination of exactly synchronizable states by 3-groups for system (II). Here, we always assume that the condition of C_3-compatibility (14.9) is satisfied and that $\text{Ker}(C_3)$ is A-marked. All the exactly synchronizable states by p-groups for general $p \geqslant 1$ can be discussed in a similar way.

By the exact boundary synchronization by 3-groups, when $t \geqslant T$, we have

$$u^{(1)} \equiv \cdots \equiv u^{(n_1)} := u_1, \tag{15.44}$$

$$u^{(n_1+1)} \equiv \cdots \equiv u^{(n_2)} := u_2, \tag{15.45}$$

$$u^{(n_2+1)} \equiv \cdots \equiv u^{(N)} := u_3. \tag{15.46}$$

Let

$$\begin{cases} e_1 = (\overbrace{1, \cdots, 1}^{n_1}, \overbrace{0, \cdots, 0}^{n_2-n_1}, \overbrace{0, \cdots, 0}^{N-n_2})^T, \\ e_2 = (\overbrace{0, \cdots, 0}^{n_1}, \overbrace{1, \cdots, 1}^{n_2-n_1}, \overbrace{0, \cdots, 0}^{N-n_2})^T, \\ e_3 = (\overbrace{0, \cdots, 0}^{n_1}, \overbrace{0, \cdots, 0}^{n_2-n_1}, \overbrace{1, \cdots, 1}^{N-n_2})^T. \end{cases} \tag{15.47}$$

We have

$$\mathrm{Ker}(C_3) = \mathrm{Span}\{e_1, e_2, e_3\} \tag{15.48}$$

and the exact boundary synchronization by 3-groups given by (15.44)–(15.46) means that

$$t \geqslant T: \quad U = u_1 e_1 + u_2 e_2 + u_3 e_3. \tag{15.49}$$

Since the invariant subspace $\mathrm{Span}\{e_1, e_2, e_3\}$ of A is of dimension 3, it may contain one, two, or three eigenvectors of A, and thus we can distinguish the following three cases.

(i) A admits three eigenvectors f_r, g_s, and h_t contained in $\mathrm{Span}\{e_1, e_2, e_3\}$, corresponding to eigenvalues λ, μ and ν, respectively. Let $f_1, \cdots, f_r$; $g_1, \cdots, g_s$ and $h_1, \cdots, h_t$ be the Jordan chains corresponding to these eigenvectors of A:

$$\begin{cases} Af_i = \lambda f_i + f_{i+1}, & 1 \leqslant i \leqslant r, \quad f_{r+1} = 0, \\ Ag_j = \mu g_j + g_{j+1}, & 1 \leqslant j \leqslant s, \quad g_{s+1} = 0, \\ Ah_k = \nu h_k + h_{k+1}, & 1 \leqslant k \leqslant t, \quad h_{t+1} = 0 \end{cases} \tag{15.50}$$

and let $\xi_1, \cdots, \xi_r$; $\eta_1, \cdots, \eta_s$ and $\zeta_1, \cdots, \zeta_t$ be the Jordan chains corresponding to the related eigenvectors of A^T:

$$\begin{cases} A^T \xi_i = \lambda \xi_i + \xi_{i-1}, & 1 \leqslant i \leqslant r, \quad \xi_0 = 0, \\ A^T \eta_j = \mu \eta_j + \eta_{j-1}, & 1 \leqslant j \leqslant s, \quad \eta_0 = 0, \\ A^T \zeta_k = \nu \zeta_k + \zeta_{k-1}, & 1 \leqslant k \leqslant t, \quad \zeta_0 = 0, \end{cases} \tag{15.51}$$

such that

$$(f_i, \xi_l) = \delta_{il}, \quad (g_j, \eta_m) = \delta_{jm}, \quad (h_k, \zeta_n) = \delta_{kn} \tag{15.52}$$

and

$$(f_i, \eta_m) = (f_i, \zeta_n) = (g_j, \xi_l) = (g_j, \zeta_n) = (h_k, \xi_l) = (h_k, \eta_m) = 0 \tag{15.53}$$

for all $i, l = 1, \cdots, r;\ j, m = 1, \cdots, s$; and $k, n = 1, \cdots, t$.

Taking the inner product with ξ_i, η_j, and ζ_k on both sides of problem (II) and (II0), respectively, and denoting

$$\phi_i = (U, \xi_i), \quad \psi_j = (U, \eta_j), \quad \theta_k = (U, \zeta_k) \tag{15.54}$$

for $i = 1, \cdots, r;\ j = 1, \cdots, s$; and $k = 1, \cdots t$, we get

$$\begin{cases} \phi_i'' - \Delta\phi_i + \lambda\phi_i + \phi_{i-1} = 0 & \text{in } (0, +\infty) \times \Omega, \\ \phi_i = 0 & \text{on } (0, +\infty) \times \Gamma_0, \\ \partial_\nu \phi_i = \xi_i^T DH & \text{on } (0, +\infty) \times \Gamma_1, \\ t = 0: \quad \phi_i = (\xi_i, \widehat{U}_0),\ \phi_i' = (\xi_i, \widehat{U}_1) & \text{in } \Omega, \end{cases} \tag{15.55}$$

$$\begin{cases} \psi_j'' - \Delta\psi_j + \mu\psi_j + \psi_{j-1} = 0 & \text{in } (0, +\infty) \times \Omega, \\ \psi_j = 0 & \text{on } (0, +\infty) \times \Gamma_0, \\ \partial_\nu \psi_j = \eta_j^T DH & \text{on } (0, +\infty) \times \Gamma_1, \\ t = 0: \quad \psi_j = (\eta_j, \widehat{U}_0),\ \psi_j' = (\eta_j, \widehat{U}_1) & \text{in } \Omega \end{cases} \tag{15.56}$$

and

$$\begin{cases} \theta_k'' - \Delta\theta_k + \nu\theta_k + \theta_{k-1} = 0 & \text{in } (0, +\infty) \times \Omega, \\ \theta_k = 0 & \text{on } (0, +\infty) \times \Gamma_0, \\ \partial_\nu \theta_k = \zeta_k^T DH & \text{on } (0, +\infty) \times \Gamma_1, \\ t = 0: \quad \theta_k = (\zeta_k, \widehat{U}_0),\ \theta_k' = (\zeta_k, \widehat{U}_1) & \text{in } \Omega \end{cases} \tag{15.57}$$

with

$$\phi_0 = \psi_0 = \theta_0 = 0. \tag{15.58}$$

Solving the reduced problems (15.55)–(15.57), we get $\phi_1, \cdots, \phi_r$; $\psi_1, \cdots, \psi_s$; and $\theta_1, \cdots, \theta_t$. Noting that f_r, g_s, and h_t are linearly independent eigenvectors and contained in $\text{Span}\{e_1, e_2, e_3\}$, we have

$$\begin{cases} e_1 = \alpha_1 f_r + \alpha_2 g_s + \alpha_3 h_t, \\ e_2 = \beta_1 f_r + \beta_2 g_s + \beta_3 h_t, \\ e_3 = \gamma_1 f_r + \gamma_2 g_s + \gamma_3 h_t. \end{cases} \tag{15.59}$$

Then, noting (15.52)–(15.53), it follows from (15.49) that

$$t \geqslant T: \quad \begin{cases} \phi_r = \alpha_1 u_1 + \beta_1 u_2 + \gamma_1 u_3, \\ \psi_s = \alpha_2 u_1 + \beta_2 u_2 + \gamma_2 u_3, \\ \theta_t = \alpha_3 u_1 + \beta_3 u_2 + \gamma_3 u_3. \end{cases} \tag{15.60}$$

Since e_1, e_2, e_3 are linearly independent, the linear system (15.59) is invertible. Then, the corresponding exactly synchronizable state by 3-groups $u = (u_1, u_2, u_3)^T$ can be uniquely determined by solving the linear system (15.60).

In particular, when $r = s = t = 1$, the invariant subspace Span$\{\xi_1, \eta_1, \zeta_1\}$ of A^T is bi-orthonormal to $\text{Ker}(C_3) = \text{Span}\{e_1, e_2, e_3\}$. By Theorem 15.2, there exists a boundary control matrix $D \in \mathcal{D}_{N-3}$, such that the exactly synchronizable state by 3-groups $u = (u_1, u_2, u_3)^T$ is independent of applied boundary controls.

(ii) A admits two eigenvectors f_r and g_s contained in Span$\{e_1, e_2, e_3\}$, corresponding to eigenvalues λ and μ, respectively. Let $f_1, f_2, \cdots, f_r$ and $g_1, g_2, \cdots, g_s$ be the Jordan chains corresponding to these eigenvectors of A:

$$\begin{cases} Af_i = \lambda f_i + f_{i+1}, & 1 \leqslant i \leqslant r, \quad f_{r+1} = 0, \\ Ag_j = \mu g_j + g_{j+1}, & 1 \leqslant j \leqslant s, \quad g_{s+1} = 0 \end{cases} \tag{15.61}$$

and let $\xi_1, \xi_2, \cdots, \xi_r$ and $\eta_1, \eta_2, \cdots, \eta_s$ be the Jordan chains corresponding to the related eigenvectors of A^T:

$$\begin{cases} A^T \xi_i = \lambda \xi_i + \xi_{i-1}, & 1 \leqslant i \leqslant r, \quad \xi_0 = 0, \\ A^T \eta_j = \mu \eta_j + \eta_{j-1}, & 1 \leqslant j \leqslant s, \quad \eta_0 = 0, \end{cases} \tag{15.62}$$

such that

$$(f_i, \xi_l) = \delta_{il}, \quad (g_j, \eta_m) = \delta_{jm} \tag{15.63}$$

and

$$(f_i, \eta_m) = (g_j, \xi_l) = 0 \tag{15.64}$$

for all $i, l = 1, \cdots, r$ and $j, m = 1, \cdots, s$.

Taking the inner product with ξ_i and η_j on both sides of problem (II) and (II0), respectively, and denoting

$$\phi_i = (U, \xi_i), \quad \psi_j = (U, \eta_j), \quad i = 1, \cdots, r; \; j = 1, \cdots, s, \tag{15.65}$$

we get again the reduced problems (15.55)–(15.56).

Since Span$\{e_1, e_2, e_3\}$ is A-marked, either f_r or g_s lies in Span$\{e_1, e_2, e_3\}$. For fixing idea, we assume that $f_{r-1} \in$ Ker$\{e_1, e_2, e_3\}$. Then we have

$$\begin{cases} e_1 = \alpha_1 f_r + \alpha_2 f_{r-1} + \alpha_3 g_s, \\ e_2 = \beta_1 f_r + \beta_2 f_{r-1} + \beta_3 g_s, \\ e_3 = \gamma_1 f_r + \gamma_2 f_{r-1} + \gamma_3 g_s. \end{cases} \tag{15.66}$$

Then, noting (15.63)–(15.64), it follows from (15.49) that

$$t \geqslant T: \quad \begin{cases} \phi_r = \alpha_1 u_1 + \beta_1 u_2 + \gamma_1 u_3, \\ \phi_{r-1} = \alpha_2 u_1 + \beta_2 u_2 + \gamma_2 u_3, \\ \psi_s = \alpha_3 u_1 + \beta_3 u_2 + \gamma_3 u_3. \end{cases} \tag{15.67}$$

Since e_1, e_2, e_3 are linearly independent, the linear system (15.66) is invertible, then the corresponding exactly synchronizable state by 3-groups $u = (u_1, u_2, u_3)^T$ can be determined by solving the linear system (15.67).

In particular, when $r = 2$, $s = 1$, the invariant subspace Span$\{\xi_1, \xi_2, \eta_1\}$ of A^T is bi-orthonormal to Span$\{e_1, e_2, e_3\}$. By Theorem 15.2, there exists a boundary control matrix $D \in \mathcal{D}_{N-3}$, such that the corresponding exactly synchronizable state by 3-groups $u = (u_1, u_2, u_3)^T$ is independent of applied boundary controls.

(iii) A admits only one eigenvector f_r contained in Span$\{e_1, e_2, e_3\}$, corresponding to the eigenvalue λ. Let $f_1, f_2, \cdots, f_r$ be the Jordan chain corresponding to this eigenvector of A:

$$Af_i = \lambda f_i + f_{i+1}, \quad 1 \leqslant i \leqslant r, \quad f_{r+1} = 0, \tag{15.68}$$

and let $\xi_1, \xi_2, \cdots, \xi_r$ be the Jordan chain corresponding to the related eigenvector of A^T:

$$A^T \xi_i = \lambda \xi_i + \xi_{i-1}, \quad 1 \leqslant i \leqslant r, \quad \xi_0 = 0, \tag{15.69}$$

such that

$$(f_i, \xi_l) = \delta_{il}, \quad i, l = 1, \cdots, r. \tag{15.70}$$

Taking the inner product with ξ_i on both sides of problem (II) and (II0), and denoting

$$\phi_i = (U, \xi_i), \quad i = 1, \cdots r, \tag{15.71}$$

we get again the reduced problem (15.55).

Since $\mathrm{Span}\{e_1, e_2, e_3\}$ is A-marked, f_{r-1} and f_{r-2} are necessarily contained in $\mathrm{Span}\{e_1, e_2, e_3\}$. We have

$$\begin{cases} e_1 = \alpha_1 f_r + \alpha_2 f_{r-1} + \alpha_3 f_{r-2}, \\ e_2 = \beta_1 f_r + \beta_2 f_{r-1} + \beta_3 f_{r-2}, \\ e_3 = \gamma_1 f_r + \gamma_2 f_{r-1} + \gamma_3 f_{r-2}. \end{cases} \tag{15.72}$$

Then, noting (15.70), it follows from (15.49) that

$$t \geqslant T: \quad \begin{cases} \phi_r = \alpha_1 u_1 + \beta_1 u_2 + \gamma_1 u_3, \\ \phi_{r-1} = \alpha_2 u_1 + \beta_2 u_2 + \gamma_2 u_3, \\ \phi_{r-2} = \alpha_3 u_1 + \beta_3 u_2 + \gamma_3 u_3. \end{cases} \tag{15.73}$$

Since e_1, e_2, e_3 are linearly independent, the linear system (15.72) is invertible, then the corresponding exactly synchronizable state by 3-groups $u = (u_1, u_2, u_3)^T$ can be uniquely determined by solving the linear system (15.73).

In particular, when $r = 3$, the invariant subspace $\{\xi_1, \xi_2, \xi_3\}$ of A^T is bi-orthonormal to $\mathrm{Span}\{e_1, e_2, e_3\}$. By Theorem 15.2, there exists a boundary control matrix $D \in \mathcal{D}_{N-3}$, such that the exactly synchronizable state by 3-groups $u = (u_1, u_2, u_3)^T$ is independent of applied boundary controls.

Part IV
Synchronization for a Coupled System of Wave Equations with Neumann Boundary Controls: Approximate Boundary Synchronization

In this part we will introduce the concept of approximate boundary null controllability, approximate boundary synchronization and approximate boundary synchronization by groups, and establish the corresponding theory for system (II) of wave equations with Neumann boundary controls in the case with fewer boundary controls. Moreover, we will show that Kalman's criterion of various kinds will play an important role in the discussion.

Chapter 16
Approximate Boundary Null Controllability

In this chapter, we will define the approximate boundary null controllability for system (II) and the D-observability for the adjoint problem, and show that these two concepts are equivalent to each other. Moreover, the corresponding Kalman's criterion is introduced and studied.

16.1 Definitions

For the coupled system (II) of wave equations with Neumann boundary controls, we still use the notations given in Chaps. 12 and 13.

Definition 16.1 Let $s > \frac{1}{2}$. System (II) is **approximately null controllable** at the time $T > 0$ if for any given initial data $(\widehat{U}_0, \widehat{U}_1) \in (\mathcal{H}_{1-s})^N \times (\mathcal{H}_{-s})^N$, there exists a sequence $\{H_n\}$ of boundary controls in $\mathcal{L}^M$ with compact support in $[0, T]$, such that the sequence $\{U_n\}$ of solutions to the corresponding problem (II) and (II0) satisfies

$$\big(U_n(T), U_n'(T)\big) \longrightarrow 0 \quad \text{in } (\mathcal{H}_{1-s})^N \times (\mathcal{H}_{-s})^N \quad \text{as } n \to +\infty \tag{16.1}$$

or equivalently,

$$\big(U_n, U_n'\big) \longrightarrow (0, 0) \quad \text{in } (\mathcal{H}_{1-s})^N \times (\mathcal{H}_{-s})^N \quad \text{as } n \to +\infty \tag{16.2}$$

in the space $C_{loc}([T, +\infty); (\mathcal{H}_{1-s})^N) \times C_{loc}([T, +\infty); (\mathcal{H}_{-s})^N)$.

Obviously, the exact boundary null controllability implies the **approximate boundary null controllability**. However, since we cannot get the convergence of the sequence $\{H_n\}$ of boundary controls from Definition 16.1, generally speaking, the approximate boundary null controllability does not lead to the exact boundary null controllability.

T. Li and B. Rao, *Boundary Synchronization for Hyperbolic Systems*,
Progress in Nonlinear Differential Equations and Their Applications 94,
https://doi.org/10.1007/978-3-030-32849-8_16

Let

$$\Phi = (\phi^{(1)}, \cdots, \phi^{(N)})^T.$$

Consider the following **adjoint problem**:

$$\begin{cases} \Phi'' - \Delta\Phi + A^T\Phi = 0 & \text{in } (0, +\infty) \times \Omega, \\ \Phi = 0 & \text{on } (0, +\infty) \times \Gamma_0, \\ \partial_\nu \Phi = 0 & \text{on } (0, +\infty) \times \Gamma_1, \\ t = 0: \quad \Phi = \widehat{\Phi}_0, \ \Phi' = \widehat{\Phi}_1 & \text{in } \Omega, \end{cases} \tag{16.3}$$

where A^T is the transpose of A.

As in the case with Dirichlet boundary controls (cf. Sect. 8.2), we give the following.

Definition 16.2 For $(\widehat{\Phi}_0, \widehat{\Phi}_1) \in (\mathcal{H}_s)^N \times (\mathcal{H}_{s-1})^N$ $(s > \frac{1}{2})$, the adjoint problem (16.3) is D-**observable** on $[0, T]$, if

$$D^T\Phi \equiv 0 \quad \text{on } [0, T] \times \Gamma_1 \Rightarrow (\widehat{\Phi}_0, \widehat{\Phi}_1) \equiv 0, \ \text{then } \Phi \equiv 0. \tag{16.4}$$

16.2 Equivalence Between the Approximate Boundary Null Controllability and the *D*-Observability

In order to find the equivalence between the approximate boundary null controllability of the original system (II) and the D-observability of the adjoint problem (16.3), let $\mathcal{C}$ be the set of all the initial states $(V(0), V'(0))$ of the following backward problem:

$$\begin{cases} V'' - \Delta V + AV = 0 & \text{in } (0, T) \times \Omega, \\ V = 0 & \text{on } (0, T) \times \Gamma_0, \\ \partial_\nu V = DH & \text{on } (0, T) \times \Gamma_1, \\ V(T) = 0, \quad V'(T) = 0 & \text{in } \Omega \end{cases} \tag{16.5}$$

with all admissible boundary controls $H \in \mathcal{L}^M$.

By Proposition 12.12, we have the following.

Lemma 16.3 *For any given $T > 0$ and any given $(V(T), V'(T)) \in (\mathcal{H}_{1-s})^N \times (\mathcal{H}_{-s})^N$ with $s > \frac{1}{2}$, and for any given boundary function H in $\mathcal{L}^M$, the backward problem (16.5) (in which the null final condition is replaced by the given final data) admits a unique weak solution such that*

$$V \in C^0([0, T]; (\mathcal{H}_{1-s})^N) \cap C^1([0, T]; (\mathcal{H}_{-s})^N). \tag{16.6}$$

Lemma 16.4 *System (II) possesses the approximate boundary null controllability if and only if*

$$\overline{\mathcal{C}} = (\mathcal{H}_{1-s})^N \times (\mathcal{H}_{-s})^N. \tag{16.7}$$

Proof Assume that $\overline{\mathcal{C}} = (\mathcal{H}_{1-s})^N \times (\mathcal{H}_{-s})^N$. By the definition of $\mathcal{C}$, for any given $(\widehat{U}_0, \widehat{U}_1) \in (\mathcal{H}_{1-s})^N \times (\mathcal{H}_{-s})^N$, there exists a sequence $\{H_n\}$ of boundary controls in $\mathcal{L}^M$ with compact support in $[0, T]$, such that the sequence $\{V_n\}$ of solutions to the corresponding backward problem (16.5) satisfies

$$\big(V_n(0), V_n'(0)\big) \to (\widehat{U}_0, \widehat{U}_1) \text{ in } (\mathcal{H}_{1-s})^N \times (\mathcal{H}_{-s})^N \text{ as } n \to +\infty. \tag{16.8}$$

Recall that

$$\mathcal{R}: \quad (\widehat{U}_0, \widehat{U}_1, H) \to (U, U') \tag{16.9}$$

is the continuous linear mapping defined by (12.73). We have

$$\begin{aligned} &\mathcal{R}(\widehat{U}_0, \widehat{U}_1, H_n) \\ =&\mathcal{R}(\widehat{U}_0 - V_n(0), \widehat{U}_1 - V_n'(0), 0) + \mathcal{R}(V_n(0), V_n'(0), H_n). \end{aligned} \tag{16.10}$$

On the other hand, by the definition of V_n, we have

$$\mathcal{R}(V_n(0), V_n'(0), H_n)(T) = 0. \tag{16.11}$$

Therefore

$$\mathcal{R}(\widehat{U}_0, \widehat{U}_1, H_n)(T) = \mathcal{R}(\widehat{U}_0 - V_n(0), \widehat{U}_1 - V_n'(0), 0)(T). \tag{16.12}$$

By Proposition 12.12 and noting (16.8), we then get

$$\begin{aligned} &\|\mathcal{R}(\widehat{U}_0, \widehat{U}_1, H_n)(T)\|_{(\mathcal{H}_{1-s})^N \times (\mathcal{H}_{-s})^N} \\ \leqslant & c\|(\widehat{U}_0 - V_n(0), \widehat{U}_1 - V_n'(0))\|_{(\mathcal{H}_{1-s})^N \times (\mathcal{H}_{-s})^N} \to 0 \end{aligned} \tag{16.13}$$

as $n \to +\infty$. Here and hereafter, c always denotes a positive constant. Thus, system (II) is approximately null controllable.

Inversely, assume that system (II) is approximately null controllable. For any given $(\widehat{U}_0, \widehat{U}_1) \in (\mathcal{H}_{1-s})^N \times (\mathcal{H}_{-s})^N$, there exists a sequence $\{H_n\}$ of boundary controls in $\mathcal{L}^M$ with compact support in $[0, T]$, such that the sequence $\{U_n\}$ of solutions to the corresponding problem (II) and (II0) satisfies

$$\big(U_n(T), U_n'(T)\big) = \mathcal{R}(\widehat{U}_0, \widehat{U}_1, H_n)(T) \to (0, 0) \tag{16.14}$$

in $(\mathcal{H}_{1-s})^N \times (\mathcal{H}_{-s})^N$ as $n \to +\infty$. Taking such H_n as the boundary control, we solve the backward problem (16.5) and get the corresponding solution V_n. By the linearity of the mapping $\mathcal{R}$, we have

$$\begin{aligned}&\mathcal{R}(\widehat{U}_0, \widehat{U}_1, H_n) - \mathcal{R}(V_n(0), V_n'(0), H_n)\\=&\mathcal{R}(\widehat{U}_0 - V_n(0), \widehat{U}_1 - V_n'(0), 0).\end{aligned} \tag{16.15}$$

By Lemma 16.3 and noting (16.14), we have

$$\begin{aligned}&\|\mathcal{R}(\widehat{U}_0 - V_n(0), \widehat{U}_1 - V_n'(0), 0)(0)\|_{(\mathcal{H}_{1-s})^N \times (\mathcal{H}_{-s})^N}\\ \leqslant& c\|(U_n(T) - V_n(T), U_n'(T) - V_n'(T)\|_{(\mathcal{H}_{1-s})^N \times (\mathcal{H}_{-s})^N}\\ =&c\|(U_n(T), U_n'(T))\|_{(\mathcal{H}_{1-s})^N \times (\mathcal{H}_{-s})^N} \to 0 \quad \text{as } n \to +\infty.\end{aligned} \tag{16.16}$$

Noting (16.15), we then get

$$\begin{aligned}&\|(\widehat{U}_0, \widehat{U}_1) - (V_n(0), V_n'(0))\|_{(\mathcal{H}_{1-s})^N \times (\mathcal{H}_{-s})^N}\\ =&\|\mathcal{R}(\widehat{U}_0, \widehat{U}_1, H_n)(0) - \mathcal{R}(V_n(0), V_n'(0), H_n)(0)\|_{(\mathcal{H}_{1-s})^N \times (\mathcal{H}_{-s})^N}\\ \leqslant& c\|\mathcal{R}(\widehat{U}_0 - V_n(0), \widehat{U}_1 - V_n'(0), 0)(0)\|_{(\mathcal{H}_{1-s})^N \times (\mathcal{H}_{-s})^N} \to 0\end{aligned} \tag{16.17}$$

as $n \to +\infty$, which shows that $\overline{\mathcal{C}} = (\mathcal{H}_{1-s})^N \times (\mathcal{H}_{-s})^N$. □

Theorem 16.5 *System (II) is approximately null controllable at the time $T > 0$ if and only if the adjoint problem* (16.3) *is D-observable on* $[0, T]$.

Proof Assume that system (II) is not approximately null controllable at the time $T > 0$. By Lemma 16.4, there is a nontrivial vector $(-\widehat{\Phi}_1, \widehat{\Phi}_0) \in \mathcal{C}^\perp$. Here and hereafter, the orthogonal complement space is always defined in the sense of duality. Thus, $(-\widehat{\Phi}_1, \widehat{\Phi}_0) \in (\mathcal{H}_{s-1})^N \times (\mathcal{H}_s)^N$. Taking $(\widehat{\Phi}_0, \widehat{\Phi}_1)$ as the initial data, we solve the adjoint problem (16.3) and get the solution $\Phi \not\equiv 0$. Multiplying Φ on both sides of the backward problem (16.5) and integrating by parts, we get

$$\begin{aligned}&\langle V(0), \widehat{\Phi}_1\rangle_{(\mathcal{H}_{1-s})^N;(\mathcal{H}_{s-1})^N} - \langle V'(0), \widehat{\Phi}_0\rangle_{(\mathcal{H}_{-s})^N;(\mathcal{H}_s)^N}\\ =&\int_0^T \int_{\Gamma_1} (DH, \Phi) d\Gamma dt.\end{aligned} \tag{16.18}$$

The right-hand side of (16.18) is meaningful due to

$$\Phi \in \left(C^0([0, T]; \mathcal{H}_s)\right)^N \subset \left(L^2(0, T; L^2(\Gamma_1))\right)^N, \quad s > \frac{1}{2}. \tag{16.19}$$

Noticing $(V(0), V'(0)) \in \mathcal{C}$ and $(-\widehat{\Phi}_1, \widehat{\Phi}_0) \in \mathcal{C}^\perp$, it is easy to see from (16.18) that for any given H in $\mathcal{L}^M$, we have

$$\int_0^T \int_{\Gamma_1} (DH, \Phi) d\Gamma dt = 0.$$

Then, it follows that

$$D^T \Phi \equiv 0 \quad \text{on } [0, T] \times \Gamma_1. \tag{16.20}$$

But $\Phi \not\equiv 0$, which implies that the adjoint problem (16.3) is not D-observable on $[0, T]$.

Inversely, assume that the adjoint problem (16.3) is not D-observable on $[0, T]$, then there exists a nontrivial initial data $(\widehat{\Phi}_0, \widehat{\Phi}_1) \in (\mathcal{H}_s)^N \times (\mathcal{H}_{s-1})^N$, such that the solution Φ to the corresponding adjoint problem (16.3) satisfies (16.20). For any given $(\widehat{U}_0, \widehat{U}_1) \in \overline{\mathcal{C}}$, there exists a sequence $\{H_n\}$ of boundary controls in $\mathcal{L}^M$, such that the solution V_n to the corresponding backward problem (16.5) satisfies

$$(V_n(0), V_n'(0)) \to (\widehat{U}_0, \widehat{U}_1) \quad \text{in } (\mathcal{H}_{1-s})^N \times (\mathcal{H}_{-s})^N \tag{16.21}$$

as $n \to +\infty$. Similarly to (16.18), multiplying Φ on both sides of the backward problem (16.5) and noting (16.20), we get

$$\langle V_n(0), \widehat{\Phi}_1 \rangle_{(\mathcal{H}_{1-s})^N;(\mathcal{H}_{s-1})^N} - \langle V_n'(0), \widehat{\Phi}_0 \rangle_{(\mathcal{H}_{-s})^N;(\mathcal{H}_s)^N} = 0. \tag{16.22}$$

Then, taking $n \to +\infty$ and noting (16.21), it follows from (16.22) that

$$\langle (\widehat{U}_0, \widehat{U}_1), (-\widehat{\Phi}_1, \widehat{\Phi}_0) \rangle_{(\mathcal{H}_{1-s})^N \times (\mathcal{H}_{-s})^N;(\mathcal{H}_{s-1})^N \times (\mathcal{H}_s)^N} = 0 \tag{16.23}$$

for all $(\widehat{U}_0, \widehat{U}_1) \in \overline{\mathcal{C}}$, which indicates that $(-\widehat{\Phi}_1, \widehat{\Phi}_0) \in \overline{\mathcal{C}}^\perp$, thus $\overline{\mathcal{C}} \neq (\mathcal{H}_{1-s})^N \times (\mathcal{H}_{-s})^N$. □

Theorem 16.6 *If for any given initial data $(\widehat{U}_0, \widehat{U}_1) \in (\mathcal{H}_1)^N \times (\mathcal{H}_0)^N$, system (II) is approximately null controllable for some s $(> \frac{1}{2})$, then for any given initial data $(\widehat{U}_0, \widehat{U}_1) \in (\mathcal{H}_{1-s})^N \times (\mathcal{H}_{-s})^N$, system (II) possesses the same approximate boundary null controllability, too.*

Proof For any given initial data $(\widehat{U}_0, \widehat{U}_1) \in (\mathcal{H}_{1-s})^N \times (\mathcal{H}_{-s})^N$ $(s > \frac{1}{2})$, by the density of $(\mathcal{H}_1)^N \times (\mathcal{H}_0)^N$ in $(\mathcal{H}_{1-s})^N \times (\mathcal{H}_{-s})^N$, we can find a sequence $\{(\widehat{U}_0^n, \widehat{U}_1^n)\}_{n\in\mathbb{N}}$ in $(\mathcal{H}_1)^N \times (\mathcal{H}_0)^N$, satisfying

$$(\widehat{U}_0^n, \widehat{U}_1^n) \to (\widehat{U}_0, \widehat{U}_1) \quad \text{in} \quad (\mathcal{H}_{1-s})^N \times (\mathcal{H}_{-s})^N \tag{16.24}$$

as $n \to +\infty$. By the assumption, for any fixed $n \geqslant 1$, there exists a sequence $\{H_k^n\}_{k\in\mathbb{N}}$ of boundary controls in $\mathcal{L}^M$ with compact support in $[0, T]$, such that the sequence of solutions $\{U_k^n\}$ to the corresponding problem (II) and (II0) satisfies

$$(U_k^n(T), (U_k^n)'(T)) \to (0, 0) \quad \text{in} \quad (\mathcal{H}_{1-s})^N \times (\mathcal{H}_{-s})^N \tag{16.25}$$

as $k \to +\infty$. For any given $n \geqslant 1$, let k_n be an integer such that

$$\begin{aligned} &\|\mathcal{R}(\widehat{U}_0^n, \widehat{U}_1^n, H_{k_n}^n)(T)\|_{(\mathcal{H}_{1-s})^N \times (\mathcal{H}_{-s})^N} \\ =&\|(U_{k_n}^n(T), (U_{k_n}^n)'(T))\|_{(\mathcal{H}_{1-s})^N \times (\mathcal{H}_{-s})^N} \leqslant \frac{1}{2^n}. \end{aligned} \tag{16.26}$$

Thus, we get a sequence $\{k_n\}$ with $k_n \to +\infty$ as $n \to +\infty$. For the sequence $\{H_{k_n}^n\}$ of boundary controls in $\mathcal{L}^M$, we have

$$\mathcal{R}(\widehat{U}_0^n, \widehat{U}_1^n, H_{k_n}^n)(T) \to 0 \text{ in } (\mathcal{H}_{1-s})^N \times (\mathcal{H}_{-s})^N \tag{16.27}$$

as $n \to +\infty$. Therefore, by the linearity of $\mathcal{R}$, the combination of (16.24) and (16.27) gives

$$\begin{aligned} &\mathcal{R}(\widehat{U}_0, \widehat{U}_1, H_{k_n}^n) \\ =&\mathcal{R}(\widehat{U}_0 - \widehat{U}_0^n, \widehat{U}_1 - \widehat{U}_1^n, 0) + \mathcal{R}(\widehat{U}_0^n, \widehat{U}_1^n, H_{k_n}^n) \to (0, 0) \end{aligned} \tag{16.28}$$

in $(\mathcal{H}_{1-s})^N \times (\mathcal{H}_{-s})^N$ as $n \to +\infty$, which indicates that the sequence $\{H_{k_n}^n\}$ of boundary controls realizes the approximate boundary null controllability for any given initial data $(\widehat{U}_0, \widehat{U}_1) \in (\mathcal{H}_{1-s})^N \times (\mathcal{H}_{-s})^N$. □

Remark 16.7 Since $(\mathcal{H}_1)^N \times (\mathcal{H}_0)^N \subset (\mathcal{H}_{1-s})^N \times (\mathcal{H}_{-s})^N$ $(s > \frac{1}{2})$ if system (II) is approximately null controllable for any given initial data $(\widehat{U}_0, \widehat{U}_1) \in (\mathcal{H}_{1-s})^N \times (\mathcal{H}_{-s})^N$ $(s > \frac{1}{2})$, then it possesses the same approximate boundary null controllability for any given initial data $(\widehat{U}_0, \widehat{U}_1) \in (\mathcal{H}_1)^N \times (\mathcal{H}_0)^N$, too.

Remark 16.8 Theorem 16.6 and Remark 16.7 indicate that for system (II), the approximate boundary null controllability for the initial data in $(\mathcal{H}_{1-s})^N \times (\mathcal{H}_{-s})^N$ $(s > \frac{1}{2})$ is equivalent to the approximate boundary null controllability for the initial data in $(\mathcal{H}_1)^N \times (\mathcal{H}_0)^N$ with the same convergence space $(\mathcal{H}_{1-s})^N \times (\mathcal{H}_{-s})^N$.

Remark 16.9 In a similar way, we can prove that if for any given initial data $(\widehat{U}_0, \widehat{U}_1) \in (\mathcal{H}_{1-s})^N \times (\mathcal{H}_{-s})^N$ $(s > \frac{1}{2})$, system (II) is approximately null controllable, then for any given initial data $(\widehat{U}_0, \widehat{U}_1) \in (\mathcal{H}_{1-s'})^N \times (\mathcal{H}_{-s'})^N$ $(s' > s)$, system (II) is still approximately null controllable.

Corollary 16.10 *If $M = N$, then, no matter whether the multiplier geometrical condition (12.3) is satisfied or not, system (II) is always approximately null controllable.*

Proof Since $M = N$, the observations given by (16.4) become

$$\Phi \equiv 0 \quad \text{on } [0, T] \times \Gamma_1. \tag{16.29}$$

By Holmgren's uniqueness theorem (cf. Theorem 8.2 in [62]), we then get the D-observability of the adjoint problem (16.3). Hence, by Theorem 16.5, we get the approximate boundary null controllability of system (II). □

16.3 Kalman's Criterion. Total (Direct and Indirect) Controls

By Theorem 16.5, similarly to Theorem 8.9, we can give the following necessary condition for the approximate boundary null controllability.

Theorem 16.11 *If system (II) is approximately null controllable at the time $T > 0$, then we have necessarily the following* **Kalman's criterion***:*

$$rank(D, AD, \cdots, A^{N-1}D) = N. \tag{16.30}$$

By Theorem 16.11, as in the case with Dirichlet boundary controls (cf. Sect. 8.3), for Neumann boundary controls, in Part IV, we should consider not only the number of direct boundary controls acting on Γ_1, which is equal to the rank of D, but also the number of total controls, given by the rank of the enlarged matrix $(D, AD, \cdots, A^{N-1}D)$, composed of the coupling matrix A and the boundary control matrix D, which should be equal to N in the case of approximate boundary null controllability. Therefore, the number of indirect controls should be equal to $\text{rank}(D, AD, \cdots, A^{N-1}D) - \text{rank}(D) = N - M$.

Generally speaking, similarly to the approximate boundary null controllability for a coupled system of wave equations with Dirichlet boundary controls, Kalman's criterion is not sufficient in general. The reason is that Kalman's criterion does not depend on T, then if it is sufficient, the approximate boundary null controllability of the original system (II), or the D-observability of the adjoint problem (16.3) could be immediately realized; however, this is impossible since the wave propagates with a finite speed.

First of all, we will give an example to show the insufficiency of Kalman's criterion.

Theorem 16.12 *Let μ_n^2 and e_n be defined by*

$$\begin{cases} -\Delta e_n = \mu_n^2 e_n & in\ \Omega, \\ e_n = 0 & on\ \Gamma_0, \\ \partial_\nu e_n = 0 & on\ \Gamma_1. \end{cases} \tag{16.31}$$

Assume that the set

$$\Lambda = \{(m, n): \quad \mu_n \neq \mu_m, \quad e_m = e_n \ \ on\ \Gamma_1\} \tag{16.32}$$

is not empty. For any given $(m, n) \in \Lambda$, setting

$$\epsilon = \frac{\mu_m^2 - \mu_n^2}{2}, \tag{16.33}$$

the adjoint problem

$$\begin{cases} \phi'' - \Delta\phi + \epsilon\psi = 0 & in\ (0, +\infty) \times \Omega, \\ \psi'' - \Delta\psi + \epsilon\phi = 0 & in\ (0, +\infty) \times \Omega, \\ \phi = \psi = 0 & on\ (0, +\infty) \times \Gamma_0, \\ \partial_\nu\phi = \partial_\nu\psi = 0 & on\ (0, +\infty) \times \Gamma_1 \end{cases} \tag{16.34}$$

admits a nontrivial solution $(\phi, \psi) \not\equiv (0, 0)$ *with the observation on the infinite time interval*

$$\phi \equiv 0 \quad on\ [0 + \infty) \times \Gamma_1, \tag{16.35}$$

hence the corresponding D-observability of the adjoint problem (16.34) *fails*

Proof Let

$$\phi = (e_n - e_m), \quad \psi = (e_n + e_m), \quad \lambda^2 = \frac{\mu_m^2 + \mu_n^2}{2}. \tag{16.36}$$

It is easy to check that (ϕ, ψ) satisfies the following system:

$$\begin{cases} \lambda^2\phi + \Delta\phi - \epsilon\psi = 0 & \text{in } (0, +\infty) \times \Omega, \\ \lambda^2\psi + \Delta\psi - \epsilon\phi = 0 & \text{in } (0, +\infty) \times \Omega, \\ \phi = \psi = 0 & \text{on } (0, +\infty) \times \Gamma_0, \\ \partial_\nu\phi = \partial_\nu\psi = 0 & \text{on } (0, +\infty) \times \Gamma_1. \end{cases} \tag{16.37}$$

Moreover, noting the definition (16.32) of Λ, we have

$$\phi = 0 \quad \text{on } \Gamma_1. \tag{16.38}$$

Then, let

$$\phi_\lambda = e^{i\lambda t}\phi, \quad \psi_\lambda = e^{i\lambda t}\psi. \tag{16.39}$$

It is easy to see that $(\phi_\lambda, \psi_\lambda)$ is a nontrivial solution to the adjoint system (16.34), which satisfies condition (16.35). □

In order to illustrate the validity of the assumptions given in Theorem 16.12, we may examine the following situations, in which the set Λ is indeed not empty.

(1) $\Omega = (0, \pi)$, $\Gamma_1 = \{\pi\}$. In this case, we have

$$\mu_n = n + \frac{1}{2}, \quad e_n = (-1)^n \sin(n + \frac{1}{2})x, \quad e_n(\pi) = e_m(\pi) = 1. \tag{16.40}$$

Thus, $(m, n) \in \Lambda$ for all $m \neq n$.

(2) $\Omega = (0, \pi) \times (0, \pi)$, $\Gamma_1 = \{\pi\} \times [0, \pi]$. Let

$$\mu_{m,n} = \sqrt{(m + \frac{1}{2})^2 + n^2}, \quad e_{m,n} = (-1)^m \sin(m + \frac{1}{2})x \sin ny. \tag{16.41}$$

We have

$$e_{m,n}(\pi, y) = e_{m',n}(\pi, y) = \sin ny, \quad 0 \leqslant y \leqslant \pi. \tag{16.42}$$

Thus, $(\{m, n\}, \{m', n\}) \in \Lambda$ for all $m \neq m'$ and $n \geqslant 1$.

Remark 16.13 Theorem 16.12 implies that Kalman's criterion (16.30) is not sufficient in general. As a matter of fact, for the adjoint system (16.34) which satisfies the condition of observation (16.35), we have $N = 2$,

$$A = \begin{pmatrix} 0 & \epsilon \\ \epsilon & 0 \end{pmatrix}, \quad D = \begin{pmatrix} 1 \\ 0 \end{pmatrix}, \quad (D, AD) = \begin{pmatrix} 1 & 0 \\ 0 & \epsilon \end{pmatrix}, \tag{16.43}$$

and the corresponding Kalman's criterion (16.30) is satisfied. Theorem 16.12 shows that Kalman's criterion cannot guarantee the D-observability of the adjoint system (16.34) even the observation is given on the infinite time interval $[0, +\infty)$.

16.4 Sufficiency of Kalman's Criterion for $T > 0$ Large Enough in the One-Space-Dimensional Case

Similarly to the situation with Dirichlet boundary controls in the one-space-dimensional case, we will show that Kalman's criterion (16.30) is also sufficient in certain cases for the approximate boundary null controllability of the original system.

Now consider the following one-space-dimensional adjoint problem:

$$\begin{cases} \Phi'' - \Delta\Phi + \epsilon A^T \Phi = 0, & t > 0, \quad 0 < x < \pi, \\ \Phi(t, 0) = 0, & t > 0, \\ \partial_\nu \Phi(t, \pi) = 0, & t > 0, \\ t = 0: \quad \Phi = \widehat{\Phi}_0, \ \Phi' = \widehat{\Phi}_1, & 0 < x < \pi \end{cases} \tag{16.44}$$

with the observation on $x = \pi$:

$$D^T \Phi(t, \pi) = 0 \quad \text{on } [0, T], \tag{16.45}$$

where $|\epsilon| > 0$ is small enough.

Assume that the $N \times N$ matrix A^T is diagonalizable with m ($\leqslant N$) distinct real eigenvalues:

$$\delta_1 < \delta_2 < \cdots < \delta_m \tag{16.46}$$

and the corresponding eigenvectors $w^{(l,\mu)}$:

$$A^T w^{(l,\mu)} = \delta_l w^{(l,\mu)}, \quad 1 \leqslant l \leqslant m, \quad 1 \leqslant \mu \leqslant \mu_l, \tag{16.47}$$

we have

$$\sum_{l=1}^{m} \mu_l = N. \tag{16.48}$$

Let

$$e_n = (-1)^n \sin(n + \frac{1}{2})x, \quad n \geqslant 0. \tag{16.49}$$

e_n is an eigenfunction of $-\Delta$ in $\mathcal{H}_1$, satisfying

$$\begin{cases} -\Delta e_n = \mu_n^2 e_n, & \text{in } 0 < x < \pi, \\ e_n = 0 & \text{on } x = 0, \\ \partial_\nu e_n = 0 & \text{on } x = \pi, \end{cases} \tag{16.50}$$

in which $\mu_n = (n + \frac{1}{2})$. Thus, $e_n w^{(l,\mu)}$ is an eigenvector of $(-\Delta + \epsilon A^T)$, corresponding to the eigenvalue $(n + \frac{1}{2})^2 + \delta_l \epsilon$.

Define

$$\begin{cases} \beta_n^{(l)} = \sqrt{(n + \frac{1}{2})^2 + \delta_l \epsilon}, & l = 1, 2, \cdots, m, \quad n \geqslant 0, \\ \beta_{-n}^{(l)} = -\beta_n^{(l)}, & l = 1, 2, \cdots, m, \quad n \geqslant 1. \end{cases} \tag{16.51}$$

The eigenvectors associated with system (16.44) are given by

$$E_n^{(l,\mu)} = \begin{pmatrix} \frac{e_n w^{(l,\mu)}}{i\beta_n^{(l)}} \\ e_n w^{(l,\mu)} \end{pmatrix}, \quad 1 \leqslant l \leqslant m, \quad 1 \leqslant \mu \leqslant \mu_l, \quad n \in \mathbb{Z}, \tag{16.52}$$

where we define $e_{-n} = e_n$ for all $n \geqslant 0$. Similarly to Sect. 8.7, we can show that $\{E_n^{(l,\mu)}\}_{1 \leqslant l \leqslant m, 1 \leqslant \mu \leqslant \mu_l, n \in \mathbb{Z}}$ forms a Riesz basis in $(\mathcal{H}_1)^N \times (\mathcal{H}_0)^N$. Then, for any given initial data in $(\mathcal{H}_1)^N \times (\mathcal{H}_0)^N$:

$$\begin{pmatrix} \Phi_0 \\ \Phi_1 \end{pmatrix} = \sum_{n \in \mathbb{Z}} \sum_{l=1}^{m} \sum_{\mu=1}^{\mu_l} \alpha_n^{(l,\mu)} E_n^{(l,\mu)}, \tag{16.53}$$

the solution to the joint problem (16.44) is given by

$$\begin{pmatrix} \Phi \\ \Phi' \end{pmatrix} = \sum_{n\in\mathbb{Z}} \sum_{l=1}^{m} \sum_{\mu=1}^{\mu_l} \alpha_n^{(l,\mu)} e^{i\beta_n^{(l)}t} E_n^{(l,\mu)}. \tag{16.54}$$

In particular, we have

$$\Phi = \sum_{n\in\mathbb{Z}} \sum_{l=1}^{m} \sum_{\mu=1}^{\mu_l} \frac{\alpha_n^{(l,\mu)}}{i\beta_n^{(l)}} e^{i\beta_n^{(l)}t} e_n w^{(l,\mu)}, \tag{16.55}$$

and the condition of D-observation (16.45) leads to

$$\sum_{n\in\mathbb{Z}} \sum_{l=1}^{m} D^T \Big(\sum_{\mu=1}^{\mu_l} \frac{\alpha_n^{(l,\mu)}}{i\beta_n^{(l)}} w^{(l,\mu)} \Big) e^{i\beta_n^{(l)}t} = 0 \quad \text{on } [0, T]. \tag{16.56}$$

Now let us examine the sequence $\{\beta_n^{(l)}\}_{1\leqslant l\leqslant m; n\in\mathbb{Z}}$ defined by (16.51):

$$\cdots \beta_{-1}^{(1)} < \cdots < \beta_{-1}^{(m)} < \beta_0^{(1)} < \cdots < \beta_0^{(m)} < \beta_1^{(1)} < \cdots < \beta_1^{(m)} < \cdots. \tag{16.57}$$

First, for $|\epsilon| > 0$ small enough, the sequence $\{\beta_n^{(l)}\}_{1\leqslant l\leqslant m; n\in\mathbb{Z}}$ is strictly increasing. On the other hand, for $|\epsilon| > 0$ small enough and for $|n| > 0$ large enough, similarly to (8.106) and (8.107), we check easily that

$$\beta_{n+1}^{(l)} - \beta_n^{(l)} = O(1) \tag{16.58}$$

and

$$\beta_n^{(l+1)} - \beta_n^{(l)} = o\Big(\Big|\frac{\epsilon}{n}\Big|\Big). \tag{16.59}$$

Thus, the sequence $\{\beta_n^{(l)}\}_{1\leqslant l\leqslant m; n\in\mathbb{Z}}$ satisfies all the assumptions given in Theorem 8.22 (in which $s = 1$). Moreover, by definition given in (8.93), a computation shows $D^+ = m$. Then the sequence $\{e^{i\beta_n^{(l)}t}\}_{1\leqslant l\leqslant m; n\in\mathbb{Z}}$ is ω-linearly independent in $L^2(0, T)$, provided that $T > 2m\pi$.

Theorem 16.14 *Assume that A and D satisfy Kalman's criterion* (16.30). *Assume furthermore that A^T is diagonalizable with* (16.46)–(16.47). *Then the adjoint problem* (16.44) *is D-observable for $|\epsilon| > 0$ small enough, provided that $T > 2m\pi$.*

Proof Since the sequence $\{e^{i\beta_n^{(l)}t}\}_{1\leqslant l\leqslant m; n\in\mathbb{Z}}$ is ω-linearly independent in $L^2(0, T)$ as $T > 2m\pi$, it follows from (16.56) that

$$D^T \Big(\sum_{\mu=1}^{\mu_l} \frac{\alpha_n^{(l,\mu)}}{i\beta_n^{(l)}} w^{(l,\mu)} \Big) = 0, \quad 1 \leqslant l \leqslant m, \quad n \in \mathbb{Z}. \tag{16.60}$$

Noting Kalman's criterion (16.30), by Proposition 2.12 (ii) with $d = 0$, $\text{Ker}(D^T)$ does not contain any nontrivial invariant subspace of A^T, then it follows that

$$\sum_{\mu=1}^{\mu_l} \frac{\alpha_n^{(l,\mu)}}{i\beta_n^{(l)}} w^{(l,\mu)} = 0, \quad 1 \leqslant l \leqslant m, \quad n \in \mathbb{Z}. \tag{16.61}$$

Then

$$\alpha_n^{(l,\mu)} = 0, \quad 1 \leqslant \mu \leqslant \mu_l, \quad 1 \leqslant l \leqslant m, \quad n \in \mathbb{Z}, \tag{16.62}$$

hence, $\Phi \equiv 0$, namely, the adjoint problem (16.44) is D-observable. □

Moreover, similarly to Theorem 8.25, we have

Theorem 16.15 *Under the assumptions of Theorem 16.14, assume furthermore that A^T possesses N distinct real eigenvalues:*

$$\delta_1 < \delta_2 < \cdots < \delta_N. \tag{16.63}$$

Then the adjoint problem (16.44) is D-observable for $|\epsilon| > 0$ small enough, provided that $T > 2\pi(N - M + 1)$, where $M = rank(D)$.

16.5 Unique Continuation for a Cascade System of Two Wave Equations

Consider the following problem for the cascade system of two wave equations:

$$\begin{cases} \phi'' - \Delta\phi = 0 & \text{in } (0, T) \times \Omega, \\ \psi'' - \Delta\psi + \phi = 0 & \text{in } (0, T) \times \Omega, \\ \phi = \psi = 0 & \text{on } (0, T) \times \Gamma_0, \\ \partial_\nu\phi = \partial_\nu\psi = 0 & \text{on } (0, T) \times \Gamma_1, \\ t = 0: \quad (\phi, \psi, \phi', \psi') = (\phi_0, \psi_0, \phi_1, \psi_1) & \text{in } \Omega \end{cases} \tag{16.64}$$

with the Dirichlet observation

$$a\phi + b\psi = 0 \quad \text{on } [0, T] \times \Gamma_1, \tag{16.65}$$

where a and b are constants.

Let

$$A = \begin{pmatrix} 0 & 1 \\ 0 & 0 \end{pmatrix} \quad \text{and} \quad D = \begin{pmatrix} a \\ b \end{pmatrix}. \tag{16.66}$$

It is easy to see that Kalman's criterion (16.30) holds if and only if $b \neq 0$. Moreover, we have (cf. [2])

Theorem 16.16 *Let $b \neq 0$. If the solution (ϕ, ψ) to adjoint problem (16.64) with the initial data $(\phi_0, \psi_0, \phi_1, \psi_1) \in H^1_{\Gamma_0}(\Omega) \times H^1_{\Gamma_0}(\Omega) \times L^2(\Omega) \times L^2(\Omega)$ satisfies the Dirichlet observation (16.65), then $\phi \equiv \psi \equiv 0$, provided that $T > 0$ is large enough.*

Theorem 16.16 can be generalized to the following system of n blocks of cascade systems of two wave equations:

$$\begin{cases} \phi_j'' - \Delta\phi_j = 0 & \text{in } (0, T) \times \Omega, \\ \psi_j'' - \Delta\psi_j + \phi_j = 0 & \text{in } (0, T) \times \Omega, \\ \phi_j = \psi_j = 0 & \text{on } (0, T) \times \Gamma_0, \\ \partial_\nu\phi_j = \partial_\nu\psi_j = 0 & \text{on } (0, T) \times \Gamma_1 \end{cases} \tag{16.67}$$

for $j = 1, \ldots, n$ with the Dirichlet observations

$$\sum_{j=1}^{n} (a_{ji}\phi_j + b_{ji}\psi_j) = 0 \quad \text{on } [0, T] \times \Gamma_1, \quad i = 1, \cdots, n. \tag{16.68}$$

Let $N = 2n$ and

$$\Phi = (\phi_1, \psi_1, \cdots \phi_n, \psi_n)^T, \tag{16.69}$$

and let

$$A = \begin{pmatrix} \begin{pmatrix} 0 & 1 \\ 0 & 0 \end{pmatrix} & & \\ & \ddots & \\ & & \begin{pmatrix} 0 & 1 \\ 0 & 0 \end{pmatrix} \end{pmatrix} \tag{16.70}$$

and

$$D = \begin{pmatrix} a_{11} & a_{12} & \cdots & a_{1n} \\ b_{11} & b_{12} & \cdots & b_{1n} \\ \vdots & \vdots & \cdots & \vdots \\ a_{n1} & a_{n2} & \cdots & a_{nn} \\ b_{n1} & b_{n2} & \cdots & b_{nn} \end{pmatrix}. \tag{16.71}$$

Setting (16.67) and (16.68) into the forms of (16.44) and (16.45), a straightforward computation shows that Kalman's criterion (16.30) holds if and only if the matrix $B = (b_{ij})$ is invertible. Moreover, we have the following

Corollary 16.17 *Assume that the matrix $B = (b_{ij})$ is invertible. If the solution $\{(\phi_i, \psi_i)\}$ to system (16.67) with $(\phi_{i0}, \psi_{i0}) \in H^1_{\Gamma_0} \times L^2(\Omega)$ for $i = 1, \ldots, n$ satisfies the observations (16.68), then $\phi_i \equiv \psi_i \equiv 0$ for $i = 1, \ldots, n$, provided that $T > 0$ is large enough.*

Proof For $i = 1, \cdots, n$, multiplying the first j-th equation of (16.67) by a_{ji} and the second j-th equation of (16.67) by b_{ji}, respectively, then summing up for j from 1 to n, we get

$$\sum_{j=1}^{N}(a_{ji}\phi_j + b_{ji}\psi_j)'' - \sum_{j=1}^{N}\Delta(a_{ji}\phi_j + b_{ji}\psi_j) + \sum_{j=1}^{N} b_{ji}\phi_j = 0 \tag{16.72}$$

for $i = 1, \cdots, n$.

Let

$$u_i = \sum_{j=1}^{N} b_{ji}\phi_j, \quad v_i = \sum_{j=1}^{N}(a_{ji}\phi_j + b_{ji}\psi_j), \quad i = 1, \cdots, n. \tag{16.73}$$

It follows from (16.72) that for each $i = 1, \cdots, n$, we have

$$\begin{cases} u_i'' - \Delta u_i = 0 & \text{in } (0, T) \times \Omega, \\ v_i'' - \Delta v_i + u_i = 0 & \text{in } (0, T) \times \Omega, \\ u_i = v_i = 0 & \text{on } (0, T) \times \Gamma_0, \\ \partial_\nu u_i = \partial_\nu v_i = 0 & \text{on } (0, T) \times \Gamma_1 \end{cases} \tag{16.74}$$

with the Dirichlet observations

$$v_i \equiv 0 \quad \text{on } [0, T] \times \Gamma_1. \tag{16.75}$$

Applying Theorem 16.16 to each sub-system (16.74) with (16.75), it follows that

$$u_i \equiv v_i \equiv 0 \quad \text{in } [0, T] \times \Omega, \quad i = 1, \cdots, n. \tag{16.76}$$

Finally, since the matrix B is invertible, it follows from (16.76) that

$$\phi_i \equiv \psi_i \equiv 0 \quad \text{in } [0, T] \times \Omega, \quad i = 1, \cdots, n. \tag{16.77}$$

The proof is thus complete. □

Chapter 17
Approximate Boundary Synchronization

The approximate boundary synchronization is defined and studied in this chapter for system (II) with Neumann boundary controls.

17.1 Definition

Definition 17.1 Let $s > \frac{1}{2}$. System (II) is **approximately synchronizable** at the time $T > 0$ if for any given initial data $(\widehat{U}_0, \widehat{U}_1) \in (\mathcal{H}_{1-s})^N \times (\mathcal{H}_{-s})^N$, there exists a sequence $\{H_n\}$ of boundary controls in $\mathcal{L}^M$ with compact support in $[0, T]$, such that the sequence $\{U_n\}$ of solutions to problem (II) and (II0) satisfies

$$u_n^{(k)} - u_n^{(l)} \to 0 \quad \text{as } n \to +\infty \tag{17.1}$$

for $1 \leqslant k, l \leqslant N$ in the space

$$C^0_{loc}([T, +\infty); \mathcal{H}_{1-s}) \cap C^1_{loc}([T, +\infty); \mathcal{H}_{-s}). \tag{17.2}$$

Let C_1 be the **synchronization matrix** of order $(N-1) \times N$, defined by

$$C_1 = \begin{pmatrix} 1 & -1 & & & \\ & 1 & -1 & & \\ & & \ddots & \ddots & \\ & & & 1 & -1 \end{pmatrix}. \tag{17.3}$$

C_1 is a full row-rank matrix, and

$$\text{Ker}(C_1) = \text{Span}\{e_1\} \tag{17.4}$$

T. Li and B. Rao, *Boundary Synchronization for Hyperbolic Systems*, Progress in Nonlinear Differential Equations and Their Applications 94,
https://doi.org/10.1007/978-3-030-32849-8_17

with

$$e_1 = (1, \cdots, 1)^T. \tag{17.5}$$

Obviously, the **approximate boundary synchronization** (17.1) can be equivalently rewritten as

$$C_1 U_n \to 0 \quad \text{as } n \to +\infty \tag{17.6}$$

in the space

$$C^0_{loc}([T, +\infty); (\mathcal{H}_{1-s})^{N-1}) \cap C^1_{loc}([T, +\infty); (\mathcal{H}_{-s})^{N-1}). \tag{17.7}$$

17.2 Condition of C_1-Compatibility

Similarly to Theorem 9.2, we have

Theorem 17.2 *Assume that system (II) is approximately synchronizable at the time $T > 0$, but not approximately null controllable. Then the coupling matrix $A = (a_{ij})$ should satisfy the following* **row-sum condition***:*

$$\sum_{j=1}^{N} a_{ij} := a \quad (i = 1, \cdots, N), \tag{17.8}$$

where a is a constant independent of $i = 1, \cdots, N$. This condition is called the **condition of C_1-compatibility**.

Remark 17.3 The condition of C_1-compatibility (17.8) for the approximate boundary synchronization is just the same as that given in (13.6) for the exact boundary synchronization. In particular, Lemma 13.3 remains valid in the present situation.

17.3 Fundamental Properties

Under the condition of C_1-compatibility (17.8), setting

$$W_1 = (w^{(1)}, \cdots, w^{(N-1)})^T = C_1 U \tag{17.9}$$

and noting (13.11), the original problem (II) and (II0) for U can be transformed into following **reduced system** for W_1:

$$\begin{cases} W_1'' - \Delta W_1 + \overline{A}_1 W_1 = 0 & \text{in } (0, +\infty) \times \Omega, \\ W_1 = 0 & \text{on } (0, +\infty) \times \Gamma_0, \\ \partial_\nu W_1 = C_1 D H & \text{on } (0, +\infty) \times \Gamma_1 \end{cases} \tag{17.10}$$

with the initial data

$$t = 0: \quad W_1 = C_1\widehat{U}_0, \; W_1' = C_1\widehat{U}_1 \quad \text{in } \Omega. \tag{17.11}$$

Accordingly, let

$$\Psi_1 = (\psi^{(1)}, \cdots, \psi^{(N-1)})^T. \tag{17.12}$$

Consider the following adjoint problem of the reduced system (17.10):

$$\begin{cases} \Psi_1'' - \Delta\Psi_1 + \overline{A}_1^T \Psi_1 = 0 & \text{in } (0, +\infty) \times \Omega, \\ \Psi_1 = 0 & \text{on } (0, +\infty) \times \Gamma_0, \\ \partial_\nu \Psi_1 = 0 & \text{on } (0, +\infty) \times \Gamma_1, \\ t = 0: \quad \Psi_1 = \widehat{\Psi}_0, \; \Psi_1' = \widehat{\Psi}_1 & \text{in } \Omega, \end{cases} \tag{17.13}$$

which is called the **reduced adjoint problem** of system (II).

From Definition 16.1 and Definition 17.1, we immediately get

Lemma 17.4 *Assume that the coupling matrix A satisfies the condition of C_1-compatibility (17.8). Then system (II) is approximately synchronizable at the time $T > 0$ if and only if the reduced system given by (17.10) is approximately null controllable at the time $T > 0$, or equivalently if and only if the reduced adjoint problem (17.13) is C_1D-observable on the time interval $[0, T]$ (cf. Definition 16.2).*

Corollary 17.5 *Assume that the condition of C_1-compatibility (17.8) holds. If rank$(C_1D) = N - 1$, then system (II) is always approximately synchronizable.*

Proof Since $(C_1D)^T$ is an invertible matrix of order $(N - 1)$, the condition of observation similarly given in Definition 16.2 becomes

$$\Psi_1 \equiv 0 \quad \text{on } [0, T] \times \Gamma_1. \tag{17.14}$$

Thus, by means of Holmgren's uniqueness theorem (*cf.* Theorem 8.2 in [62]), we get the C_1D-observability of the reduced adjoint problem (17.13) on $[0, T]$, then by Lemma 17.4, the approximate boundary synchronization of system (II). □

Similarly to Lemma 9.5, we have

Lemma 17.6 *Under the condition of C_1-compatibility (17.8), if system (II) is approximately synchronizable at $T > 0$, then we have the following* **criterion of Kalman's type***:*

$$rank(C_1D, C_1AD, \cdots, C_1A^{N-1}D) = N - 1. \tag{17.15}$$

17.4 Properties Related to the Number of Total Controls

Since the rank of the enlarged matrix $(D, AD, \cdots, A^{N-1}D)$ measures the number of total (direct and indirect) controls, we now determine the minimal number of total controls, which is necessary to the approximate boundary synchronization of system (II), no matter whether the condition of C_1-compatibility (17.8) is satisfied or not.

Similarly to Theorem 9.6, we have

Theorem 17.7 *Assume that system (II) is approximately synchronizable under the action of a boundary control matrix D. Then we necessarily have*

$$\text{rank}(D, AD, \cdots, A^{N-1}D) \geqslant N - 1. \tag{17.16}$$

In other words, at least $(N-1)$ total controls are needed in order to realize the approximate boundary synchronization of system (II).

When system (II) is approximately synchronizable under the minimal rank condition

$$\text{rank}(D, AD, \cdots, A^{N-1}D) = N - 1, \tag{17.17}$$

the coupling matrix A should possess some fundamental properties related to the synchronization matrix C_1. Similarly to Theorem 9.7, we have

Theorem 17.8 *Assume that system (II) is approximately synchronizable under the minimal rank condition (17.17). Then we have the following assertions:*

(i) The coupling matrix A satisfies the condition of C_1-compatibility (17.8).

(ii) There exists a scalar function u as the **approximately synchronizable state***, such that*

$$u_n^{(k)} \to u \quad as\ n \to +\infty \tag{17.18}$$

for all $1 \leqslant k \leqslant n$ in the space

$$C^0_{loc}([T, +\infty); \mathcal{H}_{1-s}) \cap C^1_{loc}([T, +\infty); \mathcal{H}_{-s}) \tag{17.19}$$

with $s > 1/2$. Moreover, the approximately synchronizable state u is independent of the sequence $\{H_n\}$ of applied boundary controls.

(iii) The transpose A^T of the coupling matrix A admits an eigenvector E_1 such that $(E_1, e_1) = 1$, where $e_1 = (1, \cdots, 1)^T$ is the eigenvector of A, associated with the eigenvalue a given by (17.8).

Remark 17.9 Under the hypothesis of Theorem 17.8, system (II) is approximately synchronizable in the **pinning sense**, while that originally given by Definition 17.1 is in the **consensus sense**.

On the other hand, similarly to Theorem 9.12, we have

Theorem 17.10 *Let A satisfy the condition of C_1-compatibility (17.8). Assume that A^T admits an eigenvector E_1 such that $(E_1, e_1) = 1$ with $e_1 = (1, \cdots, 1)^T$. Then there exists a boundary control matrix D satisfying the minimal rank condition (17.17), which realizes the approximate boundary synchronization of system (II). Moreover, the approximately synchronizable state u is independent of applied boundary controls.*

17.5 An Example

In this subsection, we will examine the approximate boundary synchronization for a coupled system of three wave equations, the reduced adjoint system of which is given by (16.64).

For this purpose, we first look for all the matrices A of order 3, such that $C_1 A = \bar{A}_1 C_1$. More precisely, let

$$N = 3, \quad \bar{A}_1 = \begin{pmatrix} 0 & 1 \\ 0 & 0 \end{pmatrix}, \quad C_1 = \begin{pmatrix} 1 & -1 & 0 \\ 0 & 1 & -1 \end{pmatrix}. \tag{17.20}$$

For getting A, we solve the linear system:

$$C_1 A = \bar{A}_1 C_1 = \begin{pmatrix} 0 & 1 & -1 \\ 0 & 0 & 0 \end{pmatrix}. \tag{17.21}$$

Noting that the matrix

$$A_0 = \begin{pmatrix} 0 & 1 & -1 \\ 0 & 0 & 0 \\ 0 & 0 & 0 \end{pmatrix} \tag{17.22}$$

satisfies (17.21), it is easy to see that

$$A = A_0 + \begin{pmatrix} \alpha & \beta & \gamma \\ \alpha & \beta & \gamma \\ \alpha & \beta & \gamma \end{pmatrix} = \begin{pmatrix} \alpha & \beta+1 & \gamma-1 \\ \alpha & \beta & \gamma \\ \alpha & \beta & \gamma \end{pmatrix}, \tag{17.23}$$

where α, β, δ are arbitrarily given real numbers. Par construction, A satisfies the condition of C_1-compatibility (17.8) with the row-sum $a = \alpha + \beta + \delta$.

Proposition 17.11 *In the case*

$$\alpha + \beta + \gamma \neq 0 \tag{17.24}$$

or

$$\alpha + \beta + \gamma = 0 \text{ and } \alpha = 0, \tag{17.25}$$

$\lambda = \alpha + \beta + \gamma$ is an eigenvalue of A associated with the eigenvector $e_1 = (1, 1, 1)^T$, and the corresponding eigenvector E_1 of A^T can be chosen such that $(E_1, e_1) = 1$. While, in the case

$$\alpha + \beta + \gamma = 0 \text{ but } \alpha \neq 0, \tag{17.26}$$

$\lambda = 0$ is the only eigenvalue of A, and all the corresponding eigenvectors E_1 of A^T necessarily satisfy $(E_1, e_1) = 0$.

Proof First, a straightforward computation gives

$$det(\lambda I - A) = \lambda^3 - (\alpha + \beta + \gamma)\lambda^2. \tag{17.27}$$

(i) If $\alpha + \beta + \gamma \neq 0$, then $\lambda = \alpha + \beta + \gamma$ is a simple eigenvalue of A with the eigenvector e_1. So, A^T admits an eigenvector E_1 such that $(E_1, e_1) = 1$. More precisely, the vector E_1 is given by

$$E_1 = \frac{1}{a^2}\begin{pmatrix} a\alpha \\ a\beta + \alpha \\ a\gamma - \alpha \end{pmatrix} \quad \text{with } a = \alpha + \beta + \gamma. \tag{17.28}$$

(ii) If $\alpha + \beta + \gamma = 0$ and $\alpha = 0$, then $\lambda = 0$ is the eigenvalue of A with multiplicity 3 and dim Ker $(A) = 2$. Moreover, the eigenvector E_1 of A^T given by

$$E_1 = \frac{1}{2}\begin{pmatrix} -2\beta \\ \beta + 1 \\ \beta + 1 \end{pmatrix} \tag{17.29}$$

satisfies $(E_1, e_1) = 1$.

(iii) If $\alpha + \beta + \gamma = 0$ and $\alpha \neq 0$, then $\lambda = 0$ is also the eigenvalue of A with multiplicity 3 and dim Ker $(A) = 1$. Then, the eigenvector E_1 of A^T given by

$$E_1 = \begin{pmatrix} 0 \\ 1 \\ -1 \end{pmatrix} \tag{17.30}$$

necessarily satisfies $(E_1, e_1) = 0$. The proof is complete. □

Now let us consider the corresponding problem (II) and (II0) with the boundary control matrix $D = (d_1, d_2, d_3)^T$ and the coupling matrix A given by (17.23):

$$\begin{cases} u'' - \Delta u + \alpha u + (\beta + 1)v + (\gamma - 1)w = 0 & \text{in } (0, +\infty) \times \Omega, \\ v'' - \Delta v + \alpha u + \beta v + \gamma w = 0 & \text{in } (0, +\infty) \times \Omega, \\ w'' - \Delta w + \alpha u + \beta v + \gamma w = 0 & \text{in } (0, +\infty) \times \Omega, \\ u = v = w = 0 & \text{on } (0, +\infty) \times \Gamma_0, \\ \partial_\nu u = d_1 h, \quad \partial_\nu v = d_2 h, \quad \partial_\nu w = d_3 h & \text{on } (0, +\infty) \times \Gamma_1 \end{cases} \tag{17.31}$$

with the initial data

$$t = 0: \quad (u, v, w) = (u_0, v_0, w_0), \ (u', v', w') = (u_1, v_1, w_1) \quad \text{in } \Omega. \qquad (17.32)$$

Since A satisfies the condition of C_1-compatibility, the reduced system (17.10) with $\bar{A}_1$ given by (17.20) is written as

$$\begin{cases} y'' - \Delta y + z = 0 & \text{in } (0, +\infty) \times \Omega, \\ z'' - \Delta z = 0 & \text{in } (0, +\infty) \times \Omega, \\ y = z = 0 & \text{on } (0, +\infty) \times \Gamma_0, \\ \partial_\nu y = (d_1 - d_2)h, \quad \partial_\nu z = (d_2 - d_3)h & \text{on } (0, +\infty) \times \Gamma_1, \end{cases} \qquad (17.33)$$

and the corresponding reduced adjoint system (17.13) becomes

$$\begin{cases} \psi_1'' - \Delta \psi_1 = 0 & \text{in } (0, +\infty) \times \Omega, \\ \psi_2'' - \Delta \psi_2 + \psi_1 = 0 & \text{in } (0, +\infty) \times \Omega, \\ \psi_1 = \psi_2 = 0 & \text{on } (0, +\infty) \times \Gamma_0, \\ \partial_\nu \psi_1 = \partial_\nu \psi_2 = 0 & \text{on } (0, +\infty) \times \Gamma_1. \end{cases} \qquad (17.34)$$

Moreover, the $C_1 D$-observation becomes

$$(d_1 - d_2)\psi_1 + (d_2 - d_3)\psi_2 = 0 \quad \text{on } [0, T] \times \Gamma_1. \qquad (17.35)$$

Theorem 17.12 *Let Ω satisfy the usual multiplier geometrical condition. Then, there exists a boundary control matrix $D = (d_1, d_2, d_3)^T$ such that system (17.31) is approximately synchronizable. Moreover, the approximately synchronizable state u is independent of applied boundary controls in two cases (17.24) and (17.25).*

Proof By Lemma 17.4, the approximate boundary synchronization of system (17.31) is equivalent to the $C_1 D$-observability of the reduced adjoint system (17.34). By Theorem 16.16, under the multiplier geometrical condition, the reduced adjoint system (17.34), together with the $C_1 D$-observation (17.35), has only the trivial solution $\psi_1 \equiv \psi_2 \equiv 0$, provided that $d_2 - d_3 \neq 0$ and that $T > 0$ is large enough. We get thus the approximate boundary synchronization of system (17.31).

More precisely, in case (17.24), we can choose

$$\begin{cases} d_1 = \frac{1}{\sigma} - \frac{\gamma}{\alpha}, \quad d_2 = 0, \quad d_3 = 1, & \text{if } \alpha \neq 0, \\ d_1 = 0, \quad d_2 = 1, \quad d_3 = -\frac{\beta}{\gamma}, & \text{if } \alpha = 0 \text{ and } \gamma \neq 0, \\ d_1 = 0, \quad d_2 = -\frac{\gamma}{\beta}, \quad d_3 = 1, & \text{if } \alpha = 0 \text{ and } \beta \neq 0, \end{cases} \qquad (17.36)$$

while, in case (17.25), we can choose

$$d_1 = 0, \quad d_2 = -1, \quad d_3 = 1, \qquad (17.37)$$

such that $E_1^T D = 0$ in both cases, where E_1 is given by (17.28) and (17.29), respectively. Then, applying E_1 to system (17.31), and setting $\phi = (E_1, U_n)$ with $U_n = (u_n, v_n, w_n)^T$, it follows that

$$\begin{cases} \phi'' - \Delta\phi + a\phi = 0 & \text{in } (0, +\infty) \times \Omega, \\ \phi = 0 & \text{on } (0, +\infty) \times \Gamma_0, \\ \partial_\nu \phi = 0, & \text{on } (0, +\infty) \times \Gamma_1, \\ t = 0: \quad \phi = (E_1, \widehat{U}_0), \ \phi' = (E_1, \widehat{U}_1) & \text{in } \Omega. \end{cases} \tag{17.38}$$

Since ϕ is independent of applied boundary controls, and $(E_1, e_1) = 1$, the convergence

$$U_n \to u e_1 \quad \text{as } n \to +\infty \tag{17.39}$$

in the space

$$C^0([T, +\infty); (\mathcal{H}_0)^3) \cap C^1([T, +\infty); (\mathcal{H}_{-1})^3) \tag{17.40}$$

will imply that

$$t \geqslant T: \quad \phi = (E_1, U_n) \to (E_1, e_1)u = u. \tag{17.41}$$

Therefore, the approximately synchronizable state u is indeed independent of applied boundary controls h. □

Remark 17.13 In two cases (17.24) and (17.25), under the minimal rank condition

$$\text{rank}(D, AD, AD^2) = 2, \tag{17.42}$$

the approximately synchronizable state u is independent of applied boundary controls. While, in case (17.26), A is similar to a Jordan block of order 3, we must have the rank condition:

$$\text{rank}(D, AD, AD^2) = 3, \tag{17.43}$$

which is necessary for the approximate boundary null controllability of the nilpotent system (17.31). We hope, but we do not know up to now if it is also sufficient.

Chapter 18
Approximate Boundary Synchronization by p-Groups

The approximate boundary synchronization by p-groups is introduced and studied in this chapter for system (II) with Neumann boundary controls.

18.1 Definition

When $\text{rank}(D, AD, \cdots, A^{N-1}D)$, the number of total controls, is further reduced, we can consider the **approximate boundary synchronization by p-groups** ($p \geqslant 1$). In the special case $p = 1$, it is just the approximate boundary synchronization considered in Chap. 17.

Let $p \geqslant 1$ be an integer and let

$$0 = n_0 < n_1 < n_2 < \cdots < n_p = N \tag{18.1}$$

be integers such that $n_r - n_{r-1} \geqslant 2$ for all $1 \leqslant r \leqslant p$. The approximate boundary synchronization by p-groups means that the components of U are divided into p groups:

$$(u^{(1)}, \cdots, u^{(n_1)}),\ (u^{(n_1+1)}, \cdots, u^{(n_2)}), \cdots, (u^{(n_{p-1}+1)}, \cdots, u^{(n_p)}), \tag{18.2}$$

and each group possesses the corresponding approximate boundary synchronization, respectively.

Definition 18.1 Let $s > \frac{1}{2}$. System (II) is approximately synchronizable by p-groups at $T > 0$ if for any given initial data $(\widehat{U}_0, \widehat{U}_1) \in (\mathcal{H}_{1-s})^N \times (\mathcal{H}_{-s})^N$, there exists a sequence $\{H_n\}$ of boundary controls in $\mathcal{L}^M$ with compact support in $[0, T]$, such that the corresponding sequence $\{U_n\}$ of solutions to problem (II) and (II0) satisfies

T. Li and B. Rao, *Boundary Synchronization for Hyperbolic Systems*, Progress in Nonlinear Differential Equations and Their Applications 94,
https://doi.org/10.1007/978-3-030-32849-8_18

$$u_n^{(k)} - u_n^{(l)} \to 0 \quad \text{in } C^0_{loc}([T,+\infty); \mathcal{H}_{1-s}) \cap C^1_{loc}([T,+\infty); \mathcal{H}_{-s}) \tag{18.3}$$

as $n \to +\infty$ for all $n_{r-1}+1 \leqslant k, l \leqslant n_r$ and $1 \leqslant r \leqslant p$.

Let

$$S_r = \begin{pmatrix} 1 & -1 & & & \\ & 1 & -1 & & \\ & & \ddots & \ddots & \\ & & & 1 & -1 \end{pmatrix}, \quad 1 \leqslant r \leqslant p \tag{18.4}$$

be an $(n_r - n_{r-1} - 1) \times (n_r - n_{r-1})$ matrix with full row-rank, and

$$C_p = \begin{pmatrix} S_1 & & & \\ & S_2 & & \\ & & \ddots & \\ & & & S_p \end{pmatrix} \tag{18.5}$$

be the $(N-p) \times N$ **matrix of synchronization by p-groups**. Clearly,

$$\text{Ker}(C_p) = \text{Span}\{e_1, \cdots, e_p\}, \tag{18.6}$$

where

$$(e_r)_i = \begin{cases} 1, & n_{r-1}+1 \leqslant i \leqslant n_r, \\ 0, & \text{otherwise} \end{cases} \tag{18.7}$$

for $1 \leqslant r \leqslant p$. Thus, the approximate boundary synchronization by p-groups (18.3) can be written as

$$C_p U_n \to 0 \quad \text{as } n \to +\infty \tag{18.8}$$

in the space

$$C^0_{loc}([T,+\infty); (\mathcal{H}_{1-s})^{N-p}) \cap C^1_{loc}([T,+\infty); (\mathcal{H}_{-s})^{N-p}). \tag{18.9}$$

18.2 Fundamental Properties

Similarly to Lemma 13.3, the coupling matrix A satisfies the condition of C_p-compatibility if $\text{Ker}(C_p)$ is an invariant subspace of A:

$$A\text{Ker}(C_p) \subseteq \text{Ker}(C_p), \tag{18.10}$$

or equivalently, there exists a unique matrix $\overline{A}_p$ of order $(N-p)$, such that

$$C_p A = \overline{A}_p C_p. \tag{18.11}$$

$\overline{A}_p$ is called the **reduced matrix of** A **by** C_p .

Under the condition of C_p-compatibility (18.10), setting

$$W_p = C_p U = (w^{(1)}, \cdots, w^{(N-p)})^T \tag{18.12}$$

and noting (18.11), from problem (II) and (II0) for U we get the following **reduced system** for W_p:

$$\begin{cases} W_p'' - \Delta W_p + \overline{A}_p W_p = 0 & \text{in } (0, +\infty) \times \Omega, \\ W_p = 0 & \text{on } (0, +\infty) \times \Gamma_0, \\ \partial_\nu W_p = C_p D H & \text{on } (0, +\infty) \times \Gamma_1 \end{cases} \tag{18.13}$$

with the initial data

$$t = 0: \quad W_p = C_p \widehat{U}_0, \; W_p' = C_p \widehat{U}_1 \quad \text{in } \Omega. \tag{18.14}$$

Accordingly, let

$$\Psi_p = (\psi^{(1)}, \cdots, \psi^{(N-p)})^T. \tag{18.15}$$

Consider the following **reduced adjoint problem**:

$$\begin{cases} \Psi_p'' - \Delta \Psi_p + \overline{A}_p^T \Psi_p = 0 & \text{in } (0, +\infty) \times \Omega, \\ \Psi_p = 0 & \text{on } (0, +\infty) \times \Gamma_0, \\ \partial_\nu \Psi_p = 0 & \text{on } (0, +\infty) \times \Gamma_1, \\ t = 0: \quad \Psi_p = \widehat{\Psi}_{p0}, \; \Psi_p' = \widehat{\Psi}_{p1} & \text{in } \Omega. \end{cases} \tag{18.16}$$

By Definitions 16.1 and 18.1, and using Theorem 16.5, we get

Lemma 18.2 *Assume that the coupling matrix A satisfies the condition of C_p-compatibility (18.10). Then system (II) is approximately synchronizable by p-groups at the time $T > 0$ if and only if the reduced system (18.13) is approximately null controllable at the time $T > 0$, or equivalently, the reduced adjoint problem (18.16) is $C_p D$-observable on the time interval $[0, T]$, namely, the partial observation*

$$(C_p D)^T \partial_\nu \Psi_p \equiv 0 \quad on \ [0, T] \times \Gamma_1 \tag{18.17}$$

implies $\widehat{\Psi}_{p0} = \widehat{\Psi}_{p1} \equiv 0$, then $\Psi_p \equiv 0$.

Thus, by Theorem 16.11 and noting (18.11), it is easy to get

Theorem 18.3 *Under the condition of C_p-compatibility (18.10), assume that system (II) is approximately synchronizable by p-groups, we necessarily have the following* **criterion of Kalman's type***:*

$$rank(C_pD, C_pAD, \cdots, C_pA^{N-1}D) = N - p. \tag{18.18}$$

18.3 Properties Related to the Number of Total Controls

We now consider rank$(D, AD, \cdots, A^{N-1}D)$, the number of total (direct and indirect) controls, and determine the minimal number of total controls, which is necessary to the approximate boundary synchronization by p-groups for system (II), no matter whether the condition of C_p-compatibility (18.10) is satisfied or not.

Similarly to Theorem 10.4, we have

Theorem 18.4 *Assume that system (II) is approximately synchronizable by p-groups under the action of a boundary control matrix D. Then, we necessarily have*

$$rank(D, AD, \cdots, A^{N-1}D) \geqslant N - p. \tag{18.19}$$

In other words, at least $(N - p)$ total controls are needed in order to realize the approximate boundary synchronization by p-groups for system (II).

Based on Theorem 18.4, we consider the approximate boundary synchronization by p-groups for system (II) under the minimal rank condition

$$\text{rank}(D, AD, \cdots, A^{N-1}D) = N - p. \tag{18.20}$$

In this case, the coupling matrix A should possess some fundamental properties related to C_p, the matrix of synchronization by p-groups.

Similarly to Theorem 10.5, we have

Theorem 18.5 *Assume that system (II) is approximately synchronizable by p-groups under the minimal rank condition* (18.20). *Then we have the following assertions:*

(i) The coupling matrix A satisfies the condition of C_p-compatibility (18.10).

(ii) There exists some linearly independent scalar functions $u_1, \cdots, u_p$ such that

$$u_n^{(k)} \to u_r \quad as\ n \to +\infty \tag{18.21}$$

for all $n_{r-1} + 1 \leqslant k \leqslant n_r$ and $1 \leqslant r \leqslant p$ in the space

$$C^0_{loc}([T, +\infty); \mathcal{H}_{1-s}) \cap C^1_{loc}([T, +\infty); \mathcal{H}_{-s}) \tag{18.22}$$

with $s > 1/2$. Moreover, the **approximately synchronizable state by p-groups** $u = (u_1, \cdots, u_p)^T$ *is independent of applied boundary controls H_n.*

(iii) A^T admits an invariant subspace Span$\{E_1, \cdots, E_p\}$, which is contained in Ker(D^T) and bi-orthonormal to Ker(C_p)=Span$\{e_1, \cdots, e_p\}$.

Remark 18.6 Under hypothesis (18.20), system (II) is approximately synchronizable by p-groups in the **pinning sense**, while that originally given by Definition 18.1 is in the **consensus sense**.

Remark 18.7 Since the invariant subspace Span$\{E_1, \cdots, E_p\}$ of A^T is bi-orthonormal to the invariant subspace Span$\{e_1, \cdots, e_p\}$ of A, by Proposition 2.8, the invariant subspace Span$\{E_1, \cdots, E_p\}^\perp$ of A is a supplement of Span$\{e_1, \cdots, e_p\}$

Moreover, similarly to Theorem 10.10, we have

Theorem 18.8 *Let A satisfy the condition of C_p-compatibility (18.10). Assume that Ker(C_p) admits a supplement which is also invariant for A. Then there exists a boundary control matrix D satisfying the minimal rank condition* (18.20), *which realizes the approximate boundary synchronization by p-groups for system (II) in the pinning sense.*

with $T > T/2$. Moreover, the approximately synchronizable state by p-groups $(u_1, \cdots, u_p)^T$ is independent of applied boundary controls H.

Ker $\mathcal{D}^T$ admits an invariant subspace $\mathrm{Span}\{E_1, \cdots, E_p\}$ which is contained in $\mathrm{Ker}(D^T)$ and bi-orthonormal to $\mathrm{Ker}(C_p) = \mathrm{Span}\{e_1, \cdots, e_p\}$.

Remark 18.6 Under hypothesis (18.20), system (I) is approximately synchronizable by p-groups in the pinning sense, while that originally given by Definition 18.1 is in the consensus sense.

Remark 18.7 Since the invariant subspace $\mathrm{Span}\{E_1, \cdots, E_p\}$ of A^T is bi-orthonormal to the invariant subspace $\mathrm{Span}\{e_1, \cdots, e_p\}$ of A, by Proposition 2.8, the invariant subspace $\mathrm{Span}\{E_1, \cdots, E_p\}^{\perp}$ of A is a supplement of $\mathrm{Span}\{e_1, \cdots, e_p\}$.

Moreover, similarly to Theorem 10.10, we have

Theorem 18.8 *Let A satisfy the condition of C_p-compatibility (18.11). Assume that A^T admits a supplement which is invariant for A. Then there exists a boundary control matrix D satisfying the rank condition (18.20), which realizes the approximate boundary synchronization by p-groups for system (I) in the pinning sense.*

Part V
Synchronization for a Coupled System of Wave Equations with Coupled Robin Boundary Controls: Exact Boundary Synchronization

We consider the following coupled system of wave equations with coupled Robin boundary controls:

$$\begin{cases} U'' - \Delta U + AU = 0 & in \quad (0, +\infty) \times \Omega, \\ U = 0 & on \quad (0, +\infty) \times \Gamma_0, \\ \partial_\nu U + BU = DH & on \quad (0, +\infty) \times \Gamma_1 \end{cases} \tag{III}$$

with the initial condition

$$t = 0: \quad U = \widehat{U}_0, \ U' = \widehat{U}_1 \quad in \quad \Omega, \tag{III0}$$

where $\Omega \subset \mathbb{R}^n$ *is a bounded domain with smooth boundary* $\Gamma = \Gamma_1 \cup \Gamma_0$ *such that* $\overline{\Gamma}_1 \cap \overline{\Gamma}_0 = \emptyset$ and $\text{mes}(\Gamma_1) > 0$; *"′" stands for the time derivative;* $\Delta = \sum_{k=1}^{n} \frac{\partial^2}{\partial x_k^2}$ *is the Laplacian operator;* $U = \left(u^{(1)}, \cdots, u^{(N)}\right)^T$ *and* $H = \left(h^{(1)}, \cdots, h^{(M)}\right)^T (M \leqslant N)$ *denote the state variables and the boundary controls, respectively; the* **internal coupling matrix** $A = (a_{ij})$ *and the* **boundary coupling matrix** B *are of order N, and D as the boundary control matrix is a full column-rank matrix of order* $N \times M$, *all with constant elements.*

The exact boundary synchronization and the exact boundary synchronization by groups for system (III) will be discussed in this part, while, the approximate boundary synchronization and the approximate boundary synchronization by groups for system (III) will be considered in the next part (Part VI).

Chapter 19
Preliminaries on Problem (III) and (III0)

In order to consider the exact boundary controllability and the exact boundary synchronization of system (III), we first give some necessary results on problem (III) and (III0) in this chapter.

In order to consider the exact boundary controllability and the exact boundary synchronization of system (III), we first give some necessary results on problem (III) and (III0) in this chapter.

19.1 Regularity of Solutions with Neumann Boundary Conditions

Similarly to the problem of wave equations with Neumann boundary conditions, a problem with Robin boundary conditions no longer enjoys the hidden regularity as in the case with Dirichlet boundary conditions. As a result, the solution to problem (III) and (III0) with coupled Robin boundary conditions is not smooth enough in general for the proof of the non-exact boundary controllability of the system.

In order to overcome this difficulty, we should deeply study the regularity of solutions to wave equations with Neumann boundary conditions. For this purpose, consider the following second-order hyperbolic problem on a bounded domain $\Omega \subset R^n$ $(n \geqslant 2)$ with boundary Γ:

$$\begin{cases} y'' + A(x, \partial)y = f & \text{in } (0, T) \times \Omega = Q, \\ \frac{\partial y}{\partial \nu_A} = g & \text{on } (0, T) \times \Gamma = \Sigma, \\ t = 0: \quad y = y_0, \quad y' = y_1 & \text{in } \Omega, \end{cases} \tag{19.1}$$

T. Li and B. Rao, *Boundary Synchronization for Hyperbolic Systems*, Progress in Nonlinear Differential Equations and Their Applications 94, https://doi.org/10.1007/978-3-030-32849-8_19

where

$$A(x,\partial) = -\sum_{i,j=1}^{n} a_{ij}(x)\frac{\partial^2}{\partial x_i \partial x_j} + \sum_{i=1}^{n} b_i(x)\frac{\partial}{\partial x_i} + c_0(x), \tag{19.2}$$

in which $a_{ij}(x)$ with $a_{ij}(x) = a_{ji}(x)$, $b_j(x)$ and $c_0(x)$ $(i, j = 1, \cdots, n)$ are smooth real coefficients, and the principal part of $A(x, \partial)$ is supposed to be uniformly strong elliptic in Ω:

$$\sum_{i,j=1}^{n} a_{ij}(x)\eta_i\eta_j \geqslant c\sum_{j=1}^{n}\eta_j^2 \tag{19.3}$$

for all $x \in \Omega$ and for all $\eta = (\eta_1, \cdots, \eta_n) \in \mathbb{R}^n$, where $c > 0$ is a positive constant; moreover, $\frac{\partial y}{\partial \nu_A}$ is the outward normal derivative associated with A:

$$\frac{\partial y}{\partial \nu_A} = \sum_{i=1}^{N}\sum_{j=1}^{N} a_{ij}(x)\frac{\partial y}{\partial x_i}\nu_j, \tag{19.4}$$

$\nu = (\nu_1, \cdots, \nu_n)^T$ being the unit outward normal vector on the boundary Γ.

Define the operator $\mathcal{A}$ by

$$\mathcal{A} = A(x,\partial), \quad \mathcal{D}(\mathcal{A}) = \left\{y \in H^2(\Omega) : \frac{\partial y}{\partial \nu_A} = 0 \text{ on } \Gamma\right\}. \tag{19.5}$$

We know that if $-\mathcal{A}$ is closed, and if $-\mathcal{A}$ is maximal dissipative, namely, if $-\mathcal{A}$ is dissipative: $\mathrm{Re}(\mathcal{A}u, u) \geqslant 0$, $\forall u \in D(\mathcal{A})$, and it does not possess any proper dissipative extension, then for any given α with $0 < \alpha < 1$, the fractional power $\mathcal{A}^\alpha$ can be defined in a natural way (cf. [6, 24, 70]), for example, we have

$$\mathcal{A}^\alpha u = \frac{\sin \pi\alpha}{\pi}\int_0^\infty \lambda^{\alpha-1}\mathcal{A}(\lambda + \mathcal{A})^{-1}u\, d\lambda, \quad u \in D(\mathcal{A}). \tag{19.6}$$

We can verify that $-\mathcal{A}$ given by (19.5) is closed, and there exists a constant $c > 0$ so large that $-(cI + \mathcal{A})$ is maximal dissipative, thus we can define the fractional powers of $(cI + \mathcal{A})$. Since a suitable translation of the operator does not change the regularity of the solution in $[0, T]$ with $T < +\infty$, for any α with $0 < \alpha < 1$, the fractional powers $\mathcal{A}^\alpha$ can be well defined, moreover, we have

$$\|y\|_{D(\mathcal{A}^\alpha)} = \|\mathcal{A}^\alpha y\|_{L^2(\Omega)}. \tag{19.7}$$

In [30] (cf. also [29, 31]), Lasiecka and Triggiani got the optimal regularity for the solution to problem (19.1) by means of the theory of cosine operator. In particular, more regularity results can be obtained when the domain is a parallelepiped. For conciseness and clarity, we list only those results which are needed in what follows.

Let $\epsilon > 0$ be an arbitrarily given small number. Here and hereafter, we always assume that α and β are given, respectively, as follows:

$$\begin{cases} \alpha = 3/5 - \epsilon,\ \beta = 3/5, & \Omega \text{ is a smooth bounded domain} \\ & \text{and } A(x, \partial) \text{ is defined by (19.2)}, \\ \alpha = \beta = 2/3, & \Omega \text{ is a sphere and } A(x, \partial) = -\Delta, \\ \alpha = \beta = 3/4 - \epsilon, & \Omega \text{ is a parallelepiped and } A(x, \partial) = -\Delta. \end{cases} \tag{19.8}$$

Lemma 19.1 *Assume that* $y_0 \equiv y_1 \equiv 0$ *and* $f \equiv 0$. *For any given* $g \in L^2(0, T; L^2(\Gamma))$, *problem* (19.1) *admits a unique solution* y *such that*

$$(y, y') \in C^0([0, T]; H^\alpha(\Omega) \times H^{\alpha-1}(\Omega)) \tag{19.9}$$

and

$$y|_\Sigma \in H^{2\alpha-1}(\Sigma) = L^2(0, T; H^{2\alpha-1}(\Gamma)) \cap H^{2\alpha-1}(0, T; L^2(\Gamma)), \tag{19.10}$$

where $H^\alpha(\Omega)$ *denotes the usual Sobolev space of order* α *and* $\Sigma = (0, T) \times \Gamma$.

Lemma 19.2 *Assume that* $y_0 \equiv y_1 \equiv 0$ *and* $g \equiv 0$. *For any given* $f \in L^2(0, T; L^2(\Omega))$, *the unique solution* y *to problem* (19.1) *satisfies*

$$(y, y') \in C^0([0, T]; H^1(\Omega) \times L^2(\Omega)) \tag{19.11}$$

and

$$y|_\Sigma \in H^\beta(\Sigma). \tag{19.12}$$

Lemma 19.3 *Assume that* $f \equiv 0$ *and* $g \equiv 0$.

(1) *If* $(y_0, y_1) \in H^1(\Omega) \times L^2(\Omega)$, *then problem* (19.1) *admits a unique solution* y *such that*

$$(y, y') \in C^0([0, T]; H^1(\Omega) \times L^2(\Omega)) \tag{19.13}$$

and

$$y|_\Sigma \in H^\beta(\Sigma). \tag{19.14}$$

(2) *If* $(y_0, y_1) \in L^2(\Omega) \times (H^1(\Omega))'$, *where* $(H^1(\Omega))'$ *denotes the dual space of* $H^1(\Omega)$ *with respect to* $L^2(\Omega)$, *then problem* (19.1) *admits a unique solution* y *such that*

$$(y, y') \in C^0([0, T]; L^2(\Omega) \times (H^1(\Omega))') \tag{19.15}$$

and

$$y|_\Sigma \in H^{\alpha-1}(\Sigma). \tag{19.16}$$

Remark 19.4 The regularities given in Lemmas 19.1, 19.2, and 19.3 remain true when we replace the Neumann boundary condition on the whole boundary Γ by the homogeneous Dirichlet boundary condition on Γ_0 and the Neumann boundary condition on Γ_1 with $\Gamma_0 \cup \Gamma_1 = \Gamma$ and $\overline{\Gamma}_0 \cap \overline{\Gamma}_1 = \emptyset$.

Remark 19.5 In the results mentioned above, the mappings from the given data to the solution are all continuous with respect to the corresponding topologies.

19.2 Well-Posedness of a Coupled System of Wave Equations with Coupled Robin Boundary Conditions

We now prove the well-posedness of problem (III) and (III0). Throughout this part and the next part, when $\text{mes}(\Gamma_0) > 0$, we define

$$\mathcal{H}_0 = L^2(\Omega), \quad \mathcal{H}_1 = H^1_{\Gamma_0}(\Omega), \tag{19.17}$$

in which $H^1_{\Gamma_0}(\Omega)$ is the subspace of $H^1(\Omega)$, composed of all the functions with the null trace on Γ_0, while, when $\text{mes}(\Gamma_0) = 0$, instead of (19.17), we define

$$\mathcal{H}_0 = \left\{u : u \in L^2(\Omega), \int_\Omega u dx = 0\right\}, \quad \mathcal{H}_1 = H^1(\Omega) \cap \mathcal{H}_0. \tag{19.18}$$

For simplifying the statement, we often assume that $\text{mes}(\Gamma_0) \neq 0$; however, all the conclusions are still valid when $\text{mes}(\Gamma_0) = 0$.

Let $\Phi = (\phi^{(1)}, \cdots, \phi^{(N)})^T$. We first consider the following adjoint system:

$$\begin{cases} \Phi'' - \Delta\Phi + A^T\Phi = 0 & \text{in } (0, +\infty) \times \Omega, \\ \Phi = 0 & \text{on } (0, +\infty) \times \Gamma_0, \\ \partial_\nu\Phi + B^T\Phi = 0 & \text{on } (0, +\infty) \times \Gamma_1 \end{cases} \tag{19.19}$$

with the initial data

$$t = 0: \quad \Phi = \widehat{\Phi}_0, \ \Phi' = \widehat{\Phi}_1 \quad \text{in } \Omega, \tag{19.20}$$

where A^T and B^T denote the transposes of A and B, respectively.

Theorem 19.6 *Assume that B is similar to a real symmetric matrix. Then for any given $(\widehat{\Phi}_0, \widehat{\Phi}_1) \in (\mathcal{H}_1)^N \times (\mathcal{H}_0)^N$, the adjoint problem* (19.19), (19.20) *admits a unique weak solution*

$$(\Phi, \Phi') \in C^0_{loc}([0, +\infty); (\mathcal{H}_1)^N \times (\mathcal{H}_0)^N) \tag{19.21}$$

in the sense of C_0-semigroup, where $\mathcal{H}_1$ and $\mathcal{H}_0$ are defined by (19.17).

Proof Without loss of generality, we assume that B is a real symmetric matrix.
We first formulate system (19.19) into the following variational form:

$$\int_\Omega (\Phi'', \widehat{\Phi})dx + \int_\Omega \langle \nabla\Phi, \nabla\widehat{\Phi}\rangle dx + \int_{\Gamma_1} (\Phi, B\widehat{\Phi})d\Gamma + \int_\Omega (\Phi, A\widehat{\Phi})dx = 0 \tag{19.22}$$

for any given test function $\widehat{\Phi} \in (\mathcal{H}_1)^N$, where $(\cdot, \cdot)$ denotes the inner product of $\mathbb{R}^N$, while $\langle \cdot, \cdot \rangle$ denotes the inner product of $\mathbb{M}^{N\times N}(\mathbb{R})$.

Recalling the following interpolation inequality [66]:

$$\int_{\Gamma_1} |\phi|^2 d\Gamma \leqslant c\|\phi\|_{H^1(\Omega)}\|\phi\|_{L^2(\Omega)}, \quad \forall \phi \in H^1(\Omega),$$

we have

$$\int_{\Gamma_1} (\Phi, B\Phi)d\Gamma \leqslant \|B\| \int_{\Gamma_1} |\Phi|^2 d\Gamma \leqslant c\|B\|\|\Phi\|_{(\mathcal{H}_1)^N}\|\Phi\|_{(\mathcal{H}_0)^N},$$

then it is easy to see that

$$\int_\Omega \langle \nabla\Phi, \nabla\Phi\rangle dx + \int_{\Gamma_1} (\Phi, B\Phi)d\Gamma + \lambda\|\Phi\|^2_{(\mathcal{H}_0)^N} \geqslant c'\|\Phi\|^2_{(\mathcal{H}_1)^N}$$

for some suitable constants $\lambda > 0$ and $c' > 0$. Therefore, the bilinear symmetric form

$$\int_\Omega \langle \nabla\Phi, \nabla\widehat{\Phi}\rangle dx + \int_{\Gamma_1} (\Phi, B\widehat{\Phi})d\Gamma$$

is coercive in $(\mathcal{H}_1)^N \times (\mathcal{H}_0)^N$. Moreover, the nonsymmetric part in (19.22) satisfies

$$\int_\Omega (\Phi, A\widehat{\Phi})dx \leqslant \|A\|\|\Phi\|_{(\mathcal{H}_0)^N}\|\widehat{\Phi}\|_{(\mathcal{H}_0)^N}.$$

By Theorem 1.1 of Chap. 8 in [60], the variational problem (19.22) with the initial data (19.20) admits a unique solution Φ with the smoothness (19.21). The proof is complete. □

Remark 19.7 From now on, in order to guarantee the well-posedness of problem (III) and (III0), we always assume that B is similar to a real symmetric matrix.

Definition 19.8 U is a weak solution to problem (III) and (III0) if

$$U \in C^0_{loc}([0, +\infty); (\mathcal{H}_0)^N) \cap C^1_{loc}([0, +\infty); (\mathcal{H}_{-1})^N), \tag{19.23}$$

where $\mathcal{H}_{-1}$ denotes the dual space of $\mathcal{H}_1$, such that for any given $(\widehat{\Phi}_0, \widehat{\Phi}_1) \in (\mathcal{H}_1)^N \times (\mathcal{H}_0)^N$ and for all given $t \geqslant 0$, we have

$$\langle\!\langle (U'(t), -U(t)), (\Phi(t), \Phi'(t)) \rangle\!\rangle \tag{19.24}$$
$$= \langle\!\langle (\widehat{U}_1, -\widehat{U}_0), (\widehat{\Phi}_0, \widehat{\Phi}_1) \rangle\!\rangle + \int_0^t \int_{\Gamma_1} (DH(\tau), \Phi(\tau)) dx dt,$$

in which $\Phi(t)$ is the solution to the adjoint problem (19.19) and (19.20), and $\langle\!\langle \cdot, \cdot \rangle\!\rangle$ denotes the duality between the spaces $(\mathcal{H}_{-1})^N \times (\mathcal{H}_0)^N$ and $(\mathcal{H}_1)^N \times (\mathcal{H}_0)^N$.

Theorem 19.9 *Assume that B is similar to a real symmetric matrix. For any given $H \in L^2_{loc}(0, +\infty; (L^2(\Gamma_1))^M)$ and $(\widehat{U}_0, \widehat{U}_1) \in (\mathcal{H}_0)^N \times (\mathcal{H}_{-1})^N$, problem (III) and (III0) admits a unique weak solution U. Moreover, the mapping*

$$(\widehat{U}_0, \widehat{U}_1, H) \to (U, U') \tag{19.25}$$

is continuous with respect to the corresponding topologies.

Proof Let Φ be the solution to the adjoint problem (19.19) and (19.20). Define a linear functional as follows:

$$L_t(\widehat{\Phi}_0, \widehat{\Phi}_1) \tag{19.26}$$
$$= \langle\!\langle (\widehat{U}_1, -\widehat{U}_0), (\widehat{\Phi}_0, \widehat{\Phi}_1) \rangle\!\rangle + \int_0^t \int_{\Gamma_1} (DH(\tau), \Phi(\tau)) dx dt.$$

Clearly, L_t is bounded in $(\mathcal{H}_1)^N \times (\mathcal{H}_0)^N$. Let S_t be the semigroup in $(\mathcal{H}_1)^N \times (\mathcal{H}_0)^N$, corresponding to the adjoint problem (19.19), (19.20). $L_t \circ S_t^{-1}$ is bounded in $(\mathcal{H}_1)^N \times (\mathcal{H}_0)^N$. Then, by Riesz–Fréchet representation theorem, for any given $(\widehat{\Phi}_0, \widehat{\Phi}_1) \in (\mathcal{H}_1)^N \times (\mathcal{H}_0)^N$, there exists a unique $(U'(t), -U(t)) \in (\mathcal{H}_{-1})^N \times (\mathcal{H}_0)^N$, such that

$$L_t \circ S_t^{-1}(\Phi(t), \Phi'(t)) = \langle\!\langle (U'(t), -U(t)), (\Phi(t), \Phi'(t)) \rangle\!\rangle. \tag{19.27}$$

By

$$L_t \circ S_t^{-1}(\Phi(t), \Phi'(t)) = L_t(\widehat{\Phi}_0, \widehat{\Phi}_1) \tag{19.28}$$

for any given $(\widehat{\Phi}_0, \widehat{\Phi}_1) \in (\mathcal{H}_1)^N \times (\mathcal{H}_0)^N$, (19.24) holds, then (U, U') is the unique weak solution to problem (III) and (III0). Moreover, we have

$$\|(U'(t), -U(t))\|_{(\mathcal{H}_{-1})^N \times (\mathcal{H}_0)^N} = \|L_t \circ S_t^{-1}\| \tag{19.29}$$
$$\leqslant c(\|(\widehat{U}_0, \widehat{U}_1)\|_{(\mathcal{H}_0)^N \times (\mathcal{H}_{-1})^N} + \|H\|_{L^2(0,T;(L^2(\Gamma_1))^M)})$$

for all $t \in [0, T]$.

At last, by a classic argument of density, we obtain the regularity desired by (19.23). □

19.3 Regularity of Solutions to Problem (III) and (III0)

We have

Theorem 19.10 *Assume that B is similar to a real symmetric matrix. For any given $H \in L^2(0, T; (L^2(\Gamma_1))^M)$ and any given $(\widehat{U}_0, \widehat{U}_1) \in (\mathcal{H}_1)^N \times (\mathcal{H}_0)^N$, the weak solution U to problem (III) and (III0) satisfies*

$$(U, U') \in C^0([0, T]; (H^\alpha(\Omega))^N \times (H^{\alpha-1}(\Omega))^N) \tag{19.30}$$

and

$$U|_{\Sigma_1} \in (H^{2\alpha-1}(\Sigma_1))^N, \tag{19.31}$$

where $\Sigma_1 = (0, T) \times \Gamma_1$ and α is defined by (19.8). Moreover, the linear mapping

$$(\widehat{U}_0, \widehat{U}_1, H) \to (U, U')$$

is continuous with respect to the corresponding topologies.

Proof Noting (19.13) and (19.14) in Lemma 19.3, it is easy to see that we need only to consider the case that $\widehat{U}_0 \equiv \widehat{U}_1 \equiv 0$. Assume that Ω is sufficiently smooth, for example, with C^3 boundary, there exists a function $h \in C^2(\overline{\Omega})$, such that

$$\nabla h = \nu \quad \text{on } \Sigma_1, \tag{19.32}$$

where ν is the unit outward normal vector on the boundary Σ_1 (cf. [62]). Let λ be an eigenvalue of B^T and let e be the corresponding eigenvector:

$$B^T e = \lambda e.$$

Defining

$$\phi = (e, U), \tag{19.33}$$

we have

$$\begin{cases} \phi'' - \Delta\phi = -(e, AU) & \text{in } (0, T) \times \Omega, \\ \phi = 0 & \text{on } (0, T) \times \Gamma_0, \\ \partial_\nu\phi + \lambda\phi = (e, DH) & \text{on } (0, T) \times \Gamma_1, \\ t = 0: \quad \phi = 0, \ \phi' = 0 \text{ in } \Omega. \end{cases} \tag{19.34}$$

Let

$$\psi = e^{\lambda h}\phi. \tag{19.35}$$

Problem (III) and (III0) can be rewritten to the following problem with Neumann boundary conditions:

$$\begin{cases} \psi'' - \Delta\psi + b(\psi) = -e^{\lambda h}(e, AU) & \text{in } (0,T)\times\Omega, \\ \psi = 0 & \text{on } (0,T)\times\Gamma_0, \\ \partial_\nu\psi = e^{\lambda h}(e, DH) & \text{on } (0,T)\times\Gamma_1, \\ t=0: \quad \psi = 0,\ \psi' = 0 & \text{in } \Omega, \end{cases} \tag{19.36}$$

where $b(\psi) = 2\lambda\nabla h \cdot \nabla\psi + \lambda(\Delta h - \lambda|\nabla h|^2)\psi$ is a first-order linear form of ψ with smooth coefficients.

By Theorem 19.9, $U \in C^0([0,T]; (\mathcal{H}_0)^N)$. By (19.11) in Lemma 19.2 and Remark 19.4, the solution ψ to the following problem with homogeneous Neumann boundary conditions:

$$\begin{cases} \psi'' - \Delta\psi + b(\psi) = -e^{\lambda h}(e, AU) & \text{in } (0,T)\times\Omega, \\ \psi = 0 & \text{on } (0,T)\times\Gamma_0, \\ \partial_\nu\psi = 0 & \text{on } (0,T)\times\Gamma_1, \\ t=0: \quad \psi = 0,\ \psi' = 0 & \text{in } \Omega \end{cases} \tag{19.37}$$

satisfies

$$(\psi, \psi') \in C^0([0,T]; H^1(\Omega) \times L^2(\Omega)). \tag{19.38}$$

Next, we consider the following problem with inhomogeneous Neumann boundary conditions but without internal force terms:

$$\begin{cases} \psi'' - \Delta\psi + b(\psi) = 0 & \text{in } (0,T)\times\Omega, \\ \psi = 0 & \text{on } (0,T)\times\Gamma_0, \\ \partial_\nu\psi = e^{\lambda h}(e, DH) & \text{on } (0,T)\times\Gamma_1, \\ t=0: \quad \psi = 0,\ \psi' = 0 \text{ in } \Omega. \end{cases} \tag{19.39}$$

By (19.9) and (19.10) in Lemma 19.1 and Remark 19.4, we have

$$(\psi, \psi') \in C^0([0,T]; H^\alpha(\Omega) \times H^{\alpha-1}(\Omega)) \tag{19.40}$$

and

$$\psi|_{\Sigma_1} \in H^{2\alpha-1}(\Sigma_1), \tag{19.41}$$

where α is given by (19.8). Since this regularity result holds for all the eigenvectors of B^T, and all the eigenvectors of B^T constitute a set of basis in $\mathbb{R}^N$, we get the desired (19.30) and (19.31). □

Chapter 20
Exact Boundary Controllability and Non-exact Boundary Controllability

In this chapter, we will study the exact boundary controllability and the non-exact boundary controllability for the coupled system (III) of wave equations with coupled Robin boundary controls.

In this chapter, we will study the exact boundary controllability and the non-exact boundary controllability for the coupled system (III) of wave equations with coupled Robin boundary controls. We will prove that when the number of boundary controls is equal to N, the number of state variables, system (III) is exactly controllable for any given initial data $(\widehat{U}_0, \widehat{U}_1) \in (\mathcal{H}_1)^N \times (\mathcal{H}_0)^N$, while, when $M < N$ and Ω is a parallelepiped, system (III) is not exactly boundary controllable in $(\mathcal{H}_1)^N \times (\mathcal{H}_0)^N$.

20.1 Exact Boundary Controllability

Definition 20.1 *System (III) is* **exactly null controllable** *in the space* $(\mathcal{H}_1)^N \times (\mathcal{H}_0)^N$, *if there exists a positive constant* $T > 0$, *such that for any given* $(\widehat{U}_1, \widehat{U}_0) \in (\mathcal{H}_1)^N \times (\mathcal{H}_0)^N$, *there exists a boundary control* $H \in L^2(0, T; (L^2(\Gamma_1))^M)$, *such that problem (III) and (III0) admits a unique weak solution* U *satisfying the final condition*

$$t = T: \quad U = U' = 0. \tag{20.1}$$

Theorem 20.2 *Assume that* $M = N$. *Assume furthermore that* B *is similar to a real symmetric matrix. Then there exists a time* $T > 0$, *such that for any given initial data* $(\widehat{U}_0, \widehat{U}_1) \in (\mathcal{H}_1)^N \times (\mathcal{H}_0)^N$, *there exists a boundary control* $H \in L^2(0, T; (L^2(\Gamma_1))^N)$, *such that system (III) is exactly null controllable at the time* T *and the boundary control continuously depends on the initial data:*

$$\|H\|_{L^2(0,T;(L^2(\Gamma_1))^N)} \leqslant c\|(\widehat{U}_0, \widehat{U}_1)\|_{(\mathcal{H}_1)^N \times (\mathcal{H}_0)^N}, \tag{20.2}$$

where $c > 0$ *is a positive constant.*

T. Li and B. Rao, *Boundary Synchronization for Hyperbolic Systems*, Progress in Nonlinear Differential Equations and Their Applications 94, https://doi.org/10.1007/978-3-030-32849-8_20

Proof We first consider the corresponding problem (II) and (II0). By Theorem 12.14, for any given initial data $(\widehat{U}_0, \widehat{U}_1) \in (\mathcal{H}_1)^N \times (\mathcal{H}_0)^N$, there exists a boundary control $\widehat{H} \in L^2_{loc}(0, +\infty; (L^2(\Gamma_1))^N)$ with compact support in $[0, T]$, such that system (II) with Neumann boundary controls is exactly controllable at the time T, and the boundary control $\widehat{H}$ continuously depends on the initial data:

$$\|\widehat{H}\|_{L^2(0,T;(L^2(\Gamma_1))^N)} \leqslant c_1 \|(\widehat{U}_0, \widehat{U}_1)\|_{(\mathcal{H}_1)^N \times (\mathcal{H}_0)^N}, \tag{20.3}$$

where $c_1 > 0$ is a positive constant.

Noting that $M = N$, D is invertible and the boundary condition in system (II)

$$\partial_\nu U = D\widehat{H} \quad \text{on } (0, T) \times \Gamma_1 \tag{20.4}$$

can be rewritten as

$$\partial_\nu U + BU = D(\widehat{H} + D^{-1}BU) := DH \quad \text{on } (0, T) \times \Gamma_1. \tag{20.5}$$

Thus, problem (II) and (II0) with (20.4) can be equivalently regarded as problem (III) and (III0) with (20.5). In other words, the boundary control H is given by

$$H = \widehat{H} + D^{-1}BU \quad \text{on } (0, T) \times \Gamma_1, \tag{20.6}$$

where U is the solution to problem (II) and (II0) with (20.4), realizes the exact boundary controllability of system (III).

It remains to check that H given in (20.6) belongs to the control space $L^2(0, T; (L^2(\Gamma_1))^N)$ with continuous dependence (20.2). By the regularity result given in Theorem 19.10 (in which we take B = 0), the trace $U|_{\Sigma_1} \in (H^{2\alpha-1}(\Sigma_1))^N$, where α is defined by (19.8). Since $2\alpha - 1 > 0$, we have $H \in L^2(0, T; (L^2(\Gamma_1))^N)$. Moreover, still by Theorem 19.9, we have

$$\|U\|_{(L^2(\Sigma_1))^N} \leqslant c_2\big(\|(\widehat{U}_0, \widehat{U}_1)\|_{(\mathcal{H}_1)^N \times (\mathcal{H}_0)^N} + \|\widehat{H}\|_{L^2(0,T;(L^2(\Gamma_1))^N)}\big), \tag{20.7}$$

where $c_2 > 0$ is another positive constant. Noting (20.6), (20.2) follows from (20.3) and (20.7). The proof is complete. □

20.2 Non-exact Boundary Controllability

Differently from the cases with Neumann boundary controls, the **non-exact boundary controllability** for a coupled system with coupled Robin boundary controls in a general domain is still an open problem. Fortunately, for some special domain, the solutions to problem (III) and (III0) may possess higher regularity. In particular, when Ω is a parallelepiped, the optimal regularity of trace $U|_{\Sigma_1}$ almost reaches

$H^{1/2}(\Gamma_1)$. This benefits a lot in the proof of the non-exact boundary controllability with fewer boundary controls.

Lemma 20.3 *Let $\mathcal{L}$ be a compact linear mapping from $L^2(\Omega)$ to $L^2(0,T;L^2(\Omega))$, and let $\mathcal{R}$ be a compact linear mapping from $L^2(\Omega)$ to $L^2(0,T;H^{1-\alpha}(\Gamma_1))$, where α is defined by (19.8). Then we cannot find a $T>0$, such that for any given $\theta \in L^2(\Omega)$, the solution to the following problem:*

$$\begin{cases} w'' - \Delta w = \mathcal{L}\theta & \text{in } (0,T)\times\Omega, \\ w = 0 & \text{on } (0,T)\times\Gamma_0, \\ \partial_\nu w = \mathcal{R}\theta & \text{on } (0,T)\times\Gamma_1, \\ t=0: \quad w=0,\ w'=\theta \text{ in } \Omega \end{cases} \tag{20.8}$$

satisfies the final condition

$$w(T) = w'(T) = 0. \tag{20.9}$$

Proof For any given $\theta \in L^2(\Omega)$, let ϕ be the solution to the following problem:

$$\begin{cases} \phi'' - \Delta\phi = 0 & \text{in } (0,T)\times\Omega, \\ \phi = 0 & \text{on } (0,T)\times\Gamma_0, \\ \partial_\nu\phi = 0 & \text{on } (0,T)\times\Gamma_1, \\ t=0: \quad \phi=\theta,\ \phi'=0 \text{ in } \Omega. \end{cases} \tag{20.10}$$

By (19.13) and (19.14) in Lemma 19.3, we have

$$\|\phi\|_{L^2(0,T;L^2(\Omega))} \leqslant c\|\theta\|_{L^2(\Omega)} \tag{20.11}$$

and

$$\|\phi\|_{L^2(0,T;H^{\alpha-1}(\Gamma_1))} \leqslant c\|\theta\|_{L^2(\Omega)}. \tag{20.12}$$

Noting (20.9) and taking the inner product with ϕ on both sides of (20.8) and integrating by parts, it is easy to get

$$\|\theta\|^2_{L^2(\Omega)} = \int_0^T\int_\Omega \mathcal{L}\theta\phi dx + \int_0^T\int_{\Gamma_1} \mathcal{R}\theta\phi d\Gamma. \tag{20.13}$$

Noting (20.11)–(20.12), we then have

$$\|\theta\|_{L^2(\Omega)} \leqslant c(\|\mathcal{L}\theta\|_{L^2(0,T;L^2(\Omega))} + \|\mathcal{R}\theta\|_{L^2(0,T;H^{1-\alpha}(\Gamma))}) \tag{20.14}$$

for all $\theta \in L^2(\Omega)$, which contradicts the compactness of $\mathcal{L}$ and $\mathcal{R}$. □

Remark 20.4 *When Ω is a parallelepiped with $\Gamma_0 = \emptyset$, by Fourier analysis, the solution to the adjoint problem (19.19) and (19.20) is sufficiently smooth. Therefore, Lemma 20.3 is still valid.*

Theorem 20.5 *Assume that rank$(D) = M < N$ and that B is similar to a real symmetric matrix. Assume furthermore that $\Omega \subset \mathbb{R}^n$ is a parallelepiped with $\Gamma_0 = \emptyset$. Then, no matter how large $T > 0$ is, system (III) is not exactly null controllable for any given initial data $(\widehat{U}_0, \widehat{U}_1) \in (\mathcal{H}_1)^N \times (\mathcal{H}_0)^N$.*

Proof Since $M < N$, there exists an $e \in \mathbb{R}^N$, such that $D^T e = 0$. Take the special initial data

$$t = 0: \quad U = 0, \; U' = e\theta \tag{20.15}$$

for system (III). If the system is exactly controllable for any given $\theta \in L^2(\Omega)$ at the time $T > 0$, then there exists a boundary control $H \in L^2(0, T; (L^2(\Gamma))^M)$, such that the corresponding solution satisfies

$$U(T) = U'(T) = 0. \tag{20.16}$$

Let

$$w = (e, U), \quad \mathcal{L}\theta = -(e, AU), \quad \mathcal{R}\theta = -(e, BU)|_\Sigma. \tag{20.17}$$

Noting that $D^T e = 0$, we see that w satisfies problem (20.8) and the final condition (20.9).

By Lemma 20.3, in order to show Theorem 20.5, it suffices to show that the linear mapping $\mathcal{L}$ is compact from $L^2(\Omega)$ to $L^2(0, T; L^2(\Omega))$, and $\mathcal{R}$ is compact from $L^2(\Omega)$ to $H^{1-\alpha}(\Sigma)$.

Assume that system (III) with special initial data (20.15) is exactly null controllable. The linear mapping $\theta \to H$ is continuous from $L^2(\Omega)$ to $L^2(0, T; (L^2(\Gamma))^M)$. By Theorem 19.10, $(\theta, H) \to (U, U')$ is a continuous mapping from $L^2(\Omega) \times L^2(0, T; (L^2(\Gamma))^M)$ to $C^0([0, T]; (H^\alpha(\Omega))^N) \cap C^1([0, T]; (H^{\alpha-1}(\Omega))^N)$. Besides, by Lions' compact embedding theorem (cf. Theorem 5.1 in [63]), the embedding $L^2(0, T; (H^\alpha(\Omega))^N) \cap H^1(0, T; (H^{\alpha-1}(\Omega))^N)\} \subset L^2(0, T; (L^2(\Omega))^N)$ is compact, hence the linear mapping $\mathcal{L}$ is compact from $L^2(\Omega)$ to $L^2(0, T; L^2(\Omega))$.

On the other hand, by (19.31), $H \to U|_\Sigma$ is a continuous mapping from $L^2(0, T; (L^2(\Gamma))^M)$ to $(H^{2\alpha-1}(\Sigma))^N$, then $\mathcal{R}: \theta \to -(e, BU)|_\Sigma$ is a continuous mapping from $L^2(\Omega)$ into $H^{2\alpha-1}(\Sigma_1)$. When Ω is a parallelepiped, $\alpha = 3/4 - \epsilon$, then $2\alpha - 1 > 1 - \alpha$, hence $H^{2\alpha-1}(\Sigma) \hookrightarrow H^{1-\alpha}(\Sigma)$ is a compact embedding; therefore, $\mathcal{R}$ is a compact mapping from $L^2(\Omega)$ to $H^{1-\alpha}(\Sigma)$. The proof is complete. □

Remark 20.6 *We obtain the non-exact boundary controllability for system (III) with coupled Robin boundary controls in a parallelepiped Ω when there is a lack of boundary controls. The main idea is to use the compact perturbation theory which has a higher requirement on the regularity of the solution to the original problem with coupled Robin boundary condition. How to generalize this result to the general domain is still an open problem.*

Chapter 21
Exact Boundary Synchronization

Based on the results of the exact boundary controllability and the non-exact boundary controllability, we study the exact boundary synchronization for system (III) with coupled Robin boundary controls.

21.1 Exact Boundary Synchronization

Based on the results of the exact boundary controllability and the non-exact boundary controllability, we continue to study the **exact boundary synchronization** for system (III) under coupled Robin boundary controls. Let

$$C_1 = \begin{pmatrix} 1 & -1 & & & \\ & 1 & -1 & & \\ & & \ddots & \ddots & \\ & & & 1 & -1 \end{pmatrix}_{(N-1)\times N} \tag{21.1}$$

be the corresponding full row-rank **matrix of synchronization**. We have

$$\mathrm{Ker}(C_1) = \mathrm{Span}\{e_1\}, \tag{21.2}$$

where $e_1 = (1, \cdots, 1)^T$.

Definition 21.1 System (III) is exactly synchronizable at the time $T > 0$ if for any given initial data $(\widehat{U}_0, \widehat{U}_1) \in (\mathcal{H}_1)^N \times (\mathcal{H}_0)^N$, there exists a boundary control $H \in (L^2_{loc}(0, +\infty; L^2(\Gamma_1)))^M$ with compact support in $[0, T]$, such that the corresponding solution $U = U(t, x)$ to the mixed problem (III) and (III0) satisfies

$$t \geqslant T: \quad C_1 U \equiv 0 \quad \text{in } \Omega. \tag{21.3}$$

T. Li and B. Rao, *Boundary Synchronization for Hyperbolic Systems*, Progress in Nonlinear Differential Equations and Their Applications 94,
https://doi.org/10.1007/978-3-030-32849-8_21

Theorem 21.2 *Let $\Omega \subset \mathbb{R}^n$ be a parallelepiped. Assume that the coupled system (III) of wave equations with coupled Robin boundary controls is exactly boundary synchronizable. Then we have*

$$rank(C_1 D) = N - 1. \tag{21.4}$$

Proof If $Ker(D^T) \cap Im(C_1^T) = \{0\}$, then, by Proposition 2.11, we get

$$\text{rank}(C_1 D) = \text{rank}(D^T C_1^T) = \text{rank}(C_1^T) = N - 1. \tag{21.5}$$

Otherwise, if $Ker(D^T) \cap Im(C_1^T) \neq \{0\}$, then there exists a vector $E \neq 0$, such that

$$D^T C_1^T E = 0. \tag{21.6}$$

Let

$$w = (E, C_1 U), \quad \mathcal{L}\theta = -(E, C_1 AU), \quad \mathcal{R}\theta = -(E, C_1 BU). \tag{21.7}$$

For any given initial data $(\widehat{U}_0, \widehat{U}_1) = (0, \theta e_1)$, where $\theta \in L^2(\Omega)$, we can still get problem (20.8) for w. By the exact boundary synchronization of system (III), for this initial data, there exists a boundary control $H \in L^2(0, T, (L^2(\Gamma_1))^M)$, such that (21.3) holds, then the corresponding solution w satisfies (20.9). Similarly to the proof of Theorem 20.5, when Ω is a parallelepiped, we can prove that $\mathcal{L}$ is a compact linear mapping from $L^2(\Omega)$ to $L^2(0, T; L^2(\Omega))$, and $\mathcal{R}$ is a compact linear mapping from $L^2(\Omega)$ to $L^2(0, T; H^{1-\alpha}(\Gamma_1))$. It contradicts Lemma 20.3. The proof is complete. □

21.2 Conditions of C_1-Compatibility

Remark 21.3 If rank$(D) = N$, by Theorem 20.2, system (III) is exactly null controllable, then exactly synchronizable. In order to exclude this trivial situation, in what follows, we always assume that rank$(D) = N - 1$.

Theorem 21.4 *Let $\Omega \subset \mathbb{R}^n$ be a parallelepiped. Assume that $rank(D) = N - 1$ and that B is similar to a real symmetric matrix. If the coupled system (III) of wave equations with coupled Robin boundary controls is exactly boundary synchronizable, then we have the following* **conditions of C_1-compatibility***:*

$$AKer(C_1) \subseteq Ker(C_1) \quad and \quad BKer(C_1) \subseteq Ker(C_1). \tag{21.8}$$

Proof Let $e_1 = (1, \cdots, 1)^T$. Then, we have

$$t \geqslant T: \quad U = ue_1, \tag{21.9}$$

where u is the corresponding **exactly synchronizable state**.

Taking the inner product with C_1 on both sides of (III), we get

$$t \geqslant T: \quad C_1 A e_1 u = 0 \quad \text{in } \Omega \quad \text{and} \quad C_1 B e_1 u = 0 \text{ on } \Gamma_1. \tag{21.10}$$

By Theorem 20.5, system (III) is not exactly boundary controllable when $\text{rank}(D) = N - 1$, hence there exists at least an initial data $(\widehat{U}_0, \widehat{U}_1)$ such that

$$t \geqslant T: \quad u \not\equiv 0 \quad \text{in } \Omega. \tag{21.11}$$

Therefore, we get $C_1 A e_1 = 0$. Noting that $\text{Ker}(C_1) = \text{Span}\{e_1\}$, we have

$$A\text{Ker}(C_1) \subseteq \text{Ker}(C_1). \tag{21.12}$$

We next prove $C_1 B e_1 = 0$. Otherwise, by the second formula in (21.10), we get $u|_{\Gamma_1} \equiv 0$. By (21.9), it is easy to see that as $t \geqslant T$, system (III) becomes

$$t \geqslant T: \quad \begin{cases} u'' - \Delta u + au = 0 \text{ in } (0, T) \times \Omega, \\ u = 0 \qquad\qquad\qquad \text{on } (0, T) \times \Gamma_0, \\ \partial_\nu u = u = 0 \qquad\quad \text{on } (0, T) \times \Gamma_1, \end{cases} \tag{21.13}$$

where a is defined by (17.8). Therefore, by Holmgren's uniqueness theorem (*cf.* Theorem 8.2 in [62]), we have $u \equiv 0$ as $t \geqslant T$. It contradicts the non-exact boundary controllability of system (III). The proof is then complete. □

Theorem 21.5 *Assume that rank$(D) = N - 1$ and that B is similar to a real symmetric matrix. Assume furthermore that both A and B satisfy the conditions of C_1-compatibility* (21.8). *Then there exists a boundary control matrix D with rank$(C_1 D) = N - 1$, such that system (III) is exactly synchronizable, and the boundary control H possesses the following continuous dependence:*

$$\|H\|_{L^2(0,T,(L^2(\Gamma_1))^{N-1})} \leqslant c\|C_1(\widehat{U}_0, \widehat{U}_1)\|_{(\mathcal{H}_1)^{N-1} \times (\mathcal{H}_0)^{N-1}}. \tag{21.14}$$

Proof Since both A and B satisfy the conditions of C_1-compatibility (21.8), by Proposition 2.15, there exist matrices $\overline{A}_1$ and $\overline{B}_1$ of order $(N - 1)$, such that

$$C_1 A = \overline{A}_1 C_1 \quad \text{and} \quad C_1 B = \overline{B}_1 C_1. \tag{21.15}$$

Let

$$W = C_1 U \quad \text{and} \quad \overline{D}_1 = C_1 D. \tag{21.16}$$

W satisfies

$$\begin{cases} W'' - \Delta W + \overline{A}_1 W = 0 \text{ in } (0, T) \times \Omega, \\ W = 0 \qquad\qquad\qquad\quad \text{on } (0, T) \times \Gamma_0, \\ \partial_\nu W + \overline{B}_1 W = \overline{D}_1 H \quad \text{on } (0, T) \times \Gamma_1 \end{cases} \tag{21.17}$$

with the initial data

$$t = 0: \quad W = C_1\widehat{U}_0, \; W' = C_1\widehat{U}_1 \quad \text{in } \Omega. \tag{21.18}$$

Noting that C_1 is a surjection from $\mathbb{R}^N$ to $\mathbb{R}^{N-1}$, any given initial data $(\widehat{U}_0, \widehat{U}_1) \in (\mathcal{H}_1)^N \times (\mathcal{H}_0)^N$ corresponds to a unique initial data $(C_1\widehat{U}_0, C_1\widehat{U}_1)$ of the reduced system (21.17). Then, it follows that the exact boundary synchronization for system (III) is equivalent to the exact boundary null controllability for the reduced system (21.17), and then the boundary control H, which realizes the exact boundary null controllability for the reduced system (21.17), is just the boundary control which realizes the exact boundary synchronization for system (III).

By Proposition 2.21, when B is similar to a real symmetric matrix, its reduced matrix $\overline{B}_1$ is also similar to a real symmetric matrix, which guarantees the well-posedness of the reduced problem (21.17) and (21.18).

Defining the boundary control matrix D by

$$D = C_1^T,$$

$$\overline{D}_1 = C_1 D = C_1 C_1^T \tag{21.19}$$

is an invertible matrix of order $(N-1)$, by Theorem 20.2, the reduced system (21.17) is exactly boundary controllable, then system (III) is exactly synchronizable. Moreover, by (20.2), we get (21.14). □

Chapter 22
Determination of Exactly Synchronizable States

When system (III) possesses the exact boundary synchronization, the corresponding exactly synchronizable states will be studied in this chapter.

Although under certain conditions, the exact boundary synchronization can be realized by fewer boundary controls; however, exactly synchronizable states are a priori unknown, which depend not only on the given initial data, but also on applied boundary controls in general. In this section, we will discuss the determination and the estimate of exactly synchronizable states.

22.1 Determination of Exactly Synchronizable States

Theorem 22.1 *Assume that both A and B satisfy the conditions of C_1-compatibility* (21.8) *and that B is similar to a real symmetric matrix. Assume furthermore that A^T and B^T have a common eigenvector $E_1 \in \mathbb{R}^N$, such that $(E_1, e_1) = 1$, where $e_1 = (1, \cdots, 1)^T$. Then there exists a boundary control matrix D with rank$(D) = N-1$, such that system (III) is exactly synchronizable, and the exactly synchronizable state is independent of applied boundary controls.*

Proof Taking the boundary control matrix D such that

$$\operatorname{Ker}(D^T) = Span\{E_1\}, \tag{22.1}$$

it is evident that $\operatorname{rank}(D) = N-1$. Noticing that $(E_1, e_1) = 1$, we have

$$\operatorname{Ker}(C_1) \cap \operatorname{Im}(D) = \operatorname{Span}(e_1) \cap \{\operatorname{Span}(E_1)\}^{\perp} = \{0\}. \tag{22.2}$$

T. Li and B. Rao, *Boundary Synchronization for Hyperbolic Systems*, Progress in Nonlinear Differential Equations and Their Applications 94, https://doi.org/10.1007/978-3-030-32849-8_22

Thus, by Proposition 2.15, we have

$$\operatorname{rank}(C_1 D) = \operatorname{rank}(D) = N - 1. \tag{22.3}$$

Then, by Theorem 21.5, system (III) is exactly synchronizable.

Next, we prove that the exactly synchronizable state is independent of boundary controls which realize the exact boundary synchronization. Noting that E_1 is a common eigenvector of A^T and B^T, and B is similar to a real symmetric matrix, there exist $\lambda \in \mathbb{C}$ and $\mu \in \mathbb{R}$, such that

$$A^T E_1 = \lambda E_1 \text{ and } B^T E_1 = \mu E_1. \tag{22.4}$$

Noting (22.1), we have $D^T E_1 = 0$, then $\phi = (E_1, U)$ satisfies

$$\begin{cases} \phi'' - \Delta\phi + \lambda\phi = 0 & \text{in } (0, +\infty) \times \Omega, \\ \phi = 0 & \text{on } (0, +\infty) \times \Gamma_0, \\ \partial_\nu \phi + \mu\phi = 0 & \text{on } (0, +\infty) \times \Gamma_1, \\ t = 0: \quad \phi = (E_1, \widehat{U}_0), \ \phi' = (E_1, \widehat{U}_1) \text{ in } \Omega. \end{cases} \tag{22.5}$$

Obviously, the solution ϕ to this problem is independent of boundary controls H.

On the other hand, noting that

$$t \geqslant T: \quad \phi = (E_1, U) = (E_1, e_1)u = u, \tag{22.6}$$

the exactly synchronizable state is determined by the solution to problem (22.5), and is independent of boundary controls H. The proof is complete. □

Remark 22.2 By Proposition 6.12, a matrix D satisfying $\operatorname{rank}(C_1 D) = \operatorname{rank}(D) = N - 1$ has the following form:

$$D = C_1^T D_1 + e_1 D_0,$$

where D_1 is an invertible matrix of order $(N-1)$ and D_0 is a matrix of order $1 \times (N-1)$. Furthermore, if $(E_1, e_1) = 1$ and $D^T E_1 = 0$, then

$$E_1^T C_1^T D_1 + D_0 = 0,$$

it follows that

$$D = (I - e_1 E_1^T) C_1^T D_1.$$

Theorem 22.3 *Assume that both A and B satisfy the conditions of C_1-compatibility (21.8). Assume furthermore that B is similar to a real symmetric matrix. Assume finally that system (III) is exactly synchronizable. Let $E_1 \neq 0$ be a vector in $\mathbb{R}^N$, such that the projection $\phi = (E_1, U)$ is independent of boundary controls H which* *realize the exact boundary synchronization. Then E_1 must be a common eigenvector of A^T and B^T, $E_1 \in Ker(D^T)$ and $(E_1, e_1) \neq 0$.*

Proof Taking $(\widehat{U}_0, \widehat{U}_1) = (0, 0)$ in problem (III) and (III0), by Theorem 19.10, the linear mapping

$$F : \; H \to (U, U')$$

is continuous from $L^2(0, T; (L^2(\Gamma_1))^M)$ to $C^0([0, T]; (H^\alpha(\Omega))^N \times (H^{\alpha-1}(\Omega))^N)$, where α is defined by (19.8). Let $\widehat{U}$ be the Fréchet derivative of U in the direction of $\widehat{H}$:

$$\widehat{U} = F'(0)\widehat{H}. \tag{22.7}$$

By linearity, $\widehat{U}$ satisfies the same problem as U:

$$\begin{cases} \widehat{U}'' - \Delta\widehat{U} + A\widehat{U} = 0 & \text{in } (0, +\infty) \times \Omega, \\ \widehat{U} = 0 & \text{on } (0, +\infty) \times \Gamma_0, \\ \partial_\nu\widehat{U} + B\widehat{U} = D\widehat{H} & \text{on } (0, +\infty) \times \Gamma_1 \\ t = 0 : \quad \widehat{U} = \widehat{U}' = 0 & \text{in } \Omega. \end{cases} \tag{22.8}$$

Since the projection $\phi = (E_1, U)$ is independent of boundary controls H which realize the exact boundary synchronization, we have

$$(E_1, \widehat{U}) \equiv 0, \quad \forall \widehat{H} \in L^2(0, T; (L^2(\Gamma_1))^M). \tag{22.9}$$

We first prove that $E_1 \notin \mathrm{Im}(C_1^T)$. Otherwise, there exists a vector $R \in \mathbb{R}^{N-1}$, such that $E_1 = C_1^T R$, then

$$(R, C_1\widehat{U}) = (E_1, \widehat{U}) = 0. \tag{22.10}$$

Noting that $C_1\widehat{U}$ is the solution to the reduced system (21.17) with the zero initial data, by the exact boundary synchronization of system (III), the reduced system (21.17) has the exact boundary controllability, then the value of $C_1\widehat{U}$ at the time T can be arbitrarily chosen, and as a result, from (22.10), we get

$$R = 0,$$

which contradicts the fact that $E_1 \neq 0$, then $E_1 \notin \mathrm{Im}(C_1^T)$. Hence, noting (21.2), we can choose E_1 such that

$$(E_1, e_1) = 1.$$

Thus, $\{E_1, C_1^T\}$ constitutes a basis in $\mathbb{R}^N$, and hence there exist a $\lambda \in \mathbb{C}$ and a vector $Q \in \mathbb{R}^{N-1}$, such that

$$A^T E_1 = \lambda E_1 + C_1^T Q. \tag{22.11}$$

Then by (22.11), taking the inner product with E_1 on both sides of the equations in (22.8), and noting (22.9), we get

$$0 = (A\widehat{U}, E_1) = (\widehat{U}, A^T E_1) = (\widehat{U}, C_1^T Q) = (C_1\widehat{U}, Q). \tag{22.12}$$

Thus, by the exact boundary controllability for the reduced system (21.17), we can get $Q = 0$, then it follows from (22.11) that

$$A^T E_1 = \lambda E_1.$$

On the other hand, by (22.10), taking the inner product with E_1 on both sides of the boundary condition on Γ_1 in (22.8), we get

$$(E_1, B\widehat{U}) = (E_1, D\widehat{H}) \quad \text{on } \Gamma_1. \tag{22.13}$$

By Theorem 19.10, we have

$$\begin{aligned} &\|(E_1, D\widehat{H})\|_{H^{2\alpha-1}((0,T)\times\Gamma_1)} \\ =&\|(E_1, B\widehat{U})\|_{H^{2\alpha-1}((0,T)\times\Gamma_1)} \leqslant c\|\widehat{H}\|_{L^2(0,T;(L^2(\Gamma_1))^M)}, \end{aligned} \tag{22.14}$$

here and hereafter, c denotes a positive constant.

We claim that $D^T E_1 = 0$, namely, $E_1 \in \text{Ker}(D^T)$. Otherwise, taking $\widehat{H} = D^T E_1 v$ in (22.14), we get a contradiction:

$$\|v\|_{H^{2\alpha-1}(\Sigma_1)} \leqslant c\|v\|_{L^2(0,T;L^2(\Gamma_1))}, \quad \forall v \in H^{2\alpha-1}(\Sigma_1), \tag{22.15}$$

because of $2\alpha - 1 > 0$.

Thus, we have

$$(E_1, B\widehat{U}) = 0 \quad \text{on } \Gamma_1. \tag{22.16}$$

Similarly, since $\{E_1, C_1^T\}$ constitutes a basis in $\mathbb{R}^N$, there exist a $\mu \in \mathbb{R}$ and a vector $P \in \mathbb{R}^{N-1}$, such that

$$B^T E_1 = \mu E_1 + C_1^T P. \tag{22.17}$$

Substituting it into (22.16) and noting (22.9), we have

$$(\mu E_1, \widehat{U}) + (C_1^T P, \widehat{U}) = (P, C_1\widehat{U}) = 0, \tag{22.18}$$

then by the exact boundary controllability for the reduced system (21.17), we get $P = 0$, and hence we have

$$B^T E_1 = \mu E_1.$$

Thus, E_1 must be a common eigenvector of A^T and B^T. The proof is complete. □

22.2 Estimation of Exactly Synchronizable States

Theorem 22.4 *Assume that both A and B satisfy the conditions of C_1-compatibility* (21.8), *namely, there exist real numbers a and b such that*

$$Ae_1 = ae_1 \quad \text{and} \quad Be_1 = be_1 \text{ with } e_1 = (1, \cdots, 1)^T. \tag{22.19}$$

Assume furthermore that B is similar to a real symmetric matrix. Then there exists a boundary control matrix D such that system (III) is exactly synchronizable, and each exactly synchronizable state u satisfies the following estimate:

$$\begin{aligned} &\|(u, u')(T) - (\phi, \phi')(T)\|_{H^{\alpha+1}(\Omega)\times H^{\alpha}(\Omega)} \\ \leqslant &c\|C_1(\widehat{U}_0, \widehat{U}_1)\|_{(\mathcal{H}_1)^{N-1}\times(\mathcal{H}_0)^{N-1}}, \end{aligned} \tag{22.20}$$

where α is defined by (19.8), *and ϕ is the solution to the following problem:*

$$\begin{cases} \phi'' - \Delta\phi + a\phi = 0 & \text{in } (0, +\infty) \times \Omega, \\ \phi = 0 & \text{on } (0, +\infty) \times \Gamma_0, \\ \partial_\nu\phi + b\phi = 0 & \text{on } (0, +\infty) \times \Gamma_1, \\ t = 0: \quad \phi = (\widehat{U}_0, E_1),\ \phi' = (\widehat{U}_1, E_1) \text{ in } \Omega. \end{cases} \tag{22.21}$$

Proof We first show that there exists an eigenvector $E_1 \in \mathbb{R}^N$ of B^T, associated with the eigenvalue b, satisfying $(E_1, e_1) = 1$.

Let P be a matrix such that PBP^{-1} is a real symmetric matrix. Setting

$$E_1 = P^T Pe_1, \tag{22.22}$$

we have

$$\begin{aligned} B^T E_1 =&P^T(PBP^{-1})^T Pe_1 = P^T PBP^{-1}Pe_1 \\ =&P^T PBe_1 = bP^T Pe_1 = bE_1 \end{aligned} \tag{22.23}$$

and

$$(E_1, e_1) = (P^T Pe_1, e_1) = \|Pe_1\|^2 > 0. \tag{22.24}$$

In particular, we can choose P such that $(E_1, e_1) = 1$.

Next, define the boundary control matrix D such that

$$Ker(D^T) = Span\{E_1\}. \tag{22.25}$$

Noting that $\text{Ker}(C_1) = \text{Span}\{e_1\}$ and $(E_1, e_1) = 1$, we have

$$\text{Ker}(C_1) \cap \text{Im}(D) = \{e_1\} \cap \{E_1\}^\perp = \{0\}, \tag{22.26}$$

then, by Proposition 2.11, we have

$$\mathrm{rank}(C_1 D) = \mathrm{rank}(D) = N - 1. \tag{22.27}$$

Thus, by Theorem 21.5, system (III) is exactly synchronizable. Taking the inner product with E_1 on both sides of problem (III) and (III0) and denoting $\psi = (E_1, U)$, we get

$$\begin{cases} \psi'' - \Delta\psi + a\psi = (aE_1 - A^T E_1, U) & \text{in } (0, +\infty) \times \Omega, \\ \psi = 0 & \text{on } (0, +\infty) \times \Gamma_0, \\ \partial_\nu \psi + b\psi = 0 & \text{on } (0, +\infty) \times \Gamma_1, \\ t = 0: \quad \psi = (\widehat{U}_0, E_1), \ \psi' = (\widehat{U}_1, E_1) \text{ in } \Omega. \end{cases} \tag{22.28}$$

Since

$$(aE_1 - A^T E_1, e_1) = (E_1, ae_1 - Ae_1) = 0, \tag{22.29}$$

we have

$$aE_1 - A^T E_1 \in \{\mathrm{Span}(e_1)\}^\perp = \mathrm{Im}(C_1^T). \tag{22.30}$$

Therefore, there exists a vector $R \in \mathbb{R}^{N-1}$, such that

$$aE_1 - A^T E_1 = C_1^T R. \tag{22.31}$$

Let $\varphi = \psi - \phi$, where ϕ is the solution to problem (22.21). By (22.21) and (22.28), we have

$$\begin{cases} \varphi'' - \Delta\varphi + a\varphi = (R, C_1 U) & \text{in } (0, +\infty) \times \Omega, \\ \varphi = 0 & \text{on } (0, +\infty) \times \Gamma_0, \\ \partial_\nu \varphi + b\varphi = 0 & \text{on } (0, +\infty) \times \Gamma_1, \\ t = 0: \quad \varphi = \varphi' = 0 & \text{in } \Omega. \end{cases} \tag{22.32}$$

Then, by the classic semigroups theory (cf. [70]), we have

$$\|(\varphi, \varphi')(T)\|_{H^{\alpha+1}(\Omega) \times H^\alpha(\Omega)} \leqslant c_1 \|(R, C_1 U)\|_{L^2(0,T; H^\alpha(\Omega))}, \tag{22.33}$$

here and hereafter c_i for $i = 1, 2, \cdots$ denote different positive constants.

Recall that $W = C_1 U$ is the solution to the reduced problem (21.17)–(21.18). Noting (20.2), it follows from Theorem 19.10 that

$$\begin{aligned} &\|C_1 U\|_{L^2(0,T; (H^\alpha(\Omega))^{N-1}} \\ \leqslant & c_2 (\|C_1(\widehat{U}_0, \widehat{U}_1)\|_{(\mathcal{H}_1)^{N-1} \times (\mathcal{H}_0)^{N-1}} + \|\overline{D}_1 H\|_{L^2(0,T; (L^2(\Gamma_1))^{N-1})}) \\ \leqslant & c_3 \|C_1(\widehat{U}_0, \widehat{U}_1)\|_{(\mathcal{H}_1)^{N-1} \times (\mathcal{H}_0)^{N-1}}. \end{aligned} \tag{22.34}$$

Thus, we get

$$\|(\varphi, \varphi')(T)\|_{H^{\alpha+1}(\Omega)\times H^{\alpha}(\Omega)} \leqslant c_4\|C_1(\widehat{U}_0, \widehat{U}_1)\|_{(\mathcal{H}_1)^{N-1}\times(\mathcal{H}_0)^{N-1}}. \tag{22.35}$$

On the other hand, we have

$$t \geqslant T: \quad \psi = (E_1, U) = (E_1, e_1)u = u, \tag{22.36}$$

thus $\varphi = u - \phi$ for $t \geqslant T$. Therefore, by (22.35) we get (22.20). □

Chapter 23
Exact Boundary Synchronization by p-Groups

The exact boundary synchronization by groups will be considered in this chapter for system (III) with further lack of coupled Robin boundary controls.

23.1 Definition

When there is a further lack of boundary controls, similarly to the case with Dirichlet or Neumann boundary controls, we consider the **exact boundary synchronization by p-groups** for system (III).

Let the components of U be divided into p groups:

$$(u^{(1)}, \cdots, u^{(n_1)}),\ (u^{(n_1+1)}, \cdots, u^{(n_2)}), \cdots, (u^{(n_{p-1}+1)}, \cdots, u^{(n_p)}), \tag{23.1}$$

where $0 = n_0 < n_1 < n_2 < \cdots < n_p = N$ are integers such that $n_r - n_{r-1} \geqslant 2$ for all $r = 1, \cdots, p$.

Definition 23.1 System (III) is exactly synchronizable by p-groups at the time $T > 0$ if for any given initial data $(\widehat{U}_0, \widehat{U}_1) \in (\mathcal{H}_1)^N \times (\mathcal{H}_0)^N$, there exists a boundary control $H \in L^2_{loc}(0, +\infty; (L^2(\Gamma_1))^M)$ with compact support in $[0, T]$, such that the corresponding solution $U = U(t, x)$ to problem (III) and (III0) satisfies

$$t \geqslant T: \quad u^{(i)} = u_r, \quad n_{r-1} + 1 \leqslant i \leqslant n_r, \quad 1 \leqslant r \leqslant p, \tag{23.2}$$

where $u = (u_1, \cdots, u_p)^T$, being unknown a priori, is called the corresponding **exactly synchronizable state by p-groups**.

T. Li and B. Rao, *Boundary Synchronization for Hyperbolic Systems*,
Progress in Nonlinear Differential Equations and Their Applications 94,
https://doi.org/10.1007/978-3-030-32849-8_23

Let S_r be an $(n_r - n_{r-1} - 1) \times (n_r - n_{r-1})$ full row-rank matrix:

$$S_r = \begin{pmatrix} 1 & -1 & & & \\ & 1 & -1 & & \\ & & \ddots & \ddots & \\ & & & 1 & -1 \end{pmatrix}, \quad 1 \leqslant r \leqslant p, \tag{23.3}$$

and let C_p be the following $(N - p) \times N$ **matrix of synchronization by p-groups**:

$$C_p = \begin{pmatrix} S_1 & & & \\ & S_2 & & \\ & & \ddots & \\ & & & S_p \end{pmatrix}. \tag{23.4}$$

Evidently, we have

$$\text{Ker}(C_p) = \text{Span}\{e_1, \cdots, e_p\}, \tag{23.5}$$

where for $1 \leqslant r \leqslant p$,

$$(e_r)_i = \begin{cases} 1, & n_{r-1} + 1 \leqslant i \leqslant n_r, \\ 0, & \text{others}. \end{cases}$$

Thus, the exact boundary synchronization by p-groups (23.2) can be equivalently written as

$$t \geqslant T: \quad C_p U \equiv 0, \tag{23.6}$$

or

$$t \geqslant T: \quad U = \sum_{r=1}^{p} u_r e_r. \tag{23.7}$$

23.2 Exact Boundary Synchronization by p-Groups

Theorem 23.2 *Let $\Omega \subset \mathbb{R}^n$ be a parallelepiped. Assume that system (III) is exactly synchronizable by p-groups. Then, we have*

$$rank(C_p D) = N - p. \tag{23.8}$$

In particular, we have

$$rank(D) \geqslant N - p. \tag{23.9}$$

Proof If $Ker(D^T) \cap Im(C_p^T) = \{0\}$, by Proposition 2.11, we have

$$\text{rank}(C_p D) = \text{rank}(D^T C_p^T) = \text{rank}(C_p^T) = N - p. \tag{23.10}$$

Next, we prove that it is impossible to have $Ker(D^T) \cap Im(C_p^T) \neq \{0\}$. Otherwise, there exists a vector $E \neq 0$, such that

$$D^T C_p^T E = 0. \tag{23.11}$$

Let

$$w = (E, C_p U), \quad \mathcal{L}\theta = -(E, C_p AU), \quad \mathcal{R}\theta = -(E, C_p BU). \tag{23.12}$$

We get again problem (20.8) for w. Besides, the exact boundary synchronization by p-groups for system (III) indicates that the final condition (20.9) holds. Similarly to the proof of Theorem 21.2, we get a contradiction to Lemma 20.3. □

Theorem 23.3 *Let C_p be the $(N-p) \times N$ matrix of synchronization by p-groups defined by (23.3)–(23.4). Assume that both A and B satisfy the following* **conditions of C_p-compatibility***:*

$$AKer(C_p) \subseteq Ker(C_p), \quad BKer(C_p) \subseteq Ker(C_p). \tag{23.13}$$

Then there exists a boundary control matrix D satisfying

$$rank(D) = rank(C_p D) = N - p, \tag{23.14}$$

such that system (III) is exactly synchronizable by p-groups, and the corresponding boundary control H possesses the following continuous dependence:

$$\|H\|_{L^2(0,T,(L^2(\Gamma_1))^{N-p})} \leqslant c\|C_p(\widehat{U}_0, \widehat{U}_1)\|_{(\mathcal{H}_1)^{N-p} \times (\mathcal{H}_0)^{N-p}}, \tag{23.15}$$

where $c > 0$ is a positive constant.

Proof Since both A and B satisfy the conditions of C_p-compatibility (23.13), by Proposition 2.15, there exist matrices $\overline{A}_p$ and $\overline{B}_p$ of order $(N-p)$, such that

$$C_p A = \overline{A}_p C_p, \quad C_p B = \overline{B}_p C_p. \tag{23.16}$$

Let

$$W = C_p U, \quad \overline{D}_p = C_p D. \tag{23.17}$$

We have

$$\begin{cases} W'' - \Delta W + \overline{A}_p W = 0 & \text{in } (0, +\infty) \times \Omega, \\ W = 0 & \text{on } (0, +\infty) \times \Gamma_0, \\ \partial_\nu W + \overline{B}_p W = \overline{D}_p H & \text{on } (0, +\infty) \times \Gamma_1 \end{cases} \tag{23.18}$$

with the initial data:

$$t = 0: \quad W = C_p \widehat{U}_0, \ W' = C_p \widehat{U}_1 \quad \text{in } \Omega. \tag{23.19}$$

Noting that C_p is a surjection from $\mathbb{R}^N$ to $\mathbb{R}^{N-p}$, the exact boundary synchronization by p-groups for system (III) is equivalent to the exact boundary null controllability for the reduced system (23.18), and the boundary control H, which realizes the exact boundary null controllability for the reduced system (23.18), must be the boundary control which realizes the exact boundary synchronization by p-groups for system (III).

Let D be defined by

$$\text{Ker}(D^T) = \text{Span}\{e_1, \cdots, e_p\} = \text{Ker}(C_p). \tag{23.20}$$

We have $\text{rank}(D) = N - P$, and

$$\text{Ker}(C_p) \cap \text{Im}(D) = \text{Ker}(C_p) \cap \{\text{Ker}(C_p)\}^\perp = \{0\}. \tag{23.21}$$

By Proposition 2.11, we get $\text{rank}(C_p D) = \text{rank}(D) = N - p$, thus $\overline{D}_p$ is an invertible matrix of order $(N - p)$. By Theorem 20.2, the reduced system (23.18) is exactly null controllable, then system (III) is exactly synchronizable by p-groups. By (20.2), we get (23.15). □

Chapter 24
Necessity of the Conditions of C_p-Compatibility

In this chapter, we will discuss the necessity of the conditions of C_p-compatibility for system (III) with coupled Robin boundary controls. This problem is closely related to the number of applied boundary controls.

In this chapter, we will discuss the necessity of the conditions of C_p-compatibility. This problem is closely related to the number of applied boundary controls.

24.1 Condition of C_p-Compatibility for the Internal Coupling Matrix

We first study the condition of C_p-compatibility for the coupling matrix A.

Theorem 24.1 *Let $\Omega \subset \mathbb{R}^n$ be a parallelepiped. Assume that $M = rank(D) = N - p$. If system (III) is exactly synchronizable by p-groups, then the coupling matrix $A = (a_{ij})$ should satisfy the following condition of C_p-compatibility:*

$$AKer(C_p) \subseteq Ker(C_p). \tag{24.1}$$

Proof It suffices to prove that

$$C_p Ae_r = 0, \quad 1 \leqslant r \leqslant p. \tag{24.2}$$

By (23.7), taking the inner product with C_p on both sides of the equations in system (III), we get

$$t \geqslant T: \quad \sum_{r=1}^{p} C_p Ae_r u_r = 0 \quad \text{in } \Omega. \tag{24.3}$$

T. Li and B. Rao, *Boundary Synchronization for Hyperbolic Systems*, Progress in Nonlinear Differential Equations and Their Applications 94,
https://doi.org/10.1007/978-3-030-32849-8_24

If (24.2) does not hold, then there exist constant coefficients $\alpha_r (1 \leqslant r \leqslant p)$, not all equal to zero, such that

$$\sum_{r=1}^{p} \alpha_r u_r = 0 \quad \text{in } \Omega. \tag{24.4}$$

Let

$$c_{p+1} = \sum_{r=1}^{p} \frac{\alpha_r e_r^T}{\|e_r\|^2}. \tag{24.5}$$

Noting $(e_r, e_s) = \|e_r\|^2 \delta_{rs}$, we have

$$t \geqslant T: \quad c_{p+1} U = \sum_{r=1}^{p} \alpha_r u_r = 0 \quad \text{in } \Omega. \tag{24.6}$$

Let

$$\widetilde{C}_{p-1} = \begin{pmatrix} C_p \\ c_{p+1} \end{pmatrix}. \tag{24.7}$$

Noting (23.5) and (24.5), it is easy to see that $c_{p+1}^T \notin \mathrm{Im}(C_p^T)$, then $\mathrm{rank}(\widetilde{C}_{p-1}) = N - p + 1$. Since $\mathrm{rank}(D) = N - p$, we have $\mathrm{Ker}(D^T) \cap \mathrm{Im}(\widetilde{C}_{p-1}^T) \neq \{0\}$, then there exists a vector $E \neq 0$, such that

$$D^T \widetilde{C}_{p-1}^T E = 0. \tag{24.8}$$

Let

$$w = (E, \widetilde{C}_{p-1} U), \quad \mathcal{L}\theta = -(E, \widetilde{C}_{p-1} A U), \quad \mathcal{R}\theta = -(E, \widetilde{C}_{p-1} B U).$$

We get again problem (20.8) for w. Noting (23.6) and (24.6), we have

$$t = T: \quad w(T) = (E, \widetilde{C}_{p-1} U) = (E, \begin{pmatrix} C_p \\ c_{p+1} \end{pmatrix} U) = 0. \tag{24.9}$$

Similarly, we have $w'(T) = 0$, then (20.9) holds. Noting that $\Omega \subset \mathbb{R}^n$ is a parallelepiped, similarly to the proof of Theorem 21.2, we can get a conclusion that contradicts Lemma 20.3. □

Remark 24.2 The condition of C_p-compatibility (24.1) is equivalent to the fact that there exist constants α_{rs} $(1 \leqslant r, s \leqslant p)$ such that

$$A e_r = \sum_{s=1}^{p} \alpha_{sr} e_s \quad 1 \leqslant r \leqslant p, \tag{24.10}$$

or A satisfies the following **row-sum condition by blocks**:

$$\sum_{j=n_{s-1}+1}^{n_s} a_{ij} = \alpha_{rs}, \quad n_{r-1}+1 \leqslant i \leqslant n_r, \quad 1 \leqslant r, s \leqslant p. \tag{24.11}$$

In particular, when

$$\alpha_{sr} = 0, \quad 1 \leqslant r, s \leqslant p, \tag{24.12}$$

we say that A satisfies the **zero-sum condition by blocks**. In this case, we have

$$Ae_r = 0, \quad 1 \leqslant r \leqslant p. \tag{24.13}$$

24.2 Condition of C_p-Compatibility for the Boundary Coupling Matrix

Comparing with the internal coupling matrix A, the study on the necessity of the condition of C_p-compatibility for the boundary coupling matrix B is more complicated. It concerns the regularity of solution to the problem with coupled Robin boundary conditions.

Let

$$\varepsilon_i = (0, \cdots, \overset{(i)}{1}, \cdots, 0)^T, \quad 1 \leqslant i \leqslant N \tag{24.14}$$

be a set of classical orthogonal basis in $\mathbb{R}^N$, and let

$$V_r = \text{Span}\{\varepsilon_{n_{r-1}+1}, \cdots, \varepsilon_{n_r}\}, \quad 1 \leqslant r \leqslant p. \tag{24.15}$$

Obviously, we have

$$e_r \in V_r, \quad 1 \leqslant r \leqslant p. \tag{24.16}$$

In what follows, we will discuss the necessity of the condition of C_p-compatibility for the boundary coupling matrix B under the assumption that $Ae_r \in V_r$ and $Be_r \in V_r$ $(1 \leqslant r \leqslant p)$.

Theorem 24.3 *Let $\Omega \subset \mathbb{R}^n$ be a parallelepiped. Assume that $M = rank(D) = N - p$ and*

$$Ae_r \in V_r, \quad Be_r \in V_r \quad (1 \leqslant r \leqslant p). \tag{24.17}$$

If system (III) is exactly synchronizable by p-groups, then the boundary coupling matrix B should satisfy the following condition of C_p-compatibility:

$$BKer(C_p) \subseteq Ker(C_p). \tag{24.18}$$

Proof Noting (23.7), we have

$$\begin{cases} \sum_{r=1}^{p}(u_r''e_r - \Delta u_r e_r + u_r A e_r) = 0 \text{ in } (T, +\infty) \times \Omega, \\ \sum_{r=1}^{p}(\partial_\nu u_r e_r + u_r B e_r) = 0 \qquad \text{on } (T, +\infty) \times \Gamma_1. \end{cases} \tag{24.19}$$

Noting (24.16)–(24.17) and the fact that subspaces $V_r (1 \leqslant r \leqslant p)$ are orthogonal to each other, for $1 \leqslant r \leqslant p$, we have

$$\begin{cases} u_r''e_r - \Delta u_r e_r + u_r A e_r = 0 \text{ in } (T, +\infty) \times \Omega, \\ \partial_\nu u_r e_r + u_r B e_r = 0 \qquad \text{on } (T, +\infty) \times \Gamma_1. \end{cases} \tag{24.20}$$

Taking the inner product with C_p on both sides of the boundary condition on Γ_1 in (24.20), and noting (23.5), we get

$$u_r C_p B e_r \equiv 0 \quad \text{on } (T, +\infty) \times \Gamma_1, \quad 1 \leqslant r \leqslant p. \tag{24.21}$$

We claim that $C_p B e_r = 0$ $(r = 1, \cdots, p)$, which just means that B satisfies the condition of C_p-compatibility (24.18). Otherwise, there exists an $\bar{r}$ $(1 \leqslant \bar{r} \leqslant p)$ such that $C_p B e_{\bar{r}} \neq 0$, and consequently, we have

$$u_{\bar{r}} \equiv 0 \quad \text{on } (T, +\infty) \times \Gamma_1. \tag{24.22}$$

Then, it follows from the boundary condition in system (24.20) that

$$\partial_\nu u_{\bar{r}} \equiv 0 \quad \text{on } (T, +\infty) \times \Gamma_1. \tag{24.23}$$

Hence, applying Holmgren's uniqueness theorem (cf. Theorem 8.2 in [62]) to (24.20) yields that

$$u_{\bar{r}} \equiv 0 \quad \text{in } (T, +\infty) \times \Omega, \tag{24.24}$$

then it is easy to check that

$$t \geqslant T: \quad e_{\bar{r}}^T U \equiv 0 \quad \text{in } \Omega. \tag{24.25}$$

Let

$$\widetilde{C}_{p-1} = \begin{pmatrix} C_p \\ e_{\bar{r}}^T \end{pmatrix}. \tag{24.26}$$

Since $e_{\bar{r}}^T \notin \text{Im}(C_p^T)$, it is easy to show that $\text{rank}(\widetilde{C}_{p-1}) = N - p + 1$. On the other hand, since $\text{rank}(\text{Ker}(D^T)) = p$, we have $\text{Ker}(D^T) \cap \text{Im}(\widetilde{C}_{p-1}^T) \neq \{0\}$, then there exists a vector $E \neq 0$, such that

$$D^T \widetilde{C}_{p-1}^T E = 0. \tag{24.27}$$

Let

$$w = (E, \widetilde{C}_{p-1} U), \quad \mathcal{L}\theta = -(E, \widetilde{C}_{p-1} AU), \quad \mathcal{R}\theta = -(E, \widetilde{C}_{p-1} BU). \tag{24.28}$$

We get again problem (20.8) for w, and (20.9) holds. Therefore, noting that $\Omega \subset \mathbb{R}^n$ is a parallelepiped, similarly to the proof of Theorem 21.2, we can get a conclusion that contradicts Lemma 20.3. □

Remark 24.4 Noting that condition (24.17) obviously holds for $p = 1$, so, the conditions of C_1-compatibility (21.8) are always satisfied for both A and B in the case $p = 1$. In the next result, we will remove the restricted conditions (24.17) in the case $p = 2$, and further study the necessity of the condition of C_2-compatibility for B.

Theorem 24.5 *Let $\Omega \subset \mathbb{R}^n$ be a parallelepiped and let the matrix B be similar to a real symmetric matrix. Assume that the coupling matrix A satisfies the zero-sum condition by blocks (24.13). Assume furthermore that system (III) is exactly synchronizable by 2-groups with $\text{rank}(D) = N - 2$. Then the coupling matrix B necessarily satisfies the condition of C_2-compatibility:*

$$BKer(C_2) \subseteq Ker(C_2). \tag{24.29}$$

Proof By the exact boundary synchronization by 2-groups, we have

$$t \geqslant T: \quad U = e_1 u_1 + e_2 u_2 \quad \text{in } \Omega. \tag{24.30}$$

Noting (24.13), as $t \geqslant T$, we have

$$AU = Ae_1 u_1 + Ae_2 u_2 = 0,$$

then it is easy to see that

$$\begin{cases} U'' - \Delta U = 0 & \text{in } (T, +\infty) \times \Omega, \\ U = 0 & \text{on } (T, +\infty) \times \Gamma_0, \\ \partial_\nu U + BU = 0 & \text{on } (T, +\infty) \times \Gamma_1. \end{cases} \tag{24.31}$$

Let P be a matrix such that $\widehat{B} = PBP^{-1}$ is a real symmetric matrix. Denote

$$u = (u_1, u_2)^T. \tag{24.32}$$

Taking the inner product on both sides of (24.31) with $P^T Pe_i$ for $i = 1, 2$, we get

$$\begin{cases} Lu'' - L\Delta u = 0 \text{ in } (T, +\infty) \times \Omega, \\ Lu = 0 \qquad\qquad \text{on } (T, +\infty) \times \Gamma_0, \\ L\partial_\nu u + \Lambda u = 0 \text{ on } (T, +\infty) \times \Gamma_1, \end{cases} \tag{24.33}$$

where the matrices L and Λ are given by

$$L = (Pe_i, Pe_j) \quad \text{and} \quad \Lambda = (\widehat{B}Pe_i, Pe_j), \quad 1 \leqslant i, j \leqslant 2. \tag{24.34}$$

Clearly, L is a symmetric and positive definite matrix and Λ is a symmetric matrix.

Taking the inner product on both sides of (24.33) with $L^{-\frac{1}{2}}$ and denoting $w = L^{\frac{1}{2}}u$, we get

$$\begin{cases} w'' - \Delta w = 0 \quad \text{in } (T, +\infty) \times \Omega, \\ w = 0 \qquad\qquad \text{on } (T, +\infty) \times \Gamma_0, \\ \partial_\nu w + \widehat{\Lambda} w = 0 \text{ on } (T, +\infty) \times \Gamma_1, \end{cases} \tag{24.35}$$

where $\widehat{\Lambda} = L^{-\frac{1}{2}} \Lambda L^{-\frac{1}{2}}$ is also a symmetric matrix.

On the other hand, taking the inner product with C_2 on both sides of the boundary condition on Γ_1 in system (III) and noting (24.30), we get

$$t \geqslant T: \quad C_2Be_1u_1 + C_2Be_2u_2 \equiv 0 \quad \text{on } \Gamma_1. \tag{24.36}$$

We claim that $C_2Be_1 = C_2Be_2 = 0$, namely, B satisfies the condition of C_2-compatibility (24.29).

Otherwise, without loss of generality, we may assume that $C_2Be_1 \neq 0$. Then it follows from (24.36) that there exists a nonzero vector $D_2 \in \mathbb{R}^2$, such that

$$t \geqslant T: \quad D_2^T u \equiv 0 \quad \text{on } \Gamma_1. \tag{24.37}$$

Denoting

$$\widehat{D}^T = D_2^T L^{-\frac{1}{2}},$$

we then have

$$t \geqslant T: \quad \widehat{D}^T w = 0 \quad \text{on } \Gamma_1, \tag{24.38}$$

in which $w = L^{\frac{1}{2}}u$. By Theorem 28.5, the following Kalman's criterion

$$\text{rank}(\widehat{D}, \widehat{\Lambda}\widehat{D}) = 2 \tag{24.39}$$

is sufficient for the unique continuation of system (24.35) under the observation (24.38) on the infinite horizon $[T, +\infty)$. Since $\text{rank}(D) = N - 2$, by Theorem 20.5, system (III) is not exactly null controllable. So, the rank condition (24.39) does not hold. Thus, by Proposition 2.12 (i), there exists a vector $E \neq 0$ in $\mathbb{R}^2$, such that

$$\widehat{\Lambda}^T E = \widehat{\Lambda} E = \mu E \quad \text{and} \quad \widehat{D}^T E = 0. \tag{24.40}$$

Noting (24.38) and the second formula of (24.40), we have both E and $w|_{\Gamma_1} \in \text{Ker}(\widehat{D})$. Since $\dim \text{Ker}(\widehat{D}) = 1$, there exists a constant α such that $w = \alpha E$ on Γ_1. Therefore, noting the first formula of (24.40), we have

$$\widehat{\Lambda} w = \widehat{\Lambda} \alpha E = \mu \alpha E = \mu w \quad \text{on } \Gamma_1. \tag{24.41}$$

Thus, (24.35) can be rewritten as

$$\begin{cases} w'' - \Delta w = 0 & \text{in } (T, +\infty) \times \Omega, \\ w = 0 & \text{on } (T, +\infty) \times \Gamma_0, \\ \partial_\nu w + \mu w = 0 & \text{on } (T, +\infty) \times \Gamma_1. \end{cases} \tag{24.42}$$

Let $z = \widehat{D}^T w$. Noting (24.38), it follows from (24.42) that

$$\begin{cases} z'' - \Delta z = 0 & \text{in } (T, +\infty) \times \Omega, \\ z = 0 & \text{on } (T, +\infty) \times \Gamma_0, \\ \partial_\nu z = z = 0 & \text{on } (T, +\infty) \times \Gamma_1. \end{cases} \tag{24.43}$$

Then, by Holmgren's uniqueness theorem, we have

$$t \geqslant T: \quad z = \widehat{D}^T w = D_2^T u \equiv 0 \quad \text{in } \Omega. \tag{24.44}$$

Let $D_2^T = (\alpha_1, \alpha_2)$. Define the following row vector:

$$c_3 = \frac{\alpha_1 e_1^T}{\|e_1\|^2} + \frac{\alpha_2 e_2^T}{\|e_2\|^2}. \tag{24.45}$$

Noting $(e_1, e_2) = 0$ and (24.44), we have

$$t \geqslant T: \quad c_3 U = \alpha_1 u_1 + \alpha_2 u_2 = D_2^T u \equiv 0 \quad \text{in } \Omega. \tag{24.46}$$

Let

$$\widetilde{C}_1 = \begin{pmatrix} C_2 \\ c_3 \end{pmatrix}. \tag{24.47}$$

Since $c_3^T \notin \mathrm{Im}(C_2^T)$, it is easy to see that $\mathrm{rank}(\widetilde{C}_1) = N - 1$, and $\mathrm{Ker}(D^T) \cap \mathrm{Im}(\widetilde{C}_1^T) \neq \{0\}$, thus there exists a vector $\widetilde{E} \neq 0$, such that

$$D^T \widetilde{C}_1^T \widetilde{E} = 0. \tag{24.48}$$

Let

$$u = (\widetilde{E}, \widetilde{C}_1 U), \quad \mathcal{L}\theta = -(\widetilde{E}, \widetilde{C}_1 AU), \quad \mathcal{R}\theta = -(\widetilde{E}, \widetilde{C}_1 BU). \tag{24.49}$$

We get again problem (20.8) for w, satisfying (20.9). Noting that Ω is a parallelepiped, similarly to the proof of Theorem 21.2, we get a contradiction to Lemma 20.3. □

Chapter 25
Determination of Exactly Synchronizable States by p-Groups

When system (III) possesses the exact boundary synchronization by p-groups, the corresponding exactly synchronizable states by p-groups will be studied in this chapter.

In general, exactly synchronizable states by p-groups depend not only on initial data, but also on applied boundary controls. However, when the coupling matrices A and B satisfy certain algebraic conditions, the exactly synchronizable state by p-groups can be independent of applied boundary controls. In this section, we first discuss the case when the exactly synchronizable state by p-groups is independent of applied boundary controls, then we present the estimate on each exactly synchronizable state by p-groups in general situation.

25.1 Determination of Exactly Synchronizable States by p-Groups

Theorem 25.1 *Assume that both A and B satisfy the conditions of C_p-compatibility (23.13) and that B is similar to a real symmetric matrix. Assume furthermore that A^T and B^T possess a common invariant subspace V, bi-orthonormal to Ker(C_p). Then there exists a boundary control matrix D with rank(D) = rank(C_pD) = $N - p$, such that system (III) is exactly synchronizable by p-groups, and the exactly synchronizable state by p-groups $u = (u_1, \cdots, u_p)^T$ is independent of applied boundary controls.*

Proof Define the boundary control matrix D by

$$\mathrm{Ker}(D^T) = V. \tag{25.1}$$

T. Li and B. Rao, *Boundary Synchronization for Hyperbolic Systems*, Progress in Nonlinear Differential Equations and Their Applications 94, https://doi.org/10.1007/978-3-030-32849-8_25

Since V is bi-orthonormal to $\text{Ker}(C_p)$, by Proposition 2.5, we have

$$\text{Ker}(C_p) \cap \text{Im}(D) = \text{Ker}(C_p) \cap V^{\perp} = \{0\}, \tag{25.2}$$

then, by Proposition 2.11, we have

$$\text{rank}(C_p D) = \text{rank}(D) = N - p. \tag{25.3}$$

Thus, by Theorem 23.3, system (III) is exactly synchronizable by p-groups.

By (25.2), noting $\text{Ker}(C_p) = \text{Span}\{e_1, \cdots, e_p\}$, we may write

$$V = \text{Span}\{E_1, \cdots, E_p\} \text{ with } (e_r, E_s) = \delta_{rs}(r, s = 1, \cdots, p). \tag{25.4}$$

Since V is a common invariant subspace of A^T and B^T, there exist constants α_{rs} and $\beta_{rs}(r, s = 1, \cdots, p)$ such that

$$A^T E_r = \sum_{s=1}^{p} \alpha_{rs} E_s \quad \text{and} \quad B^T E_r = \sum_{s=1}^{p} \beta_{rs} E_s. \tag{25.5}$$

For $r = 1, \cdots, p$, let

$$\phi_r = (E_r, U). \tag{25.6}$$

By system (III) and noting (25.1), for $r = 1, \cdots, p$, we have

$$\begin{cases} \phi_r'' - \Delta\phi_r + \sum_{s=1}^{p} \alpha_{rs}\phi_s = 0 & \text{in } (0, +\infty) \times \Omega, \\ \phi_r = 0 & \text{on } (0, +\infty) \times \Gamma_0, \\ \partial_\nu\phi_r + \sum_{s=1}^{p} \beta_{rs}\phi_s = 0 & \text{on } (0, +\infty) \times \Gamma_1, \\ t = 0: \quad \phi_r = (E_r, \widehat{U}_0), \ \phi_r' = (E_r, \widehat{U}_1) \text{ in } \Omega. \end{cases} \tag{25.7}$$

On the other hand, for $r = 1, \cdots, p$, we have

$$t \geqslant T: \quad \phi_r = (E_r, U) = \sum_{s=1}^{p} (E_r, e_s) u_s = \sum_{s=1}^{p} \delta_{rs} u_s = u_r. \tag{25.8}$$

Thus, the exactly synchronizable state by p-groups $u = (u_1, \cdots, u_p)^T$ is entirely determined by the solution to problem (25.7), which is independent of applied boundary controls H. □

The following result gives the counterpart of Theorem 25.1.

Theorem 25.2 *Assume that both A and B satisfy the conditions of C_p-compatibility (23.13) and that B is similar to a real symmetric matrix. Assume furthermore that* *system (III) is exactly synchronizable by p-groups. Let $V = Span\{E_1, \cdots, E_p\}$ be a subspace of dimension p. If the projection functions*

$$\phi_r = (E_r, U), \quad r = 1, \cdots, p \tag{25.9}$$

are independent of applied boundary controls H which realize the exact boundary synchronization by p-groups, then V is a common invariant subspace of A^T and B^T, and bi-orthonormal to $Ker(C_p)$.

Proof Let $(\widehat{U}_0, \widehat{U}_1) = (0, 0)$. By Theorem 19.10, the linear mapping

$$F: \quad H \to (U, U')$$

is continuous from $L^2(0, T; (L^2(\Gamma_1))^M)$ to $C^0([0, T]; (H^{\alpha}(\Omega))^N \times (H^{\alpha-1}(\Omega))^N)$, where α is defined by (19.8).

Let F' denote the Fréchet derivative of the application F. For any given $\widehat{H} \in L^2(0, T; (L^2(\Gamma_1))^M)$, we define

$$\widehat{U} = F'(0)\widehat{H}. \tag{25.10}$$

By linearity, $\widehat{U}$ satisfies a system similar to that of U:

$$\begin{cases} \widehat{U}'' - \Delta \widehat{U} + A\widehat{U} = 0 & \text{in } (0, +\infty) \times \Omega, \\ \widehat{U} = 0 & \text{on } (0, +\infty) \times \Gamma_0, \\ \partial_\nu \widehat{U} + B\widehat{U} = D\widehat{H} & \text{on } (0, +\infty) \times \Gamma_1 \\ t = 0: \quad \widehat{U} = \widehat{U}' = 0 & \text{in } \Omega. \end{cases} \tag{25.11}$$

Since the projection functions $\phi_r = (E_r, U)$ $(r = 1, \cdots, p)$ are independent of applied boundary controls H, we have

$$(E_r, \widehat{U}) \equiv 0, \quad \forall\, \widehat{H} \in L^2(0, T; (L^2(\Gamma_1))^M), \quad r = 1, \cdots, p. \tag{25.12}$$

First, we prove that $E_r \notin \mathrm{Im}(C_p^T)$ for $r = 1, \cdots, p$. Otherwise, there exist an $\bar{r}$ and a vector $R_{\bar{r}} \in \mathbb{R}^{N-p}$, such that $E_{\bar{r}} = C_p^T R_{\bar{r}}$, then we have

$$0 = (E_{\bar{r}}, \widehat{U}) = (R_{\bar{r}}, C_p\widehat{U}), \quad \forall \widehat{H} \in L^2(0, T; (L^2(\Gamma_1))^M). \tag{25.13}$$

Since $C_p\widehat{U}$ is the solution to the corresponding reduced problem (23.18) and (23.19), noting the equivalence between the exact boundary synchronization by p-groups for the original system and the exact boundary controllability for the reduced system, from the exact boundary synchronization by p-groups for system (III), we know that the reduced system (23.18) is exactly controllable, then the value of $C_p\widehat{U}$ at the time T can be chosen arbitrarily, thus we get $R_{\bar{r}} = 0$, which contradicts $E_{\bar{r}} \neq 0$. Then, we have $E_r \notin \mathrm{Im}(C_p^T)$ $(r = 1, \cdots, p)$. Thus, $V \cap \{\mathrm{Ker}(C_p)\}^{\perp} = V \cap \mathrm{Im}(C_p^T) =$

$\{0\}$. Hence, by Propositions 2.4 and 2.5, V is bi-orthonormal to $\text{Ker}(C_p)$, and then (V, C_p^T) constitutes a set of basis in $\mathbb{R}^N$. Therefore, there exist constant coefficients α_{rs} $(r, s = 1, \cdots, p)$ and vectors $P_r \in \mathbb{R}^{N-p}$ $(r = 1, \cdots, p)$, such that

$$A^T E_r = \sum_{s=1}^{p} \alpha_{rs} E_s + C_p^T P_r, \quad r = 1, \cdots, p. \tag{25.14}$$

Taking the inner product with E_r on both sides of the equations in (25.11) and noting (25.12), we get

$$0 = (A\widehat{U}, E_r) = (\widehat{U}, A^T E_r) = (\widehat{U}, C_p^T P_r) = (C_p\widehat{U}, P_r) \tag{25.15}$$

for $r = 1, \cdots, p$. Similarly, by the exact boundary controllability for the reduced system (23.18), we get $P_r = 0$ $(r = 1, \cdots, p)$, thus we have

$$A^T E_r = \sum_{s=1}^{p} \alpha_{rs} E_s, \quad r = 1, \cdots, p,$$

which means that V is an invariant subspace of A^T.

On the other hand, noting (25.12) and taking the inner product with E_r on both sides of the boundary condition on Γ_1 in (25.11), we get

$$(E_r, B\widehat{U}) = (E_r, D\widehat{H}) \quad \text{on } \Gamma_1, \quad r = 1, \cdots, p. \tag{25.16}$$

By Theorem 19.10, for $r = 1, \cdots, p$, we have

$$\begin{aligned} &\|(E_r, D\widehat{H})\|_{H^{2\alpha-1}(\Sigma_1)} \\ =&\|(E_r, B\widehat{U})\|_{H^{2\alpha-1}(\Sigma_1)} \leqslant c\|\widehat{H}\|_{L^2(0,T;(L^2(\Gamma_1))^M)}. \end{aligned} \tag{25.17}$$

We claim that $D^T E_r = 0$ for $r = 1, \cdots, p$. Otherwise, for $r = 1, \cdots, p$, setting $\widehat{H} = D^T E_r v$, it follows from (25.17) that

$$\|v\|_{H^{2\alpha-1}(\Sigma_1)} \leqslant c\|v\|_{L^2(0,T;L^2(\Gamma_1))}. \tag{25.18}$$

Since $2\alpha - 1 > 0$, it contradicts the compactness of $H^{2\alpha-1}(\Sigma_1) \hookrightarrow L^2(\Sigma_1)$. Thus, by (25.16) we have

$$(E_r, B\widehat{U}) = 0 \quad \text{on } (0, T) \times \Gamma_1, \quad r = 1, \cdots, p. \tag{25.19}$$

Similarly, there exist constants β_{rs} $(r, s = 1, \cdots, p)$ and vectors $Q_r \in \mathbb{R}^{N-p}$ $(r = 1, \cdots, p)$, such that

$$B^T E_r = \sum_{s=1}^{p} \beta_{rs} E_s + C_p^T Q_r, \quad r = 1, \cdots, p. \tag{25.20}$$

Substituting it into (25.19) and noting (25.12), we have

$$\sum_{s=1}^{p} \beta_{rs}(E_s, \widehat{U}) + (C_p^T Q_r, \widehat{U}) = (Q_r, C_p\widehat{U}) = 0, \quad r = 1, \cdots, p. \tag{25.21}$$

Similarly, by the exact boundary controllability for the reduced system (23.18), we get $Q_r = 0$ $(r = 1, \cdots, p)$, then we have

$$B^T E_r = \sum_{s=1}^{p} \beta_{rs} E_s, \quad r = 1, \cdots, p, \tag{25.22}$$

which indicates that V is also an invariant subspace of B^T. The proof is complete. □

25.2 Estimation of Exactly Synchronizable States by p-Groups

When A and B do not satisfy all the conditions mentioned in Theorem 25.1, exactly synchronizable states by p-groups may depend on applied boundary controls. We have the following

Theorem 25.3 *Assume that both A and B satisfy the conditions of C_p-compatibility (23.13). Then there exists a boundary control matrix D such that system (III) is exactly synchronizable by p-groups, and each exactly synchronizable state by p-groups $u = (u_1, \cdots, u_p)^T$ satisfies the following estimate:*

$$\begin{aligned} &\|(u, u')(T) - (\phi, \phi')(T)\|_{(H^{\alpha+1}(\Omega))^p \times (H^{\alpha}(\Omega))^p} \\ \leqslant & c\|C_p(\widehat{U}_0, \widehat{U}_1)\|_{(\mathcal{H}_1)^{N-p} \times (\mathcal{H}_0)^{N-p}}, \end{aligned} \tag{25.23}$$

where α is defined by (19.8), c is a positive constant, and $\phi = (\phi_1, \cdots, \phi_p)^T$ is the solution to the following problem $(1 \leqslant r \leqslant p)$:

$$\begin{cases} \phi_r'' - \Delta\phi_r + \sum_{s=1}^{p} \alpha_{rs}\phi_s = 0 & \text{in } (0, +\infty) \times \Omega, \\ \phi_r = 0 & \text{on } (0, +\infty) \times \Gamma_0, \\ \partial_\nu\phi_r + \sum_{s=1}^{p} \beta_{rs}\phi_s = 0 & \text{on } (0, +\infty) \times \Gamma_1, \\ t = 0: \quad \phi_r = (E_r, \widehat{U}_0), \ \phi_r' = (E_r, \widehat{U}_1) \text{ in } \Omega, \end{cases} \tag{25.24}$$

in which

$$Ae_r = \sum_{s=1}^{p} \alpha_{sr} e_s \quad \text{and} \quad Be_r = \sum_{s=1}^{p} \beta_{sr} e_s, \quad r = 1, \cdots, p. \tag{25.25}$$

Proof We first show that there exists a subspace V which is invariant for B^T and bi-orthonormal to $\text{Ker}(C_p)$.

Let $B = P^{-1} \Lambda P$, where P is an invertible matrix, and Λ be a symmetric matrix. Let $V = \text{Span}(E_1, \cdots, E_p\}$ in which

$$E_r = P^T P e_r, \quad r = 1, \cdots, p. \tag{25.26}$$

Noting (23.5) and the fact that $\text{Ker}(C_p)$ is an invariant subspace of B, we get

$$B^T E_r = P^T P B e_r \subseteq P^T P \text{Ker}(C_p) \subseteq V, \quad r = 1, \cdots, p, \tag{25.27}$$

then V is invariant for B^T.

We next show that $V^\perp \cap Ker(C_p) = \{0\}$. Then, noting that $\dim(V) = \dim(C_p) = p$, by Propositions 2.4 and 2.5, V is bi-orthonormal to $Ker(C_p)$. For this purpose, let $a_1, \cdots, a_p$ be coefficients such that

$$\sum_{r=1}^{p} a_r e_r \in V^\perp. \tag{25.28}$$

Then

$$(\sum_{r=1}^{p} a_r e_r, E_s) = (\sum_{r=1}^{p} a_r P e_r, P e_s) = 0, \quad s = 1, \cdots, p. \tag{25.29}$$

It follows that

$$(\sum_{r=1}^{p} a_r P e_r, \sum_{s=1}^{p} a_s P e_s) = 0, \tag{25.30}$$

then $a_1 = \cdots = a_p = 0$, namely, $V^\perp \cap Ker(C_p) = \{0\}$.

Denoting

$$Be_r = \sum_{s=1}^{p} \beta_{sr} e_s, \quad r = 1, \cdots, p, \tag{25.31}$$

a direct calculation yields that

$$B^T E_r = \sum_{s=1}^{p} \beta_{rs} E_s, \quad r = 1, \cdots, p. \tag{25.32}$$

Define the boundary control matrix D by

$$Ker(D^T) = V. \tag{25.33}$$

Noting (23.5), we have

$$\begin{aligned} &\text{Ker}(C_p) \cap \text{Im}(D) \\ =&\text{Ker}(C_p) \cap \{\text{Ker}(D^T)\}^{\perp} = \text{Ker}(C_p) \cap V^{\perp} = \{0\}, \end{aligned} \tag{25.34}$$

then, by Proposition 2.11, we have

$$\text{rank}(C_p D) = \text{rank}(D) = N - p. \tag{25.35}$$

Therefore, by Theorem 23.3, system (III) is exactly synchronizable by p-groups.

Denoting $\psi_r = (E_r, U)(r = 1, \cdots, p)$, we have

$$\begin{aligned} &(E_r, AU) = (A^T E_r, U) \\ =&(\sum_{s=1}^{p} \alpha_{rs} E_s + A^T E_r - \sum_{s=1}^{p} \alpha_{rs} E_s, U) \\ =&\sum_{s=1}^{p} \alpha_{rs}(E_s, U) + (A^T E_r - \sum_{s=1}^{p} \alpha_{rs} E_s, U) \\ =&\sum_{s=1}^{p} \alpha_{rs}\psi_s + (A^T E_r - \sum_{s=1}^{p} \alpha_{rs} E_s, U). \end{aligned} \tag{25.36}$$

By the assumption that V is bi-orthonormal to $\text{Ker}(C_p)$, without loss of generality, we may assume that

$$(E_r, e_s) = \delta_{rs} \quad (r, s = 1, \cdots, p). \tag{25.37}$$

Then, for any given $t = 1, \cdots, p$, by the first formula of (25.25), we get

$$\begin{aligned} &(A^T E_r - \sum_{s=1}^{p} \alpha_{rs} E_s, e_t) = (E_r, Ae_t) - \sum_{s=1}^{p} \alpha_{rs}(E_s, e_t) \\ =&\sum_{s=1}^{p} \alpha_{st}(E_r, e_s) - \alpha_{rt} = \alpha_{rt} - \alpha_{rt} = 0, \end{aligned}$$

hence

$$A^T E_r - \sum_{s=1}^{p} \alpha_{rs} E_s \in \{Ker(C_p)\}^{\perp} = \text{Im}(C_p^T), \quad r = 1, \cdots, p. \tag{25.38}$$

Thus, there exist $R_r \in \mathbb{R}^{N-p} (r = 1, \cdots, p)$ such that

$$A^T E_r - \sum_{s=1}^{p} \alpha_{rs} E_s = C_p^T R_r, \quad r = 1, \cdots, p. \tag{25.39}$$

Taking the inner product on both sides of problem (III) and (III0) with E_r, and noting (25.32)–(25.33), for $r = 1, \cdots, p$, we have

$$\begin{cases} \psi_r'' - \Delta \psi_r + \sum_{s=1}^{p} \alpha_{rs} \psi_s = -(R_r, C_p U) & \text{in } (0, +\infty) \times \Omega, \\ \psi_r = 0 & \text{on } (0, +\infty) \times \Gamma_0, \\ \partial_\nu \psi_r + \sum_{s=1}^{p} \beta_{rs} \psi_s = 0 & \text{on } (0, +\infty) \times \Gamma_1, \\ t = 0: \quad \psi_r = (E_r, \widehat{U}_0), \ \psi_r' = (E_r, \widehat{U}_1) \text{ in } \Omega. \end{cases} \tag{25.40}$$

Similarly to the proof of Theorem 22.4, we get

$$\begin{aligned} &\|(\psi, \psi')(T) - (\phi, \phi')(T)\|_{(H^{\alpha+1}(\Omega))^p \times (H^{\alpha}(\Omega))^p} \\ \leqslant &c\|C_p(\widehat{U}_0, \widehat{U}_1)\|_{(H^1(\Omega))^{N-p} \times (L^2(\Omega))^{N-p}}. \end{aligned} \tag{25.41}$$

On the other hand, noting (25.37), it is easy to see that

$$t \geqslant T: \quad \psi_r = (E_r, U) = \sum_{s=1}^{p} (E_r, e_s) u_s = u_r, \quad r = 1, \cdots, p. \tag{25.42}$$

Substituting it into (25.41), we get (25.23). □

Part VI
Synchronization for a Coupled System of Wave Equations with Coupled Robin Boundary Controls: Approximate Boundary Synchronization

In this part, the approximate boundary synchronization and the approximate boundary synchronization by groups will be discussed so that they can be realized by means of substantially fewer number of boundary controls than the number of state variables for system (III) with coupled Robin boundary controls under suitable hypotheses.

Chapter 26
Some Algebraic Lemmas

In order to study the approximate boundary synchronization for system (III) with coupled Robin boundary controls, some algebraic lemmas are given in this chapter.

Let A be a matrix of order N and D a full column-rank matrix of order $N \times M$ with $M \leqslant N$. By Proposition 2.12, we know the following Kalman's criterion:

$$\text{rank}(D, AD, \cdots, A^{N-1}D) \geqslant N - d \tag{26.1}$$

holds if and only if the dimension of any given subspace, contained in $\text{Ker}(D^T)$ and invariant for A^T, does not exceed d. In particular, the equality holds if and only if the dimension of the largest subspace, contained in $\text{Ker}(D^T)$ and invariant for A^T, is exactly equal to d.

Let A and B be two matrices of order N and D a full column-rank matrix of order $N \times M$ with $M \leqslant N$. For any given nonnegative integers $p, q, \cdots, r, s \geqslant 0$, we define a matrix of order $N \times M$ by

$$\mathcal{R}_{(p,q,\cdots,r,s)} = A^p B^q \cdots A^r B^s D. \tag{26.2}$$

We construct an enlarged matrix

$$\mathcal{R} = (\mathcal{R}_{(p,q,\cdots,r,s)}, \mathcal{R}_{(p',q',\cdots,r',s')}, \cdots) \tag{26.3}$$

by the matrices $\mathcal{R}_{(p,q,\cdots,r,s)}$ for all possible $(p, q, \cdots, r, s)$, which, by Cayley–Hamilton's Theorem, essentially constitute a finite set $\mathcal{M}$ with $\dim(\mathcal{M}) \leqslant MN$.

Lemma 26.1 *$Ker(\mathcal{R}^T)$ is the largest subspace of all the subspaces which are contained in $Ker(D^T)$ and invariant for A^T and B^T.*

T. Li and B. Rao, *Boundary Synchronization for Hyperbolic Systems*, Progress in Nonlinear Differential Equations and Their Applications 94,
https://doi.org/10.1007/978-3-030-32849-8_26

Proof First, noting that $\mathrm{Im}(D) \subseteq \mathrm{Im}(\mathcal{R})$, we have $\mathrm{Ker}(\mathcal{R}^T) \subseteq \mathrm{Ker}(\mathcal{D}^T)$. We now show that $\mathrm{Ker}(\mathcal{R}^T)$ is invariant for A^T and B^T. Let $x \in \mathrm{Ker}(\mathcal{R}^T)$. We have

$$D^T(B^T)^s(A^T)^r \cdots (B^T)^q(A^T)^p x = 0 \tag{26.4}$$

for any given integers $p, q, \cdots, r, s \geqslant 0$. Then, it follows that $A^T x \in \mathrm{Ker}(\mathcal{R}^T)$, namely, $\mathrm{Ker}(\mathcal{R}^T)$ is invariant for A^T. Similarly, $\mathrm{Ker}(\mathcal{R}^T)$ is invariant for B^T. Thus, the subspace $\mathrm{Ker}(\mathcal{R}^T)$ is contained in $\mathrm{Ker}(D^T)$ and invariant for both A^T and B^T.

Now let V be another subspace, contained in $\mathrm{Ker}(D^T)$ and invariant for A^T and B^T. For any given $y \in V$, we have

$$A^T y \in V, \quad B^T y \in V \quad \text{and} \quad D^T y = 0. \tag{26.5}$$

Then, it is easy to see that

$$(B^T)^s(A^T)^r \cdots (B^T)^q(A^T)^p y \in V \tag{26.6}$$

for any given integers $p, q, \cdots, r, s \geqslant 0$. Thus, by the first formula of (26.5), we have

$$D^T(B^T)^s(A^T)^r \cdots (B^T)^q(A^T)^p y = 0 \tag{26.7}$$

for any given integers $p, q, \cdots, r, s \geqslant 0$, namely, we have

$$V \subseteq \mathrm{Ker}(\mathcal{R}^T). \tag{26.8}$$

The proof is then complete. □

By the rank-nullity theorem, we have $\mathrm{rank}(\mathcal{R}) + \dim \mathrm{Ker}(\mathcal{R}^T) = N$. The following lemma is a dual version of Lemma 26.1.

Lemma 26.2 *Let $d \geqslant 0$ be an integer. Then*

(i) the rank condition

$$rank(\mathcal{R}) \geqslant N - d \tag{26.9}$$

holds true if and only if the dimension of any given subspace, contained in $Ker(D^T)$ and invariant for A^T and B^T, does not exceed d;

(ii) the rank condition

$$rank(\mathcal{R}) = N - d \tag{26.10}$$

holds true if and only if the dimension of the largest subspace, contained in $Ker(D^T)$ and invariant for A^T and B^T, is exactly equal to d.

Proof (i) Let V be a subspace which is contained in $\mathrm{Ker}(D^T)$ and invariant for A^T and B^T. By Lemma 26.1, we have

$$V \subseteq Ker(\mathcal{R}^T). \tag{26.11}$$

Assume that (26.9) holds, it follows from (26.11) that

$$N - d \leqslant \text{rank}(\mathcal{R}) = N - \dim \text{Ker}(\mathcal{R}^T) \leqslant N - \dim(V), \tag{26.12}$$

namely,

$$\dim(V) \leqslant d. \tag{26.13}$$

Conversely, assume that (26.13) holds for any given subspace V which is contained in $\text{Ker}(D^T)$ and invariant for A^T and B^T. In particular, by Lemma 26.1, we have $\dim \text{Ker}(\mathcal{R}^T) \leqslant d$. Then, it follows that

$$\text{rank}(\mathcal{R}) = N - \dim \text{Ker}(\mathcal{R}^T) \geqslant N - d. \tag{26.14}$$

(ii) Noting that (26.10) can be written as

$$\text{rank}(\mathcal{R}) \geqslant N - d \tag{26.15}$$

and

$$\text{rank}(\mathcal{R}) \leqslant N - d. \tag{26.16}$$

By (i), the rank condition (26.15) means that $\dim(V) \leqslant d$ for any given invariant subspace V of A^T and B^T, contained in $\text{Ker}(D^T)$. We claim that there exists a subspace V_0, which is contained in $\text{Ker}(D^T)$ and invariant for A^T and B^T, such that $\dim(V_0) = d$. Otherwise, all the subspaces of this kind have a dimension fewer than or equal to $(d - 1)$. By (i), we get

$$\text{rank}(\mathcal{R}) \geqslant N - d + 1, \tag{26.17}$$

which contradicts (26.16). It proves (ii). The proof is then complete. □

Remark 26.3 In the special case that $B = I$, we can write

$$\mathcal{R} = (D, AD, \cdots, A^{N-1}D). \tag{26.18}$$

Then, by Lemma 26.2, we find again that Kalman's criterion (26.1) holds if and only if the dimension of any given subspace, contained in $\text{Ker}(D^T)$ and invariant for A^T, does not exceed d. In particular, the equality holds if and only if the dimension of the largest subspace, contained in $\text{Ker}(D^T)$ and invariant for A^T, is exactly equal to d.

Chapter 27
Approximate Boundary Null Controllability

In this chapter, we will define the approximate boundary null controllability for system (III) and the D-observability for the adjoint problem, and show that these two concepts are equivalent to each other.

Definition 27.1 For $(\widehat{\Phi}_0, \widehat{\Phi}_1) \in (\mathcal{H}_1)^N \times (\mathcal{H}_0)^N$, the **adjoint system** (19.19) is **D-observable** on a finite interval $[0, T]$ if the observation

$$D^T \Phi \equiv 0 \quad \text{on } [0, T] \times \Gamma_1 \tag{27.1}$$

implies that $\widehat{\Phi}_0 = \widehat{\Phi}_1 \equiv 0$, then $\Phi \equiv 0$.

Proposition 27.2 *If the adjoint system* (19.19) *is D-observable, then we necessarily have rank($\mathcal{R}$) $= N$. In particular, if $M = N$, namely, D is invertible, then system* (19.19) *is D-observable.*

Proof Otherwise, we have $\dim \mathrm{Ker}(\mathcal{R}^T) = d \geqslant 1$. Let $\mathrm{Ker}(\mathcal{R}^T) = \mathrm{Span}\{E_1, \cdots, E_d\}$. By Lemma 26.1, $\mathrm{Ker}(\mathcal{R}^T)$ is contained in $\mathrm{Ker}(D^T)$ and invariant for A^T and B^T, namely, we have

$$D^T E_r = 0 \quad \text{for } 1 \leqslant r \leqslant d \tag{27.2}$$

and there exist coefficients α_{sr} and β_{sr} such that

$$A^T E_r = \sum_{s=1}^{d} \alpha_{rs} E_s \quad \text{and} \quad B^T E_r = \sum_{s=1}^{d} \beta_{rs} E_s \quad \text{for } 1 \leqslant r \leqslant d. \tag{27.3}$$

In what follows, we restrict system (19.19) on the subspace $\mathrm{Ker}(\mathcal{R}^T)$ and look for a solution of the form

$$\Phi = \sum_{s=1}^{d} \phi_s E_s, \tag{27.4}$$

which, because of (27.2), obviously satisfies the D-observation (27.1).

T. Li and B. Rao, *Boundary Synchronization for Hyperbolic Systems*, Progress in Nonlinear Differential Equations and Their Applications 94,
https://doi.org/10.1007/978-3-030-32849-8_27

Inserting the function (27.4) into system (19.19) and noting (27.3), it is easy to see that for $1 \leqslant r \leqslant d$, we have

$$\begin{cases} \phi_r'' - \Delta\phi_r + \sum_{s=1}^d \alpha_{sr}\phi_s = 0 & \text{in } (0, +\infty) \times \Omega, \\ \phi_r = 0 & \text{on } (0, +\infty) \times \Gamma_0, \\ \partial_\nu\phi_r + \sum_{s=1}^d \beta_{sr}\phi_s = 0 & \text{on } (0, +\infty) \times \Gamma_1. \end{cases} \tag{27.5}$$

For any nontrivial initial data

$$t = 0: \quad \phi_r' = \phi_{0r}, \ \phi_r' = \phi_{1r} \quad (1 \leqslant r \leqslant d), \tag{27.6}$$

we have $\Phi \not\equiv 0$. This contradicts the D-observability of system (19.19).

Moreover, when D is invertible, the D-observation (27.1) implies that

$$\partial_\nu\Phi \equiv \Phi \equiv 0 \quad \text{on } (0, T) \times \Gamma_1. \tag{27.7}$$

Then, Holmgren's uniqueness theorem (cf. Theorem 8.2 in [62]) implies well $\Phi \equiv 0$, provided that $T > 0$ is large enough. □

Definition 27.3 System (III) is **approximately null controllable** at the time $T > 0$ if for any given initial data $(\widehat{U}_0, \widehat{U}_1) \in (\mathcal{H}_0)^N \times (\mathcal{H}_{-1})^N$, there exists a sequence $\{H_n\}$ of boundary controls in $\mathcal{L}^M$ with compact support in $[0, T]$, such that the sequence $\{U_n\}$ of solutions to problem (III) and (III0) satisfies

$$u_n^{(k)} \longrightarrow 0 \quad \text{in} \quad C^0_{loc}([T, +\infty); \mathcal{H}_0) \cap C^1_{loc}([T, +\infty); \mathcal{H}_{-1}) \tag{27.8}$$

for all $1 \leqslant k \leqslant N$ as $n \to +\infty$.

By a similar argument as in Chaps. 8 and 16, we can prove the following

Proposition 27.4 *System (III) is approximately null controllable at the time $T > 0$ if and only if its adjoint system* (19.19) *is D-observable on the interval* $[0, T]$.

Corollary 27.5 *If system (III) is approximately null controllable, then we necessarily have rank($\mathcal{R}$) $= N$. In particular, as $M = N$, namely, D is invertible, system (III) is approximately null controllable.*

Proof This corollary follows immediately from Propositions 27.2 and 27.4. However, here we prefer to give a direct proof from the point of view of control.

Suppose that $\dim \text{Ker}(\mathcal{R}^T) = d \geqslant 1$. Let $\text{Ker}(\mathcal{R}^T) = \text{Span}\{E_1, \cdots, E_d\}$. By Lemma 26.1, $\text{Ker}(\mathcal{R}^T)$ is contained in $\text{Ker}(D^T)$ and invariant for both A^T and B^T, then we still have (27.2) and (27.3). Applying E_r to problem (III) and (III0) and setting $u_r = (E_r, U)$ for $1 \leqslant r \leqslant d$, it follows that for $1 \leqslant r \leqslant d$, we have

$$\begin{cases} u_r'' - \Delta u_r + \sum_{s=1}^d \alpha_{rs}u_s = 0 \text{ in } (0, +\infty) \times \Omega, \\ u_r = 0 & \text{on } (0, +\infty) \times \Gamma_0, \\ \partial_\nu u_r + \sum_{s=1}^d \beta_{rs}u_s = 0 & \text{on } (0, +\infty) \times \Gamma_1 \end{cases} \tag{27.9}$$

with the initial condition

$$t = 0: \quad u_r = (E_r, \widehat{U}_0), \ u_r' = (E_r, \widehat{U}_1) \quad \text{in } \Omega. \tag{27.10}$$

Thus, the projections $u_1, \cdots, u_d$ of U on the subspace $\text{Ker}(\mathcal{R}^T)$ are independent of applied boundary controls H, therefore, uncontrollable. This contradicts the approximate boundary null controllability of system (III). The proof is then complete. $\square$

Chapter 28
Unique Continuation for Robin Problems

We consider the unique continuation for Robin problems in this chapter.

28.1 General Consideration

By Proposition 27.2, rank$(\mathcal{R}) = N$ is a necessary condition for the D-observability of the adjoint system (19.19).

Proposition 28.1 *Let*

$$\mu = \sup_{\alpha,\beta\in\mathbb{C}} dim\ Ker \begin{pmatrix} A^T - \alpha I \\ B^T - \beta I \end{pmatrix}. \tag{28.1}$$

Assume that

$$Ker(\mathcal{R}^T) = \{0\}. \tag{28.2}$$

Then we have the following lower bound estimate:

$$rank(D) \geqslant \mu. \tag{28.3}$$

Proof Let $\alpha, \beta \in \mathbb{C}$, such that

$$V = \text{Ker} \begin{pmatrix} A^T - \alpha I \\ B^T - \beta I \end{pmatrix} \tag{28.4}$$

is of dimension μ. It is easy to see that any given subspace W of V is still invariant for A^T and B^T, then by Lemma 26.1, condition (28.2) implies that $\text{Ker}(D^T) \cap V = \{0\}$. Then, it follows that

T. Li and B. Rao, *Boundary Synchronization for Hyperbolic Systems*, Progress in Nonlinear Differential Equations and Their Applications 94,
https://doi.org/10.1007/978-3-030-32849-8_28

$$\dim \operatorname{Ker}(D^T) + \dim (V) \leqslant N, \tag{28.5}$$

namely,

$$\mu = \dim (V) \leqslant N - \dim \operatorname{Ker}(D^T) = \operatorname{rank}(D). \tag{28.6}$$

The proof is complete. □

In general, the condition $\dim \operatorname{Ker}(\mathcal{R}^T) = 0$ does not imply $\operatorname{rank}(D) = N$, so, the D-observation (27.1) does not imply

$$\Phi = 0 \quad \text{on } (0, T) \times \Gamma_1. \tag{28.7}$$

Therefore, the unique continuation for the adjoint system (19.19) with D-observation (27.1) is not a standard type of Holmgren's uniqueness theorem (cf. [18, 62, 75]).

Let us consider the following adjoint system (19.19) with $\Phi = (\phi, \psi)^T$:

$$\begin{cases} \phi'' - \Delta\phi + a\phi + b\psi = 0 & \text{in } (0, +\infty) \times \Omega, \\ \psi'' - \Delta\psi + c\phi + d\psi = 0 & \text{in } (0, +\infty) \times \Omega, \\ \phi = \psi = 0 & \text{on } (0, +\infty) \times \Gamma_0, \\ \partial_\nu\phi + \alpha\phi = 0 & \text{on } (0, +\infty) \times \Gamma_1, \\ \partial_\nu\psi + \beta\psi = 0 & \text{on } (0, +\infty) \times \Gamma_1 \end{cases} \tag{28.8}$$

with the observation of rank one:

$$d_1\phi + d_2\psi = 0 \quad \text{on } [0, T] \times \Gamma_1, \tag{28.9}$$

where a, b, c, d; α, β; and d_1, d_2 are constants. In system (28.8), since the boundary coupling matrix B is always assumed to be similar to a real symmetric matrix, without loss of generality, we suppose that B is a diagonal matrix.

Proposition 28.2 *The following results can be easily checked.*

(a) Assume that A^T and B^T admit only one common eigenvector E. Then $Ker(\mathcal{R}^T) = \{0\}$ if and only if $(E, D) \neq 0$.

(b) Assume that A^T and B^T admit only two common eigenvectors E_1 and E_2. Then $Ker(\mathcal{R}^T) = \{0\}$ if and only if $(E_i, D) \neq 0$ for $i = 1, 2$.

(c) Assume that A^T and B^T have no common eigenvector. Then $Ker(\mathcal{R}^T) = \{0\}$ if and only if $D \neq 0$.

The above conditions are only necessary for the unique continuation. We only know a few results about their sufficiency as follows.

28.2 Examples in Higher Dimensional Cases

The first example is a cascade system with Neumann boundary conditions:

$$\begin{cases} \phi'' - \Delta\phi = 0 & \text{in } (0, +\infty) \times \Omega, \\ \psi'' - \Delta\psi + \phi = 0 & \text{in } (0, +\infty) \times \Omega, \\ \phi = \psi = 0 & \text{on } (0, +\infty) \times \Gamma_0, \\ \partial_\nu\phi = \partial_\nu\psi = 0 & \text{on } (0, +\infty) \times \Gamma_1 \end{cases} \tag{28.10}$$

with the initial data in $H^1_{\Gamma_0}(\Omega) \times H^1_{\Gamma_0}(\Omega) \times L^2(\Omega) \times L^2(\Omega)$.

Let

$$A^T = \begin{pmatrix} 0 & 0 \\ 1 & 0 \end{pmatrix} \tag{28.11}$$

denote the corresponding coupling matrix of system (28.10). Since $(0, 1)^T$ is the only eigenvector of A^T, by Proposition 28.2, $\mathrm{Ker}(\mathcal{R}) = \{0\}$ holds true if and only if $d_2 \neq 0$. The following result shows that it is also sufficient for the unique continuation.

Theorem 28.3 ([3]) *Assume that $d_2 \neq 0$. Let (ϕ, ψ) be a solution to system* (28.10). *Then the partial observation* (28.9) *implies that $\phi \equiv \psi \equiv 0$, provided that $T > 0$ is large enough.*

Remark 28.4 Let us consider the following slightly modified system:

$$\begin{cases} \phi'' - \Delta\phi = 0 & \text{in } (0, +\infty) \times \Omega, \\ \psi'' - \Delta\psi + \phi = 0 & \text{in } (0, +\infty) \times \Omega, \\ \phi = \psi = 0 & \text{on } (0, +\infty) \times \Gamma_0, \\ \partial_\nu\phi + \alpha\phi = 0 & \text{on } (0, +\infty) \times \Gamma_1, \\ \partial_\nu\psi + \beta\psi = 0 & \text{on } (0, +\infty) \times \Gamma_1. \end{cases} \tag{28.12}$$

Since $(0, 1)^T$ is the only common eigenvector of A^T and B^T, by Proposition 28.2, $\mathrm{Ker}(\mathcal{R}^T) = \{0\}$ if and only if $d_2 \neq 0$. Unfortunately, the multiplier approach used in [3] is quite technically delicate, we don't know up to now if it can be adapted to get the unique continuation for system (28.12) with the partial observation (28.9).

The second example is a system of decoupled wave equations with Robin boundary conditions:

$$\begin{cases} \phi'' - \Delta\phi = 0 & \text{in } (0, +\infty) \times \Omega, \\ \psi'' - \Delta\psi = 0 & \text{in } (0, +\infty) \times \Omega, \\ \phi = \psi = 0 & \text{on } (0, +\infty) \times \Gamma_0, \\ \partial_\nu\phi + \alpha\phi = 0 & \text{on } (0, +\infty) \times \Gamma_1, \\ \partial_\nu\psi + \beta\psi = 0 & \text{on } (0, +\infty) \times \Gamma_1. \end{cases} \tag{28.13}$$

If $\alpha = \beta$, then dim Ker$(\mathcal{R}^T) > 0$ for any given $D = (d_1, d_2)^T$. So, we assume that $\alpha \neq \beta$, then $(1, 0)^T$ and $(0, 1)^T$ are the only common eigenvectors of A^T and B^T. By Proposition 28.2, Ker$(\mathcal{R}^T) = \{0\}$ holds if and only if $d_1 d_2 \neq 0$. Moreover, we replace the observation on the finite horizon (28.9) by the observation on the infinite horizon:

$$d_1\phi + d_2\psi = 0 \quad \text{on } (0, +\infty) \times \Gamma_1. \tag{28.14}$$

The following result confirms that Ker$(\mathcal{R}^T) = \{0\}$ or equivalently, rank$(\mathcal{R}) = 2$ is sufficient for the unique continuation of system (28.13) with the observation on the infinite horizon (28.14).

Theorem 28.5 *Assume that $\alpha > 0$, $\beta > 0$, $\alpha \neq \beta$, and $d_1 d_2 \neq 0$. Let (ϕ, ψ) be a solution to system* (28.13). *Then the partial observation on the infinite horizon* (28.14) *implies that $\phi \equiv \psi \equiv 0$.*

Proof We first recall Green's formula

$$\int_\Omega \Delta u v dx = -\int_\Omega \nabla u \cdot \nabla v dx + \int_\Gamma \partial_\nu u v d\Gamma \tag{28.15}$$

and Rellich's identity (cf. [62]):

$$\begin{aligned} 2\int_\Omega \Delta u(m \cdot \nabla u) dx &= (n-2)\int_\Omega |\nabla u|^2 dx \\ &+2\int_\Gamma \partial_\nu u(m \cdot \nabla u) d\Gamma - \int_\Gamma (m, \nu)|\nabla u|^2 d\Gamma, \end{aligned} \tag{28.16}$$

where $m = x - x_0$ and ν stands for the unit outward normal vector on the boundary Γ.

The energy of system (28.13) can be defined as

$$\begin{aligned} E(t) &= \frac{1}{2}\int_\Omega (|\phi_t|^2 + |\nabla\phi|^2 + |\psi_t|^2 + |\nabla\psi|^2) dx \\ &+ \frac{1}{2}\int_{\Gamma_1} (\alpha|\phi|^2 + \beta|\psi|^2) d\Gamma. \end{aligned} \tag{28.17}$$

It is easy to check that $E'(t) = 0$ which yields the conservation of energy

$$E(t) = E(0), \quad \forall\, t \geqslant 0. \tag{28.18}$$

Now, taking the inner product with $2m \cdot \nabla u$ on both sides of the first equation in (28.13) and integrating by parts, we get

$$\left[\int_\Omega \phi_t(m\cdot\nabla\phi)dx\right]_0^T \tag{28.19}$$
$$=\int_0^T\int_\Omega \phi_t m\cdot\nabla\phi_t dxdt+\int_0^T\int_\Omega \Delta\phi(m\cdot\nabla\phi)dxdt.$$

Then, using Green's formula in the first term and Rellich's identity in the second term on the right-hand side, it follows that

$$2\left[\int_\Omega \phi_t(m\cdot\nabla\phi)dx\right]_0^T=\int_0^T\int_\Gamma (m,\nu)|\phi_t|^2 d\Gamma dt \tag{28.20}$$
$$-n\int_0^T\int_\Omega |\phi_t|^2 dxdt+(n-2)\int_0^T\int_\Omega |\nabla\phi|^2 dxdt$$
$$+2\int_0^T\int_\Gamma \partial_\nu\phi(m\cdot\nabla\phi)d\Gamma dt-\int_\Gamma (m,\nu)|\nabla\phi|^2 d\Gamma dt.$$

Noting the conservation of energy (28.18) and using Cauchy–Schwarz inequality on the left-hand side, it follows that

$$n\int_0^T\int_\Omega |\phi_t|^2 dxdt+(2-n)\int_0^T\int_\Omega |\nabla\phi|^2 dxdt \tag{28.21}$$
$$\leqslant cE(0)+\int_0^T\int_\Gamma (m,\nu)|\phi_t|^2 d\Gamma dt$$
$$+\int_0^T\int_\Gamma (2\partial_\nu\phi(m\cdot\nabla\phi)-(m,\nu)|\nabla\phi|^2)d\Gamma,$$

here and hereafter, c denotes a positive constant independent of T.

Recall the usual multiplier geometrical condition:

$$(m,\nu)\leqslant 0,\quad \forall x\in\Gamma_0;\quad (m,\nu)\geqslant\delta,\quad \forall x\in\Gamma_1, \tag{28.22}$$

where $\delta>0$ is a positive constant. Then on Γ_1, we have

$$2\partial_\nu\phi(m\cdot\nabla\phi)-(m,\nu)|\nabla\phi|^2 \tag{28.23}$$
$$\leqslant 2\|m\|_\infty|\partial_\nu\phi|\cdot|\nabla\phi|-\delta|\nabla\phi|^2$$
$$\leqslant\frac{\|m\|_\infty^2}{\delta}|\partial_\nu\phi|^2=\frac{\|m\|_\infty^2}{\delta}\alpha^2|\phi|^2.$$

While, noting that $\phi=0$ on Γ_0, we have

$$\nabla\phi=\partial_\nu\phi\nu, \tag{28.24}$$

then

$$2\partial_\nu\phi(m\cdot\nabla\phi)-(m,\nu)|\nabla\phi|^2=(m,\nu)|\partial_\nu\phi|^2\leqslant 0. \tag{28.25}$$

Substituting (28.23) and (28.25) into (28.21), it is easy to get

$$\begin{aligned}&n\int_0^T\int_\Omega|\phi_t|^2dxdt+(2-n)\int_0^T\int_\Omega|\nabla\phi|^2dxdt\\&\leqslant cE(0)+\int_0^T\int_{\Gamma_1}(m,\nu)|\phi_t|^2d\Gamma dt+\frac{\|m\|_\infty^2\alpha^2}{\delta}\int_0^T\int_{\Gamma_1}|\phi|^2d\Gamma dt\\&\leqslant cE(0)+c\int_0^T\int_{\Gamma_1}(|\phi|^2+|\phi_t|^2)d\Gamma dt.\end{aligned} \tag{28.26}$$

Next, taking the inner product with u on both sides of the first equation of system (28.13) and integrating by parts, we have

$$\begin{aligned}\Big[\int_\Omega\phi_t\phi dx\Big]_0^T=&\int_0^T\int_\Omega|\phi_t|^2dxdt\\&+\int_0^T\int_{\Gamma_1}\phi\partial_\nu\phi d\Gamma dt-\int_0^T\int_\Omega|\nabla\phi|^2dxdt.\end{aligned} \tag{28.27}$$

Using the conservation of energy (28.18), it follows that

$$-\int_0^T\int_\Omega|\phi_t|^2dxdt+\int_0^T\int_\Omega|\nabla\phi|^2dxdt\leqslant cE(0). \tag{28.28}$$

By (28.26) $+(n-1)\times$(28.28), we get

$$\begin{aligned}&\int_0^T\int_\Omega(|\phi_t|^2+|\nabla\phi|^2)dxdt\\&\leqslant c\int_0^T\int_{\Gamma_1}(|\phi|^2+|\phi_t|^2)d\Gamma dt+cE(0).\end{aligned} \tag{28.29}$$

Similarly, we have

$$\begin{aligned}&\int_0^T\int_\Omega(|\psi_t|^2+|\nabla\psi|^2)dxdt\\&\leqslant c\int_0^T\int_{\Gamma_1}(|\psi|^2+|\psi_t|^2)d\Gamma dt+cE(0).\end{aligned} \tag{28.30}$$

Combing (28.29) and (28.30), we have

$$\int_0^T E(t)dt \leqslant c\int_0^T\int_{\Gamma_1}(|\phi|^2+|\psi|^2+|\phi_t|^2+|\psi_t|^2)d\Gamma dt + cE(0). \tag{28.31}$$

Finally, taking the inner product with ψ on both sides of the first equation of (28.13) and integrating by parts, we get

$$\begin{aligned}\Big[\int_\Omega \phi_t\psi dx\Big]_0^T = &\int_0^T\int_\Omega \phi_t\psi_t dxdt \\ -\alpha &\int_0^T\int_{\Gamma_1}\phi\psi d\Gamma dt - \int_0^T\int_\Omega \nabla\phi\cdot\nabla\psi dxdt.\end{aligned} \tag{28.32}$$

Similarly, taking the inner product with ϕ on both sides of the second equation of (28.13) and integrating by parts, we get

$$\begin{aligned}\Big[\int_\Omega \psi_t\phi dx\Big]_0^T = &\int_0^T\int_\Omega \phi_t\psi_t dxdt \\ -\beta &\int_0^T\int_{\Gamma_1}\phi\psi d\Gamma dt - \int_0^T\int_\Omega \nabla\phi\cdot\nabla\psi dxdt.\end{aligned} \tag{28.33}$$

Thus, we get

$$\Big[\int_\Omega (\phi_t\psi - \psi_t\phi)dx\Big]_0^T = (\alpha-\beta)\int_0^T\int_{\Gamma_1}\phi\psi d\Gamma dt. \tag{28.34}$$

By the boundary observation (28.14), we have $\psi = -\frac{d_1}{d_2}\phi$ on Γ_1, then, by Cauchy–Schwarz' inequality and noting (28.18), it comes from (28.34) that

$$\int_0^T\int_{\Gamma_1}(|\phi|^2+|\psi|^2)d\Gamma dt \leqslant cE(0). \tag{28.35}$$

Because of the linearity, we also have

$$\int_0^T\int_{\Gamma_1}(|\phi_t|^2+|\psi_t|^2)d\Gamma dt \leqslant c\widehat{E}(0), \tag{28.36}$$

where

$$\begin{aligned}\widehat{E}(t) = &\frac{1}{2}\int_\Omega(|\phi_{tt}|^2+|\nabla\phi_t|^2+|\psi_{tt}|^2+|\nabla\psi_t|^2)dx \\ +&\frac{1}{2}\int_{\Gamma_1}(\alpha|\phi_t|^2+\beta|\psi_t|^2)d\Gamma.\end{aligned} \tag{28.37}$$

Noting the conservation of energy (28.18), and substituting (28.35)–(28.36) into (28.31), we get

$$TE(0) \leqslant c(E(0) + \widehat{E}(0)). \tag{28.38}$$

Taking $T \to +\infty$, we have $E(0) = 0$, then by (28.18), we get $E(t) \equiv 0$ for all $t \geqslant 0$. The proof is complete. □

28.3 One-Dimensional Case

In the one-space-dimensional case, Theorem 28.5 can be further improved to the observation of finite horizon.

Theorem 28.6 *Assume that* $\alpha > 0$, $\beta > 0$, $\alpha \neq \beta$, *and* $d_1 d_2 \neq 0$. *Then the following one-dimensional system*

$$\begin{cases} \phi'' - \phi_{xx} = 0 & 0 < x < 1, \\ \psi'' - \psi_{xx} = 0 & 0 < x < 1, \\ \phi(t, 0) = \psi(t, 0) = 0, \\ \phi_x(t, 1) + \alpha\phi(t, 1) = 0, \\ \psi_x(t, 1) + \beta\psi(t, 1) = 0 \end{cases} \tag{28.39}$$

with the observation of finite horizon

$$d_1\phi(t, 1) + d_2\psi(t, 1) = 0 \quad \forall t \in [0, T] \tag{28.40}$$

only has the trivial solution, provided that $T > 0$ *is large enough.*

Proof Here we only give a sketch of the proof which is similar to that in Theorem 8.24.

Consider the following eigenvalue problem:

$$\begin{cases} \lambda^2 u + u_{xx} = 0, & 0 < x < 1, \\ \lambda^2 v + v_{xx} = 0, & 0 < x < 1, \\ u(0) = v(0) = 0, \\ u_x(1) + \alpha u(1) = 0, \\ v_x(1) + \beta v(1) = 0. \end{cases} \tag{28.41}$$

Let

$$u = \sin \lambda x, \quad v = \sin \lambda x. \tag{28.42}$$

By the last two formulas in (28.41), we have

$$\lambda \cos \lambda + \alpha \sin \lambda = 0, \quad \lambda \cos \lambda + \beta \sin \lambda = 0. \tag{28.43}$$

Rewrite the first formula of (28.43) as

$$\tan\lambda + \frac{\lambda}{\alpha} = 0. \tag{28.44}$$

Noting (28.44), we have

$$\cos^2\lambda - \sin^2\lambda = \frac{1-\tan^2\lambda}{1+\tan^2\lambda} = \frac{\alpha^2-\lambda^2}{\alpha^2+\lambda^2} \tag{28.45}$$

and

$$\sin\lambda\cos\lambda = \frac{\tan\lambda}{1+\tan^2\lambda} = -\frac{\alpha\lambda}{\alpha^2+\lambda^2}. \tag{28.46}$$

Then, by

$$e^{2i\lambda} = \cos 2\lambda + i\sin 2\lambda = \cos^2\lambda - \sin^2\lambda + 2i\sin\lambda\cos\lambda, \tag{28.47}$$

we have

$$e^{2i\lambda} = \frac{\alpha^2-\lambda^2}{\alpha^2+\lambda^2} - \frac{2\alpha\lambda i}{\alpha^2+\lambda^2} = -\frac{\lambda+\alpha i}{\lambda-\alpha i}. \tag{28.48}$$

The asymptotic expansion of $e^{2i\lambda}$ at $\lambda = \infty$ gives

$$e^{2i\lambda} = -1 - \frac{2\alpha i}{\lambda} + \frac{O(1)}{\lambda^2}. \tag{28.49}$$

Taking the logarithm on both sides, we get

$$\begin{aligned} 2i\lambda &= (2n+1)\pi i + \ln\left(1 + \frac{2\alpha i}{\lambda} + \frac{O(1)}{\lambda^2}\right) \\ &= (2n+1)\pi i + \frac{2\alpha i}{\lambda} + \frac{O(1)}{\lambda^2} i, \end{aligned}$$

then

$$\lambda = \lambda_n^\alpha := (n+\frac{1}{2})\pi + \frac{2\alpha}{\lambda} + \frac{O(1)}{\lambda^2}. \tag{28.50}$$

Noting $\lambda_n^\alpha \sim n\pi$, we get

$$\lambda_n^\alpha = (n+\frac{1}{2})\pi + \frac{\alpha}{n\pi} + \frac{O(1)}{n^2}. \tag{28.51}$$

Similarly, by the second formula of (28.43), we have

$$\lambda = \lambda_n^\beta := (n+\frac{1}{2})\pi + \frac{\beta}{n\pi} + \frac{O(1)}{n^2}. \tag{28.52}$$

Noting $\alpha \neq \beta$, it follows from (28.44) that

$$\lambda_m^\alpha \neq \lambda_n^\beta, \quad \forall m, n \in \mathbb{Z}. \tag{28.53}$$

Besides, by the monotonicity of the function $\lambda \to \tan\lambda + \frac{\lambda}{\alpha}$, we have

$$\lambda_m^\alpha \neq \lambda_n^\alpha, \quad \lambda_m^\beta \neq \lambda_n^\beta, \quad \forall m, n \in \mathbb{Z}, \quad m \neq n. \tag{28.54}$$

Without loss of generality, assuming that $\alpha > \beta > 0$, we can arrange $\{\lambda_n^\alpha\} \cup \{\lambda_n^\beta\}$ into an increasing sequence

$$\cdots < \lambda_{-n}^\alpha < \lambda_{-n}^\beta < \cdots < \lambda_1^\beta < \lambda_1^\alpha < \cdots < \lambda_n^\beta < \lambda_n^\alpha < \cdots. \tag{28.55}$$

On the other hand, noting $a > \beta$ and using the expressions (28.51)–(28.52), there exists a positive constant $\gamma > 0$, such that

$$\lambda_{n+1}^\alpha - \lambda_n^\alpha \geqslant 2\gamma, \quad \lambda_{n+1}^\beta - \lambda_n^\beta \geqslant 2\gamma \tag{28.56}$$

and

$$\frac{1}{|n|} \leqslant \lambda_n^\alpha - \lambda_n^\beta \leqslant \gamma \tag{28.57}$$

for all $n \in \mathbb{Z}$ with $|n|$ large enough.

So, the sequence $\{\lambda_n^\alpha\} \cup \{\lambda_n^\beta\}$ satisfies the conditions (8.87), (8.91), and (8.92) in Theorem 8.22 with $c = 1$, $m = 2$, and $s = 1$. Moreover, the upper density D^+ defined in (8.93) for the sequence $\{\lambda_n^\alpha\} \cup \{\lambda_n^\beta\}$ is equal to 2.

Finally, we easily check that the system of eigenvectors $\{E_n^\alpha, E_n^\beta\}_{n\in\mathbb{Z}}$ with

$$E_n^\alpha = \begin{pmatrix} \frac{\sin\lambda_n^\alpha x}{\lambda_n^\alpha} \\ \sin\lambda_n^\alpha x \end{pmatrix}, \quad E_n^\beta = \begin{pmatrix} \frac{\sin\lambda_n^\beta x}{\lambda_n^\beta} \\ \sin\lambda_n^\beta x \end{pmatrix}, \quad n \in \mathbb{Z} \tag{28.58}$$

forms a Hilbert basis in $(H^1(\Omega))^N \times (L^2(\Omega))^N$. Then any given solution to system (28.39) can be represented by

$$\begin{pmatrix} \phi \\ \phi' \end{pmatrix} = \sum_{n\in\mathbb{Z}} c_n^\alpha e^{i\lambda_n^\alpha t} E_n^\alpha, \quad \begin{pmatrix} \psi \\ \psi' \end{pmatrix} = \sum_{n\in\mathbb{Z}} c_n^\beta e^{i\lambda_n^\beta t} E_n^\beta. \tag{28.59}$$

By the boundary observation (28.40), we have

$$\sum_{n\in\mathbb{Z}} d_1 c_n^\alpha e^{i\lambda_n^\alpha t} \frac{\sin\lambda_n^\alpha}{\lambda_n^\alpha} + \sum_{n\in\mathbb{Z}} d_2 c_n^\beta e^{i\lambda_n^\beta t} \frac{\sin\lambda_n^\beta}{\lambda_n^\beta} = 0. \tag{28.60}$$

Then, all the conditions of Theorem 8.22 are verified, so the sequence $\{e^{i\lambda_n^\alpha t}, e^{i\lambda_n^\beta t}\}_{n\in\mathbb{Z}}$ is ω-linearly independent in $L^2(0, T)$, provided that $T > 4\pi$. It follows that

$$d_1 c_n^\alpha \frac{\sin \lambda_n^\alpha}{\lambda_n^\alpha} = 0, \quad d_2 c_n^\beta \frac{\sin \lambda_n^\beta}{\lambda_n^\beta} = 0, \quad \forall\, n \in \mathbb{Z}, \tag{28.61}$$

namely,

$$c_n^\alpha = 0, \quad c_n^\beta = 0, \quad \forall n \in \mathbb{Z}, \tag{28.62}$$

hence $\phi \equiv \psi \equiv 0$. The proof is complete. □

Chapter 29
Approximate Boundary Synchronization

The approximate boundary synchronization is defined and studied in this chapter for system (III) with coupled Robin boundary controls.

29.1 Definitions

Definition 29.1 System (III) is **approximately synchronizable** at the time $T > 0$ if for any given initial data $(\widehat{U}_0, \widehat{U}_1) \in (\mathcal{H}_0)^N \times (\mathcal{H}_{-1})^N$, there exists a sequence $\{H_n\}$ of boundary controls in $\mathcal{L}^M$ with compact support in $[0, T]$, such that the corresponding sequence $\{U_n\}$ of solutions to problem (III) and (III0) satisfies

$$u_n^{(k)} - u_n^{(l)} \to 0 \quad \text{in} \quad C^0_{loc}([T, +\infty); \mathcal{H}_0) \cap C^1_{loc}([T, +\infty); \mathcal{H}_{-1}) \tag{29.1}$$

for all k, l with $1 \leqslant k, l \leqslant N$ as $n \to +\infty$.

Define the synchronization matrix of order $(N-1) \times N$ by

$$C_1 = \begin{pmatrix} 1 & -1 & & & \\ & 1 & -1 & & \\ & & \ddots & \ddots & \\ & & & 1 & -1 \end{pmatrix}. \tag{29.2}$$

Clearly,

$$\text{Ker}(C_1) = \text{Span}\{e_1\} \quad \text{with} \quad e_1 = (1, \cdots, 1)^T. \tag{29.3}$$

T. Li and B. Rao, *Boundary Synchronization for Hyperbolic Systems*, Progress in Nonlinear Differential Equations and Their Applications 94,
https://doi.org/10.1007/978-3-030-32849-8_29

Then, the **approximate boundary synchronization** (29.1) can be equivalently rewritten as

$$C_1 U_n \to 0 \quad \text{as } n \to +\infty \tag{29.4}$$

in the space

$$C^0_{loc}([T, +\infty); (\mathcal{H}_0)^{N-1}) \cap C^1_{loc}([T, +\infty); (\mathcal{H}_{-1})^{N-1}). \tag{29.5}$$

Definition 29.2 The matrix A satisfies the **condition of C_1-compatibility** if there exists a unique matrix $\overline{A}_1$ of order $(N-1)$, such that

$$C_1 A = \overline{A}_1 C_1. \tag{29.6}$$

The matrix $\overline{A}_1$ is called the **reduced matrix of A by C_1**.

Remark 29.3 The condition of C_1-compatibility (29.6) is equivalent to

$$A\text{Ker}(C_1) \subseteq \text{Ker}(C_1). \tag{29.7}$$

Then, noting (29.3), the vector $e_1 = (1, \cdots, 1)^T$ is an eigenvector of A, corresponding to the eigenvalue a given by

$$a = \sum_{j=1}^{N} a_{ij}, \quad i = 1, \cdots, N, \tag{29.8}$$

where $\sum_{j=1}^{N} a_{ij}$ is independent of $i = 1, \cdots, N$. Equation (29.8) is called the **row-sum condition**, which is also equivalent to the condition of C_1-compatibility (29.6) or (29.7).

Similarly, the matrix B satisfies the condition of C_1-compatibility if there exists a unique matrix $\overline{B}_1$ of order $(N-1)$, such that

$$C_1 B = \overline{B}_1 C_1, \tag{29.9}$$

which is equivalent to the fact that

$$B\text{Ker}(C_1) \subseteq \text{Ker}(C_1). \tag{29.10}$$

Moreover, the vector $e = (1, \cdots, 1)^T$ is also an eigenvector of B, corresponding to the eigenvalue b given by

$$b = \sum_{j=1}^{N} b_{ij}, \quad i = 1, \cdots, N, \tag{29.11}$$

where the row-sum $\sum_{j=1}^{N} b_{ij}$ is independent of $i = 1, \cdots, N$.

29.2 Fundamental Properties

Theorem 29.4 *Assume that system (III) is approximately synchronizable. Then we necessarily have rank*$(\mathcal{R}) \geqslant N - 1$.

Proof Otherwise, let $\mathrm{Ker}(\mathcal{R}^T) = \mathrm{Span}\{E_1, \cdots, E_d\}$ with $d > 1$. Noting that

$$\dim \mathrm{Im}(C_1^T) + \dim \mathrm{Ker}(\mathcal{R}^T) = N - 1 + d > N, \tag{29.12}$$

there exists a unit vector $E \in Im(C_1^T) \cap \mathrm{Ker}(\mathcal{R}^T)$. Let $E = C_1^T x$ with $x \in \mathbb{R}^{N-1}$. The approximate boundary synchronization (29.4) implies that

$$(E, U_n) = (x, C_1 U_n) \to 0 \tag{29.13}$$

as $n \to +\infty$ in the space

$$C^0_{loc}([T, +\infty); \mathcal{H}_0) \cap C^1_{loc}([T, +\infty); \mathcal{H}_{-1}).$$

On the other hand, since $E \in \mathrm{Ker}(\mathcal{R}^T)$, we have

$$E = \sum_{r=1}^{d} \alpha_r E_r, \tag{29.14}$$

where the coefficients $\alpha_1, \cdots, \alpha_d$ are not all zero. By Lemma 26.1, $\mathrm{Ker}(\mathcal{R}^T)$ is contained in $\mathrm{Ker}(D^T)$ and invariant for both A^T and B^T; therefore, we still have (27.2) and (27.3). Thus, applying E_r to problem (III) and (III0) and setting $u_r = (E_r, U_n)$ for $1 \leqslant r \leqslant d$, we find again problem (27.9)–(27.10) with homogeneous boundary conditions. Noting that problem (27.9)–(27.10) is independent of n, it follows from (29.13) and (29.14) that

$$\sum_{r=1}^{d} \alpha_r u_r(T) \equiv \sum_{r=1}^{d} \alpha_r u_r'(T) \equiv 0. \tag{29.15}$$

Then, by well-posedness, it is easy to see that

$$\sum_{r=1}^{d} \alpha_r (E_r, \widehat{U}_0) \equiv \sum_{r=1}^{d} \alpha_r (E_r, \widehat{U}_1) \equiv 0 \tag{29.16}$$

for any given initial data $(\widehat{U}_0, \widehat{U}_1) \in (\mathcal{H}_0)^N \times (\mathcal{H}_{-1})^N$. This yields

$$\sum_{r=1}^{d} \alpha_r E_r = 0. \tag{29.17}$$

Because of the linear independence of the vectors $E_1, \cdots, E_d$, we get a contradiction $\alpha_1 = \cdots = \alpha_d = 0$. □

Theorem 29.5 *Assume that system (III) is approximately synchronizable under the minimal rank($\mathcal{R}$) $= N - 1$. Then, we have the following assertions:*

(i) There exists a vector $E_1 \in Ker(\mathcal{R}^T)$, such that $(E_1, e_1)=1$ with $e_1=(1, \cdots, 1)^T$.

(ii) For any given initial data $(\widehat{U}_0, \widehat{U}_1) \in (\mathcal{H}_0)^N \times (\mathcal{H}_{-1})^N$, there exists a unique scalar function u such that

$$u_n^{(k)} \to u \quad in \quad C^0_{loc}([T, +\infty); \mathcal{H}_0) \cap C^1_{loc}([T, +\infty); \mathcal{H}_{-1}) \tag{29.18}$$

for all $1 \leqslant k \leqslant N$ as $n \to +\infty$.

(iii) The matrices A and B satisfy the conditions of C_1-compatibility (29.6) and (29.9), respectively.

Proof (i) Noting that $\dim \mathrm{Ker}(\mathcal{R}^T) = 1$, by Lemma 26.1, there exists a nonzero vector $E_1 \in \mathrm{Ker}(\mathcal{R}^T)$, such that

$$D^T E_1 = 0, \quad A^T E_1 = \alpha E_1 \quad \text{and} \quad B^T E_1 = \beta E_1. \tag{29.19}$$

We claim that $E_1 \notin \mathrm{Im}(C_1^T)$. Otherwise, applying E_1 to problem (III) and (III0) with $U = U_n$ and $H = H_n$, and setting $u = (E_1, U_n)$, it follows that

$$\begin{cases} u'' - \Delta u + \alpha u = 0 \text{ in } (0, +\infty) \times \Omega, \\ u = 0 \qquad\qquad\qquad \text{on } (0, +\infty) \times \Gamma_0, \\ \partial_\nu u + \beta u = 0 \qquad \text{on } (0, +\infty) \times \Gamma_1 \end{cases} \tag{29.20}$$

with the following initial data:

$$t = 0: \quad u = (E_1, \widehat{U}_0), \; u' = (E_1, \widehat{U}_1) \quad \text{in } \Omega. \tag{29.21}$$

Suppose that $E_1 \in \mathrm{Im}(C_1^T)$, there exists a vector $x \in \mathbb{R}^{N-1}$, such that $E_1 = C_1^T x$. Then, the approximate boundary synchronization (29.4) implies

$$(u(T), u'(T)) = ((x, C_1 U_n(T)), (x, C_1 U_n'(T))) \to (0, 0) \tag{29.22}$$

in the space $\mathcal{H}_0 \times \mathcal{H}_{-1}$ as $n \to +\infty$. Then, since u is independent of n, we have

$$u(T) \equiv u'(T) \equiv 0. \tag{29.23}$$

Thus, because of the well-posedness of problem (29.20)–(29.21), it follows that

$$(E_1, \widehat{U}_0) = (E_1, \widehat{U}_1) = 0 \tag{29.24}$$

for any given initial data $(\widehat{U}_0, \widehat{U}_1) \in (\mathcal{H}_0)^N \times (\mathcal{H}_{-1})^N$. This yields a contradiction $E_1 = 0$.

Since $E_1 \notin \text{Im}(C_1^T)$, noting that $\text{Im}(C_1^T) = (\text{Span}\{e_1\})^\perp$, we have $(E_1, e_1) \neq 0$. Without loss of generality, we can take E_1 such that $(E_1, e_1) = 1$.

(ii) Since $E_1 \notin Im(C_1^T)$, the matrix $\begin{pmatrix} C_1 \\ E_1^T \end{pmatrix}$ is invertible. Moreover, we have

$$\begin{pmatrix} C_1 \\ E_1^T \end{pmatrix} e_1 = \begin{pmatrix} 0 \\ 1 \end{pmatrix}. \tag{29.25}$$

Noting (29.4), we have

$$\begin{pmatrix} C_1 \\ E_1^T \end{pmatrix} U_n = \begin{pmatrix} C_1 U_n \\ (E_1, U_n) \end{pmatrix} \to \begin{pmatrix} 0 \\ u \end{pmatrix} = u \begin{pmatrix} 0 \\ 1 \end{pmatrix} \tag{29.26}$$

as $n \to +\infty$ in the space

$$C^0_{loc}([T, +\infty); (\mathcal{H}_0)^N) \cap C^1_{loc}([T, +\infty); (\mathcal{H}_{-1})^N). \tag{29.27}$$

Then, noting (29.25), it follows that

$$U_n = \begin{pmatrix} C_1 \\ E_1^T \end{pmatrix}^{-1} \begin{pmatrix} C_1 U_n \\ E_1^T U_n \end{pmatrix} \to u \begin{pmatrix} C_1 \\ E_1^T \end{pmatrix}^{-1} \begin{pmatrix} 0 \\ 1 \end{pmatrix} = u e_1 \tag{29.28}$$

in the space (29.27), namely, (29.18) holds.

(iii) Applying C_1 to system (III) with $U = U_n$ and $H = H_n$, and passing to the limit as $n \to +\infty$, it follows from (29.4) and (29.28) that

$$C_1 A e_1 u = 0 \quad \text{in } [T, +\infty) \times \Omega \tag{29.29}$$

and

$$C_1 B e_1 u = 0 \quad \text{on } [T, +\infty) \times \Gamma_1. \tag{29.30}$$

We claim that at least for an initial data $(\widehat{U}_0, \widehat{U}_1)$, we have

$$u \not\equiv 0 \quad \text{on } [T, +\infty) \times \Gamma_1. \tag{29.31}$$

Otherwise, it follows from system (29.20) that

$$\partial_\nu u \equiv u \equiv 0 \quad \text{on } [T, +\infty) \times \Gamma_1, \tag{29.32}$$

then, by Holmgren's uniqueness theorem (cf. Theorem 8.2 in [62]), we get $u \equiv 0$ for all the initial data $(\widehat{U}_0, \widehat{U}_1)$, namely, system (III) is approximately null controllable under the condition $\dim \text{Ker}(\mathcal{R}^T) = 1$. This contradicts Corollary 27.5. Then, it follows from (29.29) and (29.30) that $C_1 A e = 0$ and $C_1 B e = 0$, which give the conditions of C_1-compatibility for A and B, respectively. The proof is complete. □

Remark 29.6 Theorem 29.5 (ii) indicates that under the minimal rank condition $\text{rank}(D) = N - 1$, the approximate boundary synchronization in the **consensus sense** (29.1) is actually that in the **pinning sense** (29.18) with the **approximately synchronizable state** u.

Assume that both A and B satisfy the corresponding conditions of C_1-compatibility, namely, there exist two matrices $\overline{A}_1$ and $\overline{B}_1$ such that $C_1 A = \overline{A}_1 C_1$ and $C_1 B = \overline{B}_1 C_1$, respectively. Setting $W = C_1 U$ in problem (III) and (III0), we get the following **reduced system**:

$$\begin{cases} W'' - \Delta W + \overline{A}_1 W = 0 \text{ in } (0, +\infty) \times \Omega, \\ W = 0 \qquad\qquad\qquad\quad \text{on } (0, +\infty) \times \Gamma_0, \\ \partial_\nu W + \overline{B}_1 W = C_1 D H \text{ on } (0, +\infty) \times \Gamma_1 \end{cases} \tag{29.33}$$

with the initial condition

$$t = 0: \quad W = C_1 \widehat{U}_0, \ W' = C_1 \widehat{U}_1 \quad \text{in } \Omega. \tag{29.34}$$

Since B is similar to a real symmetric matrix, by Proposition 2.21, so is its reduced matrix $\overline{B}_1$. Then, by Theorem 19.9, the reduced problem (29.33)–(29.34) is well-posed in the space $(\mathcal{H}_0)^{N-1} \times (\mathcal{H}_{-1})^{N-1}$.

Accordingly, consider the **reduced adjoint system**

$$\begin{cases} \Psi'' - \Delta \Psi + \overline{A}_1^T \Psi = 0 & \text{in } (0, T) \times \Omega, \\ \Psi = 0 & \text{on } (0, T) \times \Gamma_0, \\ \partial_\nu \Psi + \overline{B}_1^T \Psi = 0 & \text{on } (0, T) \times \Gamma_1 \end{cases} \tag{29.35}$$

with the $C_1 D$-observation

$$(C_1 D)^T \Psi \equiv 0 \quad \text{on } (0, T) \times \Gamma_1. \tag{29.36}$$

Obviously, we have

Proposition 29.7 *Under the conditions of C_1-compatibility for A and B, system (III) is approximately synchronizable if and only if the reduced system (29.33) is approximately null controllable, or equivalently if and only if the reduced adjoint system (29.35) is $C_1 D$-observable.*

Theorem 29.8 *Assume that A and B satisfy the conditions of C_1-compatibility (29.6) and (29.9), respectively. Assume furthermore that A^T and B^T admit a common eigenvector E_1 such that $(E_1, e_1) = 1$ with $e_1 = (1, \cdots, 1)^T$. Let D be defined by*

$$Im(D) = (Span\{E_1\})^\perp. \tag{29.37}$$

Then system (III) is approximate synchronizable. Moreover, we have $\text{rank}(\mathcal{R}) = N - 1$.

Proof Since $(E_1, e_1) = 1$, noting (29.37), we have $e_1 \notin \text{Im}(D)$ and $\text{Ker}(C_1) \cap \text{Im}(D) = \{0\}$. Therefore, by Proposition 2.11, we have

$$\text{rank}(C_1 D) = \text{rank}(D) = N - 1. \tag{29.38}$$

Thus, the reduced adjoint system (29.35) is $C_1 D$-observable because of Holmgren's uniqueness theorem. By Proposition 29.7, system (III) is approximate synchronizable.

Noting (29.37), we have $E_1 \in \text{Ker}(D^T)$. Moreover, since E_1 is a common eigenvector of A^T and B^T, we have $E_1 \in \text{Ker}(\mathcal{R}^T)$, hence $\dim \text{Ker}(\mathcal{R}^T) \geqslant 1$, namely, $\text{rank}(\mathcal{R}) \leqslant N - 1$. On the other hand, since $\text{rank}(\mathcal{R}) \geqslant \text{rank}(D) = N - 1$, we get $\text{rank}(\mathcal{R}) = N - 1$. The proof is complete. □

Chapter 30
Approximate Boundary Synchronization by p-Groups

The approximate boundary synchronization by p-groups is introduced and studied in this chapter for system (III) with coupled Robin boundary controls.

30.1 Definitions

Let $p \geqslant 1$ be an integer and let

$$0 = n_0 < n_1 < n_2 < \cdots < n_p = N \tag{30.1}$$

be integers such that $n_r - n_{r-1} \geqslant 2$ for all $1 \leqslant r \leqslant p$. We rearrange the components of the state variable U into p-groups:

$$(u^{(1)}, \cdots, u^{(n_1)}), (u^{(n_1+1)}, \cdots, u^{(n_2)}), \cdots, (u^{(n_{p-1}+1)}, \cdots, u^{(n_p)}). \tag{30.2}$$

Definition 30.1 System (III) is **approximately synchronizable by p-groups** at the time $T > 0$ if for any given initial data $(\widehat{U}_0, \widehat{U}_1) \in (\mathcal{H}_0)^N \times (\mathcal{H}_{-1})^N$, there exists a sequence $\{H_n\}$ of boundary controls in $\mathcal{L}^M$ with compact support in $[0, T]$, such that the corresponding sequence $\{U_n\}$ of solutions to problem (III) and (III0) satisfies

$$u_n^{(k)} - u_n^{(l)} \to 0 \quad \text{in} \quad C^0_{loc}([T, +\infty); \mathcal{H}_0) \cap C^1_{loc}([T, +\infty); \mathcal{H}_{-1}) \tag{30.3}$$

for $n_{r-1} + 1 \leqslant k, l \leqslant n_r$ and $1 \leqslant r \leqslant p$ as $n \to +\infty$.

T. Li and B. Rao, *Boundary Synchronization for Hyperbolic Systems*,
Progress in Nonlinear Differential Equations and Their Applications 94,
https://doi.org/10.1007/978-3-030-32849-8_30

Let S_r be the following $(n_r - n_{r-1} - 1) \times (n_r - n_{r-1})$ matrix:

$$S_r = \begin{pmatrix} 1 & -1 & 0 & \cdots & 0 \\ 0 & 1 & -1 & \cdots & 0 \\ \vdots & \vdots & \ddots & \ddots & \vdots \\ 0 & 0 & \cdots & 1 & -1 \end{pmatrix}. \tag{30.4}$$

Let C_p be the following $(N - p) \times N$ full row-rank **matrix of synchronization by p-groups**:

$$C_p = \begin{pmatrix} S_1 & & & \\ & S_2 & & \\ & & \ddots & \\ & & & S_p \end{pmatrix}. \tag{30.5}$$

For $1 \leqslant r \leqslant p$, setting

$$(e_r)_i = \begin{cases} 1, & n_{r-1} + 1 \leqslant i \leqslant n_r, \\ 0, & \text{otherwise}, \end{cases} \tag{30.6}$$

it is clear that

$$\text{Ker}(C_p) = \text{Span}\{e_1, e_2, \cdots, e_p\}. \tag{30.7}$$

Moreover, the approximate boundary synchronization by p-groups (30.3) can be equivalently rewritten as

$$C_p U_n \to 0 \quad \text{as } n \to +\infty \tag{30.8}$$

in the space

$$C^0_{loc}([T, +\infty); (\mathcal{H}_0)^{N-p}) \cap C^1_{loc}([T, +\infty); (\mathcal{H}_{-1})^{N-p}). \tag{30.9}$$

Definition 30.2 The matrix A satisfies the **condition of C_p-compatibility** if there exists a unique matrix $\overline{A}_p$ of order $(N - p)$, such that

$$C_p A = \overline{A}_p C_p. \tag{30.10}$$

The matrix $\overline{A}_p$ is called the **reduced matrix of A by C_p**.

Remark 30.3 By Proposition 2.15, the condition of C_p-compatibility (30.10) is equivalent to

$$A\text{Ker}(C_p) \subseteq \text{Ker}(C_p). \tag{30.11}$$

Moreover, the reduced matrix $\overline{A}_p$ is given by

$$\overline{A}_p = C_p A C_p^T (C_p C_p^T)^{-1}. \tag{30.12}$$

Similarly, the matrix B satisfies the condition of C_p-compatibility if there exists a unique matrix $\overline{B}_p$ of order $(N-p)$, such that

$$C_p B = \overline{B}_p C_p, \tag{30.13}$$

which is equivalent to

$$B\text{Ker}(C_p) \subseteq \text{Ker}(C_p). \tag{30.14}$$

Assume that A and B satisfy the conditions of C_p-compatibility (30.10) and (30.13), respectively. Setting $W = C_p U$ in problem (III) and (III0), we get the following **reduced system**:

$$\begin{cases} W'' - \Delta W + \overline{A}_p W = 0 \text{ in } (0, T) \times \Omega, \\ W = 0 \qquad\qquad\qquad\quad \text{on } (0, T) \times \Gamma_0, \\ \partial_\nu W + \overline{B}_p W = C_p D H \text{ on } (0, T) \times \Gamma_1 \end{cases} \tag{30.15}$$

with the initial condition

$$t = 0: \quad W = C_p \widehat{U}_0, \ W' = C_p \widehat{U}_1 \quad \text{in } \Omega. \tag{30.16}$$

Since B is similar to a real symmetric matrix, by Proposition 2.21, the reduced matrix $\overline{B}_p$ is also similar to a real symmetric matrix. Then, by Theorem 19.19, the reduced problem (30.15)–(30.16) is well-posed in the space $(\mathcal{H}_0)^{N-p} \times (\mathcal{H}_{-1})^{N-p}$.

Accordingly, consider the **reduced adjoint system**

$$\begin{cases} \Psi'' - \Delta \Psi + \overline{A}_p^T \Psi = 0 & \text{in } (0, +\infty) \times \Omega, \\ \Psi = 0 & \text{on } (0, +\infty) \times \Gamma_0, \\ \partial_\nu \Psi + \overline{B}_p^T \Psi = 0 & \text{on } (0, +\infty) \times \Gamma_1 \end{cases} \tag{30.17}$$

together with the $C_p D$-observation

$$(C_p D)^T \Psi \equiv 0 \quad \text{on } (0, T) \times \Gamma_1. \tag{30.18}$$

We have

Proposition 30.4 *Assume that A and B satisfy the conditions of C_p-compatibility (30.10) and (30.13), respectively. Then system (III) is approximately synchronizable by p-groups if and only if the reduced system (30.15) is approximately null controllable, or equivalently if and only if the reduced adjoint system (30.17) is $C_p D$-observable.*

Corollary 30.5 *Under the conditions of C_p-compatibility (30.10) and (30.13), if system (III) is approximately synchronizable by p-groups, we necessarily have the following rank condition:*

$$\text{rank}(C_p \mathcal{R}) = N - p. \tag{30.19}$$

Proof Let $\overline{\mathcal{R}}$ be the matrix defined by (26.2)–(26.3) corresponding to the reduced matrices $\overline{A}_p$, $\overline{B}_p$ and $\overline{D} = C_p D$. Noting (30.10) and (30.13), we have

$$\overline{A}_p^r \overline{B}_p^s \overline{D} = \overline{A}_p^r \overline{B}_p^s C_p D = C_p A^r B^s D, \tag{30.20}$$

then

$$\overline{\mathcal{R}} = C_p \mathcal{R}. \tag{30.21}$$

Under the assumption that system (III) is approximately synchronizable by p-groups, by Proposition 30.4, the reduced system (30.15) is approximately null controllable, then by Corollary 27.5, we have $\text{rank}(\overline{\mathcal{R}}) = N - p$, which together with (30.21) implies (30.19). □

30.2 Fundamental Properties

Proposition 30.6 *Assume that system (III) is approximately synchronizable by p-groups. Then, we necessarily have $rank(\mathcal{R}) \geqslant N - p$.*

Proof Assume that $\dim \text{Ker}(\mathcal{R}^T)=d$ with $d > p$. Let $\text{Ker}(\mathcal{R}^T)=\text{Span}\{E_1, \cdots, E_d\}$. Since

$$\dim \text{Ker}(\mathcal{R}^T) + \dim \text{Im}(C_p^T) = d + N - p > N,$$

we have $\text{Ker}(\mathcal{R}^T) \cap \text{Im}(C_p^T) \neq \{0\}$. Hence, there exists a nonzero vector $x \in \mathbb{R}^{N-d}$ and coefficients $\beta_1, \cdots, \beta_d$ not all zero, such that

$$\sum_{r=1}^{d} \beta_r E_r = C_p^T x. \tag{30.22}$$

Moreover, by Lemma 26.1, we still have (27.2) and (27.3). Then, applying E_r to problem (III) and (III0) with $U = U_n$ and $H = H_n$ and setting $u_r = (E_r, U_n)$ for $1 \leqslant r \leqslant d$, it follows that

$$\begin{cases} u_r'' - \Delta u_r + \sum_{s=1}^d \alpha_{rs} u_s = 0 \text{ in } (0, +\infty) \times \Omega, \\ u_r = 0 \qquad\qquad\qquad\qquad\quad \text{on } (0, +\infty) \times \Gamma_0, \\ \partial_\nu u_r + \sum_{s=1}^d \beta_{rs} u_s = 0 \qquad \text{on } (0, +\infty) \times \Gamma_1 \end{cases} \tag{30.23}$$

with the initial condition

$$t = 0: \quad u_r = (E_r, \widehat{U}_0), \ u_r' = (E_r, \widehat{U}_1) \ \text{ in } \Omega. \tag{30.24}$$

Noting (30.8), it follows from (30.22) that

$$\sum_{r=1}^{d} \beta_r u_r = (x, C_p U_n) \to 0 \tag{30.25}$$

as $n \to +\infty$ in the space

$$C^0_{loc}([T, +\infty); \mathcal{H}_0) \cap C^1_{loc}([T, +\infty); \mathcal{H}_{-1}).$$

Since the functions $u_1, \cdots, u_d$ are independent of n and of applied boundary controls H_n, it follows that

$$\sum_{r=1}^{d} \beta_r u_r(T) = \sum_{r=1}^{d} \beta_r u_r'(T) = 0 \quad \text{in } \Omega. \tag{30.26}$$

Then, it follows from the well-posedness of problem (30.23)–(30.24) that

$$\sum_{r=1}^{d} \beta_r (E_r, \widehat{U}_0) = \sum_{r=1}^{d} \beta_r (E_r, \widehat{U}_1) = 0 \tag{30.27}$$

for any given initial data $(\widehat{U}_0, \widehat{U}_1) \in (\mathcal{H}_0)^N \times (\mathcal{H}_{-1})^N$. In particular, we get

$$\sum_{r=1}^{d} \beta_r E_r = 0, \tag{30.28}$$

then a contradiction: $\beta_1 = \cdots = \beta_d = 0$, because of the linear independence of the vectors $E_1, \cdots, E_d$. The proof is achieved. □

Theorem 30.7 *Let A and B satisfy the conditions of C_p-compatibility (30.10) and (30.13), respectively. Assume that A^T and B^T admit a common invariant subspace V, which is bi-orthonormal to $Ker(C_p)$. Then, setting the boundary control matrix D by*

$$Im(D) = V^{\perp}, \tag{30.29}$$

system (III) is approximately synchronizable by p-groups. Moreover, we have $rank(\mathcal{R}) = N - p$.

Proof Since V is bi-orthonormal to $\text{Ker}(C_p)$, we have

$$\text{Ker}(C_p) \cap V^{\perp} = \text{Ker}(C_p) \cap \text{Im}(D) = \{0\}, \tag{30.30}$$

therefore, by Proposition 2.11, we have

$$\text{rank}(C_p D) = \text{rank}(D) = N - p. \tag{30.31}$$

Thus, the C_pD-observation (29.36) becomes the full observation

$$\Psi \equiv 0 \quad \text{on } (0, T) \times \Gamma_1. \tag{30.32}$$

By Holmgren's uniqueness theorem (*cf.* Theorem 8.2 in [62]), the reduced adjoint system (30.17) is C_pD-observable and the reduced system (30.15) is approximately null controllable. Then, by Proposition 30.4, the original system (III) is approximately synchronizable by p-groups. Noting that $\text{Ker}(D^T) = V$, by Lemma 26.1, it is easy to see that $\text{rank}(\mathcal{R}) = N - p$. The proof is then complete. □

Theorem 30.8 *Assume that system (III) is approximately synchronizable by p-groups. Assume furthermore that $\text{rank}(\mathcal{R}) = N - p$. Then, we have the following assertions:*

(i) $\text{Ker}(\mathcal{R}^T)$ is bi-orthonormal to $\text{Ker}(C_p)$.

(ii) For any given initial data $(\widehat{U}_0, \widehat{U}_1) \in (\mathcal{H}_0)^N \times (\mathcal{H}_{-1})^N$, there exist unique scalar functions $u_1, u_2, \cdots, u_p$ such that

$$u_n^{(k)} \to u_r \quad in\ C^0_{loc}([T, +\infty); \mathcal{H}_0) \cap C^1_{loc}([T, +\infty); H^{-1}(\Omega)) \tag{30.33}$$

for $n_{r-1} + 1 \leqslant k \leqslant n_r$ and $1 \leqslant r \leqslant p$ as $n \to +\infty$.

(iii) The coupling matrices A and B satisfy the conditions of C_p-compatibility (30.10) and (30.13), respectively.

Proof (i) We claim that $\text{Ker}(\mathcal{R}^T) \cap \text{Im}(C_p^T) = \{0\}$. Then, noting that $\text{Ker}(\mathcal{R}^T)$ and $\text{Ker}(C_p)$ have the same dimension p and

$$\text{Ker}(\mathcal{R}^T) \cap \{\text{Ker}(C_p)\}^{\perp} = \text{Ker}(\mathcal{R}^T) \cap \text{Im}(C_p^T) = \{0\}, \tag{30.34}$$

by Proposition 2.5, $\text{Ker}(\mathcal{R}^T)$ and $\text{Ker}(C_p)$ are bi-orthonormal. Let $\text{Ker}(\mathcal{R}^T) = \text{Span}\{E_1, \cdots, E_p\}$ and $\text{Ker}(C_P) = \text{Span}\{e_1, \cdots, e_p\}$ such that

$$(E_r, e_s) = \delta_{rs}, \quad r, s = 1, \cdots, p. \tag{30.35}$$

Now we return to check that $\text{Ker}(\mathcal{R}^T) \cap \text{Im}(C_p^T) = \{0\}$. If $\text{Ker}(\mathcal{R}^T) \cap \text{Im}(C_p^T) \neq \{0\}$, then there exists a nonzero vector $x \in \mathbb{R}^{N-p}$ and some coefficients $\beta_1, \cdots, \beta_p$ not all zero, such that

$$\sum_{r=1}^{p} \beta_r E_r = C_p^T x. \tag{30.36}$$

By Lemma 26.1, we still have (27.2) and (27.3) with $d = p$. For $1 \leqslant r \leqslant p$, applying E_r to problem (III) and (III0) with $U = U_n$ and $H = H_n$, and setting

$$u_r = (E_r, U), \tag{30.37}$$

it follows that

$$\begin{cases} u_r'' - \Delta u_r + \sum_{s=1}^{p} \alpha_{rs} u_s = 0 \text{ in } (0, +\infty) \times \Omega, \\ u_r = 0 \qquad\qquad\qquad\qquad \text{on } (0, +\infty) \times \Gamma_0, \\ \partial_\nu u_r + \sum_{s=1}^{p} \beta_{rs} u_s = 0 \qquad \text{on } (0, +\infty) \times \Gamma_1 \end{cases} \tag{30.38}$$

with the initial condition

$$t = 0: \quad u_r = (E_r, \widehat{U}_0), \; u_r' = (E_r, \widehat{U}_1). \tag{30.39}$$

Noting (30.8), we have

$$\sum_{r=1}^{p} \beta_r u_r = (x, C_p U_n) \to 0 \quad \text{as } n \to +\infty \tag{30.40}$$

in the space

$$C^0_{loc}([T, +\infty); \mathcal{H}_0) \cap C^1_{loc}([T, +\infty); \mathcal{H}_{-1}). \tag{30.41}$$

Since the functions $u_1, \cdots, u_p$ are independent of n and of applied boundary controls, we have

$$\sum_{r=1}^{p} \beta_r u_r(T) \equiv \sum_{r=1}^{p} \beta_r u_r'(T) \equiv 0 \quad \text{in } \Omega. \tag{30.42}$$

Then, it follows from the well-posedness of problem (30.38)–(30.39) that

$$\sum_{r=1}^{p} \beta_r (E_r, \widehat{U}_0) = \sum_{r=1}^{p} \beta_r (E_r, \widehat{U}_1) = 0 \quad \text{in } \Omega \tag{30.43}$$

for any given initial data $(\widehat{U}_0, \widehat{U}_1) \in (\mathcal{H}_0)^N \times (\mathcal{H}_{-1})^N$. In particular, we get

$$\sum_{r=1}^{p} \beta_r E_r = 0, \tag{30.44}$$

then, a contradiction: $\beta_1 = \cdots = \beta_p = 0$, because of the linear independence of the vectors $E_1, \cdots, E_p$.

(ii) Noting (30.8) and (30.37), we have

$$\begin{pmatrix} C_p \\ E_1^T \\ \vdots \\ E_p^T \end{pmatrix} U_n = \begin{pmatrix} C_p U_n \\ E_1^T U_n \\ \vdots \\ E_p^T U_n \end{pmatrix} \to \begin{pmatrix} 0 \\ u_1 \\ \vdots \\ u_p \end{pmatrix} \tag{30.45}$$

as $n \to +\infty$ in the space

$$C^0_{loc}([T, +\infty); (\mathcal{H}_0)^N) \cap C^1_{loc}([T, +\infty); (\mathcal{H}_{-1})^N), \tag{30.46}$$

where $u_1, \cdots, u_p$ are given by (30.37). Since $\text{Ker}(\mathcal{R}^T) \cap \text{Im}(C_p^T) = \{0\}$, the matrix $\begin{pmatrix} C_p \\ E_1^T \\ \vdots \\ E_p^T \end{pmatrix}$ is invertible. Thus, it follows from (30.45) that there exists U such that

$$U_n \to \begin{pmatrix} C_p \\ E_1^T \\ \vdots \\ E_p^T \end{pmatrix}^{-1} \begin{pmatrix} 0 \\ u_1 \\ \vdots \\ u_p \end{pmatrix} =: U \tag{30.47}$$

as $n \to +\infty$ in the space (30.46). Moreover, (30.8) implies that

$$t \geqslant T: \quad C_p U \equiv 0 \quad \text{in } \Omega.$$

Noting (30.7), (30.35) and (30.37), it follows that

$$t \geqslant T: \quad U = \sum_{r=1}^{p} (E_r, U) e_r = \sum_{r=1}^{p} u_r e_r \quad \text{in } \Omega. \tag{30.48}$$

Noting (30.6), we get then (30.33).

(iii) Applying C_p to system (III) with $U = U_n$ and $H = H_n$, and passing to the limit as $n \to +\infty$, by (30.8), (30.47) and (30.48), it is easy to get that

$$\sum_{r=1}^{p} C_p A e_r u_r(T) \equiv 0 \quad \text{in } \Omega \tag{30.49}$$

and

$$\sum_{r=1}^{p} C_p B e_r u_r(T) \equiv 0 \quad \text{on } \Gamma_1. \tag{30.50}$$

On the other hand, because of the time-invertibility, system (30.38) defines an isomorphism from $(\mathcal{H}_1)^p \times (\mathcal{H}_0)^p$ onto $(\mathcal{H}_1)^p \times (\mathcal{H}_0)^p$ for all $t \geqslant 0$. Then, it follows from (30.49) and (30.50) that

$$C_p A e_r = 0 \quad \text{and} \quad C_p B e_r = 0 \quad \text{for } 1 \leqslant r \leqslant p. \tag{30.51}$$

We get thus the conditions of C_p-compatibility for A and B, respectively. The proof is complete. □

Remark 30.9 In general, the convergence (30.8) of the sequence $\{C_pU_n\}$ does not imply the convergence of the sequence of solutions $\{U_n\}$. In fact, we even don't know if the sequence $\{U_n\}$ is bounded. However, under the rank condition $\text{rank}(\mathcal{R}) = N - p$, the convergence (30.8) actually implies the convergence (30.33). Moreover, the functions $u_1, \cdots, u_p$, called the **approximately synchronizable state by p-groups**, are independent of applied boundary controls. In this case, system (III) is approximately synchronizable by p-groups in the **pinning sense**, while that originally given by Definition 30.1 is in the **consensus sense**.

Let $\mathbb{D}_p$ be the set of all the boundary control matrices D which realize the approximate boundary synchronization by p-groups for system (III). In order to show the dependence on D, we prefer to write $\mathcal{R}_D$ instead of $\mathcal{R}$ given in (26.3). Then, we may define the minimal rank as

$$N_p = \inf_{D \in \mathbb{D}_p} \text{rank}(\mathcal{R}_D). \tag{30.52}$$

Noting that $\text{rank}(\mathcal{R}_D) = N - \dim \text{Im}(\mathcal{R}_D^T)$, because of Proposition 30.6, we have

$$N_p \geqslant N - p. \tag{30.53}$$

Moreover, we have the following

Corollary 30.10 *The equality*

$$N_p = N - p \tag{30.54}$$

holds if and only if the coupling matrices A and B satisfy the conditions of C_p-compatibility (30.10) and (30.13), respectively, and A^T and B^T possess a common invariant subspace, which is bi-orthonormal to $Ker(C_p)$. Moreover, the approximate boundary synchronization by p-groups is in the pinning sense.

Proof Assume that (30.54) holds. Then there exists a boundary control matrix $D \in D_p$, such that $\dim \text{Ker}(\mathcal{R}_D^T) = p$. By Theorem 30.8, the coupling matrices A and B satisfy the conditions of C_p-compatibility (30.10) and (30.13), respectively, and $\text{Ker}(\mathcal{R}_D^T)$ which, by Lemma 26.1, is bi-orthonormal to $\text{Ker}(C_p)$, is invariant for both A^T and B^T. Moreover, the approximate boundary synchronization by p-groups is in the pinning sense.

Conversely, let V be a subspace which is invariant for both A^T and B^T, and bi-orthonormal to $\text{Ker}(C_p)$. Noting that A and B satisfy the conditions of C_p-compatibility (30.10) and (30.13), respectively, by Theorem 30.7, there exists a boundary control matrix $D \in \mathbb{D}_p$, such that $\dim \text{Ker}(\mathcal{R}_D^T) = p$, which together with (30.53) implies (30.54). □

Remark 30.11 When $N_p > N - p$, the situation is more complicated. We don't know if the conditions of C_p-compatibility (30.10) and (30.13) are necessary, either if the approximate boundary synchronization by p-groups is in the pinning sense.

Chapter 31
Approximately Synchronizable States by p-Groups

When system (III) is approximately synchronizable by p-groups, the corresponding approximately synchronizable states by p-groups will be considered in this chapter.

In Theorem 30.8, we have shown that if system (III) is approximately synchronizable by p-groups under the condition $\dim \text{Ker}(\mathcal{R}^T) = p$, then A and B satisfy the corresponding conditions of C_p-compatibility, and $\text{Ker}(\mathcal{R}^T)$ is bi-orthonormal to $\text{Ker}(C_p)$; moreover, the approximately synchronizable state by p-groups is independent of applied boundary controls. The following is the counterpart.

Theorem 31.1 *Let A and B satisfy the conditions of C_p-compatibility (30.10) and (30.13), respectively. Assume that system (III) is approximately synchronizable by p-groups. If the projection of the solution U to problem (III) and (III0) on a subspace V of dimension p is independent of applied boundary controls, then $V = Ker(\mathcal{R}^T)$. Moreover, $Ker(\mathcal{R}^T)$ is bi-orthonormal to $Ker(C_p)$.*

Proof Fixing $\widehat{U}_0 = \widehat{U}_1 = 0$, by Theorem 19.9, the linear map

$$F: \quad H \to U$$

is continuous, then, infinitely differential from $L^2_{loc}(0, +\infty; (L^2(\Gamma_1))^M)$ to $C^0_{loc}([0, +\infty); (\mathcal{H}_0)^N) \cap C^1_{loc}([0, +\infty); (\mathcal{H}_{-1})^N)$. For any given boundary control $\widehat{H} \in L^2_{loc}(0, +\infty; (L^2(\Gamma_1))^M)$, let $\widehat{U}$ be the Fréchet derivative of U on $\widehat{H}$:

$$\widehat{U} = F'(0)\widehat{H}.$$

Then, it follows from problem (III) and (III0) that

$$\begin{cases} \widehat{U}'' - \Delta\widehat{U} + A\widehat{U} = 0 & \text{in } (0, +\infty) \times \Omega, \\ \widehat{U} = 0 & \text{on } (0, +\infty) \times \Gamma_0, \\ \partial_\nu \widehat{U} + B\widehat{U} = D\widehat{H} & \text{on } (0, +\infty) \times \Gamma_1, \\ t = 0: \quad \widehat{U} = \widehat{U}' = 0 \text{ in } \Omega. \end{cases} \tag{31.1}$$

T. Li and B. Rao, *Boundary Synchronization for Hyperbolic Systems*, Progress in Nonlinear Differential Equations and Their Applications 94,
https://doi.org/10.1007/978-3-030-32849-8_31

Let $V = \text{Span}\{E_1, \cdots, E_p\}$. Then, the independence of the projection of U on the subspace V, with respect to the boundary controls, implies that

$$(E_r, \widehat{U}) \equiv 0 \quad \text{in } (0, +\infty) \times \Omega \quad \text{for } 1 \leqslant r \leqslant p. \tag{31.2}$$

We first show that $E_r \notin \text{Im}(C_p^T)$ for any given r with $1 \leqslant r \leqslant p$. Otherwise, there exists an $\bar{r}$ with $1 \leqslant \bar{r} \leqslant p$ and a vector $x_{\bar{r}} \in \mathbb{R}^{N-p}$, such that $E_{\bar{r}} = C_p^T x_{\bar{r}}$. Then, it follows from (31.2) that

$$0 = (E_{\bar{r}}, \widehat{U}) = (x_{\bar{r}}, C_p \widehat{U}).$$

Since $W = C_p \widehat{U}$ is the solution to the reduced system (30.15) with $H = \widehat{H}$, which is approximately controllable, we get thus $x_{\bar{r}} = 0$, which contradicts $E_{\bar{r}} \neq 0$. Thus, since $\dim \text{Im}(C_p^T) = N - p$ and $\dim(V) = p$, we have $V \oplus \text{Im}(C_p^T) = \mathbb{R}^N$. Then, for any given r with $1 \leqslant r \leqslant p$, there exists a vector $y_r \in \mathbb{R}^{N-p}$, such that

$$A^T E_r = \sum_{s=1}^{p} \alpha_{rs} E_s + C_p^T y_r.$$

Noting (31.2) and applying E_r to system (31.1), it follows that

$$0 = (A\widehat{U}, E_r) = (\widehat{U}, A^T E_r) = (\widehat{U}, C_p^T y_r) = (C_p \widehat{U}, y_r).$$

Once again, the approximate controllability of the reduced system (30.15) implies that $y_r = 0$ for $1 \leqslant r \leqslant p$. Then, it follows that

$$A^T E_r = \sum_{s=1}^{p} \alpha_{rs} E_s, \quad 1 \leqslant r \leqslant p.$$

So, the subspace V is invariant for A^T.

By the sharp regularity given in [30] (cf. also [29, 31]) on the problem of Neumann, we have improved the regularity of the solution to problem (III) and (III0). In fact, setting

$$\alpha = \begin{cases} 3/5 - \epsilon, & \Omega \text{ is a bounded smooth domain,} \\ 2/3, & \Omega \text{ is a sphere,} \\ 3/4 - \epsilon, & \Omega \text{ is a parallelepiped,} \end{cases} \tag{31.3}$$

where $\epsilon > 0$ is a sufficiently small number, by Theorem 19.10, the trace

$$U|_{\Gamma_1} \in (H^{2\alpha-1}_{loc}((0, +\infty) \times \Gamma_1))^N \tag{31.4}$$

with the corresponding continuous dependence with respect to $(\widehat{U}_0, \widehat{U}_1, H)$.

Next, noting (31.2) and applying $E_r (1 \leqslant r \leqslant p)$ to the boundary condition on Γ_1 in (31.1), we get

$$(D^T E_r, \widehat{H}) = (E_r, B\widehat{U}).$$

Since $2\alpha - 1 > 0$, $H^{2\alpha-1}_{loc}((0, +\infty) \times \Gamma_1)$ is compactly embedded in $L^2_{loc}((0, +\infty) \times \Gamma_1)$, we get then $D^T E_r = 0$ for all $1 \leqslant r \leqslant p$, namely,

$$V \subseteq \mathrm{Ker}(D^T). \tag{31.5}$$

In particular, the subspace V is contained in $\mathrm{Ker}(D^T)$.

Moreover, for $1 \leqslant i \leqslant p$, we have

$$(E_r, B\widehat{U}) = 0 \quad \text{on } (0, +\infty) \times \Gamma_1. \tag{31.6}$$

Now, let $x_r \in \mathbb{R}^{N-p}$, such that

$$B^T E_r = \sum_{s=1}^{p} \beta_{rs} E_s + C_p^T x_r. \tag{31.7}$$

Noting (31.2) and inserting the expression (31.7) into (31.6), it follows that

$$(x_r, C_p\widehat{U}) = 0 \quad \text{on } (0, +\infty) \times \Gamma_1.$$

Once again, because of the approximate boundary controllability of the reduced system (30.15), we deduce that $x_r = 0$ for $1 \leqslant r \leqslant p$. Then, we get

$$B^T E_r = \sum_{s=1}^{p} \beta_{rs} E_s, \quad 1 \leqslant r \leqslant p.$$

So, the subspace V is also invariant for B^T.

Finally, since $\dim(V) = p$, by Lemma 26.1 and Proposition 30.6, we have $\mathrm{Ker}(\mathcal{R}^T) = V$. Then, by assertion (i) of Theorem 30.8, $\mathrm{Ker}(\mathcal{R}^T)$ is bi-orthonormal to $\mathrm{Ker}(C_p)$. This achieves the proof. □

Let d be a column vector of D and contained in $\mathrm{Ker}(C_p)$. Then d will be canceled in the product matrix $C_p D$, and therefore it cannot give any effect to the reduced system (30.15). However, the vectors in $\mathrm{Ker}(C_p)$ may play an important role for the approximate boundary controllability. More precisely, we have the following

Theorem 31.2 *Let A and B satisfy the conditions of C_p-compatibility (30.10) and (30.13), respectively. Assume that system (III) is approximately synchronizable by p-groups under the action of a boundary control matrix D. Assume furthermore that*

$$e_1, \cdots, e_p \in Im(D), \tag{31.8}$$

where $e_1, \cdots, e_p$ are given by (30.6). Then system (III) is actually approximately null controllable.

Proof By Proposition 27.4, it is sufficient to show that the adjoint system (19.19) is D-observable. For $1 \leqslant r \leqslant p$, applying e_r to the adjoint system (19.19) and noting $\phi_r = (e_r, \Phi)$, it follows that

$$\begin{cases} \phi_r'' - \Delta\phi_r + \sum_{s=1}^p \alpha_{sr}\phi_s = 0 & \text{in } (0, +\infty) \times \Omega, \\ \phi_r = 0 & \text{on } (0, +\infty) \times \Gamma_0, \\ \partial_\nu \phi_r + \sum_{s=1}^p \beta_{sr}\phi_s = 0 & \text{on } (0, +\infty) \times \Gamma_1, \end{cases} \tag{31.9}$$

where the constant coefficients α_{sr} and β_{sr} are given by

$$Ae_r = \sum_{s=1}^p \alpha_{sr} e_s \quad \text{and} \quad Be_r = \sum_{s=1}^p \beta_{sr} e_s, \quad 1 \leqslant r \leqslant p. \tag{31.10}$$

On the other hand, noting (31.8), the D-observation (27.1) implies that

$$\phi_r \equiv 0 \quad \text{on } (0, T) \times \Gamma_1 \tag{31.11}$$

for $1 \leqslant r \leqslant p$. Then, by Holmgren's uniqueness theorem (cf. Theorem 8.2 in [62]), we get

$$\phi_r \equiv 0 \quad \text{in } (0, +\infty) \times \Omega \tag{31.12}$$

for $1 \leqslant r \leqslant p$. Thus, $\Phi \in \text{Im}(C_p^T)$, then we can write $\Phi = C_p^T \Psi$ and the adjoint system (19.19) becomes

$$\begin{cases} C_p^T \Psi'' - C_p^T \Delta\Psi + A^T C_p^T \Psi = 0 & \text{in } (0, +\infty) \times \Omega, \\ C_p^T \Psi = 0 & \text{on } (0, +\infty) \times \Gamma_0, \\ C_p^T \partial_\nu \Psi + B^T C_p^T \Psi = 0 & \text{on } (0, +\infty) \times \Gamma_1. \end{cases} \tag{31.13}$$

Noting the conditions of C_p-compatibility (30.10) and (30.13), it follows that

$$\begin{cases} C_p^T (\Psi'' - \Delta\Psi + \overline{A}_p^T \Psi) = 0 & \text{in } (0, +\infty) \times \Omega, \\ C_p^T \Psi = 0 & \text{on } (0, +\infty) \times \Gamma_0 \\ C_p^T (\partial_\nu \Psi + \overline{B}_p^T \Psi) = 0 & \text{on } (0, +\infty) \times \Gamma_1. \end{cases} \tag{31.14}$$

Since the mapping C_p^T is injective, we find again the reduced adjoint system (30.17). Accordingly, the D-observation (27.1) implies that

$$D^T \Phi \equiv D^T C_p^T \Psi \equiv 0 \quad \text{on } [0, T] \times \Gamma_1. \tag{31.15}$$

Since system (III) is approximately synchronizable by p-groups under the action of the boundary control matrix D, by Proposition 30.4, the reduced adjoint system (30.17) for Ψ is $C_p D$-observable, therefore, $\Psi \equiv 0$, then $\Phi \equiv 0$. So, the adjoint system (19.19) is D-observable, then by Proposition 27.4, system (III) is approximately null controllable. □

Chapter 32
Closing Remarks

Some closing remarks including related literatures and prospects are given in this chapter.

32.1 Related Literatures

The main contents of this monograph have been published or will be published in a series of papers written by the authors with their collaborators (cf. [3, 35–56]).

For closing this monograph, let us comment on some related literatures. One motivation for studying the synchronization consists of establishing a weak exact boundary controllability in the case of fewer boundary controls. In order to realize the exact boundary controllability, because of its uniform character with respect to the state variables, the number of boundary controls should be equal to the degrees of freedom of the considered system. However, when the components of initial data are allowed to have different levels of energy, the exact boundary controllability by means of only one boundary control for a system of two wave equations was established in [65, 72], and for the cascade system of N wave equations in [2]. In [10], the authors established the controllability of two coupled wave equations on a compact manifold with only one local distributed control. Moreover, both the optimal time of controllability and the controllable spaces are given in the cases with the same or different wave speeds.

The approximate boundary null controllability is more flexible with respect to the number of applied boundary controls. In [37, 53, 55], for a coupled system of wave equations with Dirichlet, Neumann or Robin boundary controls, some fundamental algebraic properties on the coupling matrices are used to characterize the unique continuation for the solution to the corresponding adjoint systems. Although these

T. Li and B. Rao, *Boundary Synchronization for Hyperbolic Systems*,
Progress in Nonlinear Differential Equations and Their Applications 94,
https://doi.org/10.1007/978-3-030-32849-8_32

criteria are only necessary in general, they open an important way to the research on the unique continuation for a system of hyperbolic partial differential equations.

In contrast with hyperbolic systems, in [4] (also [5, 13] and the reference therein), it was shown that Kalman's criterion is sufficient to the exact boundary null controllability for systems of parabolic equations. Recently, in [76], the authors have established the minimal time for a control problem related to the exact synchronization for a linear parabolic system.

The average controllability proposed in [67, 86] gives another way to deal with the controllability with fewer controls. The observability inequality is particularly interesting for a trial on the decay rate of approximate controllability.

32.2 Prospects

The study of exact and approximate boundary synchronizations for systems of PDEs has just begun, and there are still many problems worthy of further research.

1. We have discussed above a coupled system of wave equations with Dirichlet, Neumann, or coupled Robin boundary controls. In the case of coupled Robin boundary controls, since the solution has less trace regularity, we can only obtain a relatively complete result when the domain Ω is a parallelepiped. It is still an open problem to get a relevant result for more general domains. Moreover, besides the original coupling matrix A, there is one more boundary coupling matrix B, which brings more difficulties and opportunities to the study.

2. The above research is limited to the linear situation, and the method employed is also linear. However, the study of synchronization should be extended to the nonlinear situation. At present, in the one-space-dimensional case, for the coupled system of quasilinear wave equations with various boundary controls, the exact boundary synchronization has been achieved in the framework of classical solutions (cf. [20, 68]) and there is still larger developing space for more general research.

3. In addition to the study of exact boundary controllability of hyperbolic systems on the whole space domain (cf. [32, 62, 73]), in the recent years, due to the demand of applications, in the one-space-dimensional situation, the research on exact boundary controllability of hyperbolic systems at a node has been developed, which is called the **exact boundary controllability of nodal profile** (cf. [16, 33, 59]). Correspondingly, the investigation on the approximate boundary controllability of nodal profile as well as on the exact and approximate boundary synchronizations of nodal profile should also be carried further.

4. We can also probe into similar problems on a complicate domain formed by networks. The corresponding results in the one-space-dimensional case on the exact boundary controllability and on the exact boundary controllability of nodal profile can be found in [32, 59], etc., and the synchronization on a tree-like network for a coupled system of wave equations should be carried out.

5. Some essentially new results can be obtained if we study the phenomena of synchronization through coupling among individuals with possibly different motion

laws (governing equations), whose nature is yet to be explored. The research on the existence of the exactly synchronizable state for a coupled system of wave equations with different wave speeds has been initiated.

6. The stability of the exactly synchronizable state or the approximately synchronizable state should be studied systematically.

7. It is worth to consider the **generalized exact boundary synchronization** (cf. [58] in the 1-D case and [77–80] in higher dimensional case) and the generalized approximate boundary synchronization.

8. The coupled system of first-order linear or quasilinear hyperbolic systems has wider implications and applications than the coupled system of wave equations, and to do the study of similar problems on it will be of great significance, though more difficult (cf. [68]).

9. To do similar research on the coupled system of other linear or nonlinear evolution equations (such as beam equations, plate equations, heat equations, etc.) will reveal many new properties and characteristics, which is also quite meaningful.

10. To extend the concept of synchronization to the case of components with different time delays will be more challenging and may expose quite different features.

11. The study above focuses on the exact or approximate synchronization for a coupled system of wave equations in finite time through boundary controls. It is also worthwhile to study the asymptotic synchronization in the linear and nonlinear situations without any boundary control when $t \to +\infty$ (cf. [56]). This should be a meaningful extension of the research on the asymptotic stability of the solution to the coupled system of wave equations.

12. To dig into more practical applications of related results will further enrich the new field of study and lead to greater promotion and influence of the subject.

References

1. F. Alabau-Boussouira, A two-level energy method for indirect boundary observability and controllability of weakly coupled hyperbolic systems. SIAM J. Control Optim. **42**, 871–904 (2003)
2. F. Alabau-Boussouira, A hierarchic multi-level energy method for the control of bidiagonal and mixed n-coupled cascade systems of PDE's by a reduced number of controls. Adv. Differ. Equ. **18**, 1005–1072 (2013)
3. F. Alabau-Boussouira, T.T. Li, B.P. Rao, Indirect observation and control for a coupled cascade system of wave equations with Neumann boundary conditions, to appear
4. F. Ammar Khodja, A. Benabdallah, C. Dupaix, Null-controllability of some reaction-diffusion systems with one control force. J. Math. Anal. Appl. **320**, 928–943 (2006)
5. F. Ammar Khodja, A. Benabdallah, M. González-Burgos, L. de Teresa, Recent results on the controllability of linear coupled parabolic problems: a survey. Math. Control Relat. Fields **1**, 267–306 (2011)
6. A.V. Balakrishnan, Fractional powers of closed operators and the semigroups generated by them. Pac. J. Math. **10**, 419–437 (1960)
7. C. Bardos, G. Lebeau, J. Rauch, Sharp sufficient conditions for the observation, control, and stabilization of waves from the boundary. SIAM J. Control Optim. **30**, 1024–1064 (1992)
8. H. Brezis, *Functional Analysis, Sobolev Spaces and Partial Differential Equations* (Springer, New York, 2011)
9. R. Bru, L. Rodman, H. Schneider, Extensions of Jordan bases for invariant subspaces of a matrix. Linear Algebra Appl. **150**, 209–225 (1991)
10. B. Dehman, J. Le Rousseau, M. Léautaud, Controllability of two coupled wave equations on a compact manifold. Arch. Ration. Mech. Anal. **211**, 113–187 (2014)
11. Th. Duyckaerts, X. Zhang, E. Zuazua, On the optimality of the observability inequalities for parabolic and hyperbolic systems with potentials. Ann. Inst. H. Poincaré Anal. Non Linéaire **25**, 1–41 (2008)
12. S. Ervedoza, E. Zuazua, A systematic method for building smooth controls for smooth data. Discret. Contin. Dyn. Syst. Ser. B **14**, 1375–1401 (2010)
13. E. Fernández-Cara, M. González-Burgos, L. de Teresa, Boundary controllability of parabolic coupled equations. J. Funct. Anal. **259**, 1720–1758 (2010)
14. N. Garofalo, F.-H. Lin, Monotonicity properties of variational integrals, A_p weights and unique continuation. Indiana Univ. Math. J. **35**, 245–268 (1986)

T. Li and B. Rao, *Boundary Synchronization for Hyperbolic Systems*,
Progress in Nonlinear Differential Equations and Their Applications 94,
https://doi.org/10.1007/978-3-030-32849-8

15. I.C. Gohberg, M.G. Krein, *Introduction to the Theory of Linear Nonselfadjoint Operators* (AMS, Providence, 1969)
16. M. Gugat, M. Hertz, V. Schleper, Flow control in gas networks: exact controllability to a given demand. Math. Methods Appl. Sci. **34**, 745–757 (2011)
17. M.L.J. Hautus, Controllability and observability conditions for linear autonomous systems. Indag. Math. (N.S.) **31**, 443–446 (1969)
18. L. Hörmander, *Linear Partial Differential Operators* (Springer, Berlin, 1976)
19. L. Hu, F.Q. Ji, K. Wang, Exact boundary controllability and exact boundary observability for a coupled system of quasilinear wave equations. Chin. Ann. Math. Ser. B **34**, 479–490 (2013)
20. L. Hu, T.T. Li, P. Qu, Exact boundary synchronization for a coupled system of 1-D quasilinear wave equations. ESAIM: Control Optim. Calc. Var. **22**, 1136–1183 (2016)
21. L. Hu, T.T. Li, B.P. Rao, Exact boundary synchronization for a coupled system of 1-D wave equations with coupled boundary conditions of dissipative type. Commun. Pure Appl. Anal. **13**, 881–901 (2014)
22. Ch. Huygens, *Oeuvres Complètes*, vol. 15 (Swets & Zeitlinger B.V., Amsterdam, 1967)
23. R.E. Kalman, Contributions to the theory of optimal control. Bol. Soc. Mat. Mex. **5**, 102–119 (1960)
24. T. Kato, Fractional powers of dissipative operators. J. Math. Soc. Jpn. **13**(3), 246–274 (1961)
25. T. Kato, *Perturbation Theory for Linear Operators*. Grundlehren der Mathematischen Wissenschaften, vol. 132 (Springer, Berlin, 1976)
26. V. Komornik, *Exact Controllability and Stabilization, the Multiplier Method* (Masson, Paris, 1994)
27. V. Komornik, P. Loreti, Observability of compactly perturbed systems. J. Math. Anal. Appl. **243**, 409–428 (2000)
28. V. Komornik, P. Loreti, *Fourier Series in Control Theory*. Springer Monographs in Mathematics (Springer, New York, 2005)
29. I. Lasiecka, R. Triggiani, Trace regularity of the solutions of the wave equation with homogeneous Neumann boundary conditions and data supported away from the boundary. J. Math. Anal. Appl. **141**, 49–71 (1989)
30. I. Lasiecka, R. Triggiani, Sharp regularity for mixed second-order hyperbolic equations of Neumann type, I. L2 non-homogeneous data. Ann. Mat. Pura Appl. **157**, 285–367 (1990)
31. I. Lasiecka, R. Triggiani, Regularity theory of hyperbolic equations with non-homogeneous Neumann boundary conditions II, general boundary data. J. Differ. Equ. **94**, 112–164 (1991)
32. T.T. Li, *Controllability and Observability for Quasilinear Hyperbolic Systems*. AIMS on Applied Mathematics, vol. 3 (American Institute of Mathematical Sciences and Higher Education Press, Springfield and Beijing, 2010)
33. T.T. Li, Exact boundary controllability of nodal profile for quasilinear hyperbolic systems. Math. Methods Appl. Sci. **33**, 2101–2106 (2010)
34. T.T. Li, Exact boundary synchronization for a coupled system of wave equations, in *Differential Geometry, Partial Differential Equations and Mathematical Physics*, ed. by M. Ge, J. Hong, T. Li, W. Zhang (World Scientific, Singapore, 2014), pp. 219–241
35. T.T. Li, From phenomena of synchronization to exact synchronization and approximate synchronization for hyperbolic systems. Sci. China Math. **59**, 1–18 (2016)
36. T.T. Li, X. Lu, B.P. Rao, Exact boundary synchronization for a coupled system of wave equations with Neumann boundary controls. Chin. Ann. Math. Ser. B **39**, 233–252 (2018)
37. T.T. Li, X. Lu, B.P. Rao, Approximate boundary null controllability and approximate boundary synchronization for a coupled system of wave equations with Neumann boundary controls, in *Contemporary Computational Mathematics–A Celebration of the 80th Birthday of Ian Sloan*, vol. II, ed. by J. Dich, F.Y. Kuo, H. Wozniakowski (Springer, Berlin, 2018), pp. 837–868
38. T.T. Li, X. Lu, B.P. Rao, Exact boundary controllability and exact boundary synchronization for a coupled system of wave equations with coupled Robin boundary controls, to appear
39. T.T. Li, B.P. Rao, Asymptotic controllability for linear hyperbolic systems. Asymptot. Anal. **72**, 169–187 (2011)

40. T.T. Li, B.P. Rao, Contrôlabilité asymptotique de systèmes hyperboliques linéaires. C. R. Acad. Sci. Paris Ser. I **349**, 663–668 (2011)
41. T.T. Li, B.P. Rao, Synchronisation exacte d'un système couplé d'équations des ondes par des contrôles frontières de Dirichlet. C. R. Acad. Sci. Paris Ser. 1 **350**, 767–772 (2012)
42. T.T. Li, B.P. Rao, Exact synchronization for a coupled system of wave equations with Dirichlet boundary controls. Chin. Ann. Math. **34B**, 139–160 (2013)
43. T.T. Li, B.P. Rao, Contrôlabilité asymptotique et synchronisation asymptotique d'un système couplé d'équations des ondes avec des contrôles frontièères de Dirichlet. C. R. Acad. Sci. Paris Ser. I **351**, 687–693 (2013)
44. T.T. Li, B.P. Rao, Asymptotic controllability and asymptotic synchronization for a coupled system of wave equations with Dirichlet boundary controls. Asymptot. Anal. **86**, 199–226 (2014)
45. T.T. Li, B.P. Rao, Sur l'état de synchronisation exacte d'un système couplé d'équations des ondes. C. R. Acad. Sci. Paris Ser. I **352**, 823–829 (2014)
46. T.T. Li, B.P. Rao, On the exactly synchronizable state to a coupled system of wave equations. Port. Math. **72** 2–3, 83–100 (2015)
47. T.T. Li, B.P. Rao, Critères du type de Kálmán pour la contrôlabilité approchée et la synchronisation approchée d'un système couplé d'équations des ondes. C. R. Acad. Sci. Paris Ser. 1 **353**, 63–68 (2015)
48. T.T. Li, B.P. Rao, A note on the exact synchronization by groups for a coupled system of wave equations. Math. Methods Appl. Sci. **38**, 241–246 (2015)
49. T.T. Li, B.P. Rao, Exact synchronization by groups for a coupled system of wave equations with Dirichlet boundary control. J. Math. Pures Appl. **105**, 86–101 (2016)
50. T.T. Li, B.P. Rao, Criteria of Kalman's type to the approximate controllability and the approximate synchronization for a coupled system of wave equations with Dirichlet boundary controls. SIAM J. Control Optim. **54**, 49–72 (2016)
51. T.T. Li, B.P. Rao, Une nouvelle approche pour la synchronisation approchée d'un système couplé d'équations des ondes: contrôles directs et indirects. C. R. Acad. Sci. Paris Ser. I **354**, 1006–1012 (2016)
52. T.T. Li, B.P. Rao, Exact boundary controllability for a coupled system of wave equations with Neumann boundary controls. Chin. Ann. Math. Ser. B **38**, 473–488 (2017)
53. T.T. Li, B.P. Rao, On the approximate boundary synchronization for a coupled system of wave equations: direct and indirect controls. ESIAM: Control Optim. Calc. Var. **24**, 1675–1704 (2018)
54. T.T. Li, B.P. Rao, Kalman's criterion on the uniqueness of continuation for the nilpotent system of wave equations. C. R. Acad. Sci. Paris Ser. I **356**, 1188–1194 (2018)
55. T.T. Li, B.P. Rao, Approximate boundary synchronization for a coupled system of wave equations with coupled Robin boundary conditions, to appear
56. T.T. Li, B.P. Rao, Unique continuation for elliptic operators and application to the asymptotic synchronization of second order evolution equations, to appear
57. T.T. Li, B.P. Rao, L. Hu, Exact boundary synchronization for a coupled system of 1-D wave equations. ESAIM: Control Optim. Calc. Var. **20**, 339–361 (2014)
58. T.T. Li, B.P. Rao, Y.M. Wei, Generalized exact boundary synchronization for second order evolution systems. Discret. Contin. Dyn. Syst. **34**, 2893–2905 (2014)
59. T.T. Li, K. Wang, Q. Gu, *Exact Boundary Controllability of Nodal Profile for Quasilinear Hyperbolic Systems*. Springer Briefs in Mathematics (Springer, Berlin, 2016)
60. J.-L. Lions, *Equations Différentielles Opérationnelles et Problèmes aux Limites*. Grundlehren, vol. 111 (Springer, Berlin, 1961)
61. J.-L. Lions, Exact controllability, stabilization and perturbations for distributed systems. SIAM Rev. **30**, 1–68 (1988)
62. J.-L. Lions, *Contrôlabilité exacte, Perturbations et Stabilisation de Systèmes Distribués*, vol. 1 (Masson, Paris, 1988)
63. J.-L. Lions, *Quelques Méthodes de Résolution des Problèmes aux Limites Non Linéaires* (Dunod, Gauthier-Villars, Paris, 1969)

64. J.-L. Lions, E. Magenes, *Problèmes aux Limites non Homogènes et Applications*, vol. 1 (Dunod, Paris, 1968)
65. Z.Y. Liu, B.P. Rao, A spectral approach to the indirect boundary control of a system of weakly coupled wave equations. Discret. Contin. Dyn. Syst. **23**, 399–414 (2009)
66. Z.Y. Liu, S.M. Zheng, *Semigroups Associated with Dissipative Systems*, vol. 398 (CRC Press, Boca Raton, 1999)
67. Q. Lü, E. Zuazua, Averaged controllability for random evolution partial differential equations. J. Math. Pures Appl. **105**, 367–414 (2016)
68. X. Lu, Local exact boundary synchronization for a kind of first order quasilinear hyperbolic systems. Chin. Ann. Math. Ser. B **40**, 79–96 (2019)
69. M. Mehrenberger, Observability of coupled systems. Acta Math. Hung. **103**, 321–348 (2004)
70. A. Pazy, *Semigroups of Linear Operators and Applications to Partial Differential Equations* (Springer, New York, 1983)
71. A. Pikovsky, M. Rosenblum, J. Kurths, *Synchronization: A Universal Concept in Nonlinear Sciences* (Cambridge University Press, Cambridge, 2001)
72. L. Rosier, L. de Teresa, Exact controllability of a cascade system of conservation equations. C. R. Acad. Sci. Paris **1**(349), 291–296 (2011)
73. D.L. Russell, Controllability and stabilization theory for linear partial differential equations: recent progress and open questions. SIAM Rev. **20**, 639–739 (1978)
74. S. Strogatz, *SYNC: The Emerging Science of Spontaneous Order* (THEIA, New York, 2003)
75. F. Trèves, *Basic Linear Partial Differential Equations*. Pure and Applied Mathematics, vol. 62 (Academic, New York, 1975)
76. L.J. Wang, Q.S. Yan, Minimal time control of exact synchronization for parabolic systems (2018). arXiv:1803.00244vl
77. Y.Y. Wang, Generalized exact boundary synchronization for a coupled system of wave equations with Dirichlet boundary controls, to appear in Chin. Ann. Math. Ser. B
78. Y.Y. Wang, On the generalized exact boundary synchronization for a coupled system of wave equations, Math. Methods Appl. Sci. **42**, 7011–7029 (2019)
79. Y.Y. Wang, Induced generalized exact boundary synchronization for a coupled system of wave equations, to appear in Appl. Math. J. Chin. Univ
80. Y.Y. Wang, Determination of the generalized exact boundary synchronization matrix for a coupled system of wave equations, to appear in Front. Math. China
81. N. Wiener, *Cybernetics, or Control and Communication in the Animal and the Machine*, 2nd edn. (The M.I.T. Press/Wiley, Cambridge/New York, 1961)
82. P.F. Yao, On the observability inequalities for exact controllability of wave equations with variable coefficients. SIAM J. Control Optim. **37**, 1568–1599 (1999)
83. R.M. Young, *An Introduction to Nonharmonic Fourier Series* (Academic, San Diego, 2001)
84. X. Zhang, E. Zuazua, A sharp observability inequality for Kirchhoff plate systems with potentials. Comput. Appl. Math. **25**, 353–373 (2006)
85. E. Zuazua, Exact controllability for the semilinear wave equation. J. Math. Pures Appl. **69**, 1–31 (1990)
86. E. Zuazua, Averaged control. Autom. J. IFAC **50**, 3077–3087 (2014)

Index

A

B

T. Li and B. Rao, *Boundary Synchronization for Hyperbolic Systems*, Progress in Nonlinear Differential Equations and Their Applications 94,
https://doi.org/10.1007/978-3-030-32849-8

Printed in the United States
By Bookmasters